LABORATORY MANUAL

INTRODUCTORY
GEOLOGY

Bradley Deline, PhD | Randa Harris, MS | Karen Tefend, PhD

UNG
UNIVERSITY of
NORTH GEORGIA
UNIVERSITY PRESS

Dahlonega, GA

Laboratory Manual for Introductory Geology is licensed under a Creative Commons Attribution-ShareAlike 4.0 International License.

This license allows you to remix, tweak, and build upon this work, even commercially, as long as you credit this original source for the creation and license the new creation under identical terms.

If you reuse this content elsewhere, in order to comply with the attribution requirements of the license please attribute the original source to the University System of Georgia.

NOTE: The above copyright license which University System of Georgia uses for their original content does not extend to or include content which was accessed and incorporated, and which is licensed under various other CC Licenses, such as ND licenses. Nor does it extend to or include any Special Permissions which were granted to us by the rightsholders for our use of their content.

Image Disclaimer: All images and figures in this book are believed to be (after a reasonable investigation) either public domain or carry a compatible Creative Commons license. If you are the copyright owner of images in this book and you have not authorized the use of your work under these terms, please contact the University of North Georgia Press at ungpress@ung.edu to have the content removed.

ISBN: 978-1-940771-36-6

Produced by:
University System of Georgia

Published by:
University of North Georgia Press
Dahlonega, Georgia

Cover Design and Layout Design:
Corey Parson

For more information, please visit http://ung.edu/university-press
Or email ungpress@ung.edu

TABLE OF CONTENTS

CHAPTER 1: INTRODUCTION TO PHYSICAL GEOLOGY — 1
Bradley Deline

CHAPTER 2: EARTH'S INTERIOR — 22
Randa Harris and Bradley Deline

CHAPTER 3: TOPOGRAPHIC MAPS — 41
Karen Tefend and Bradley Deline

CHAPTER 4: PLATE TECTONICS — 65
Bradley Deline

CHAPTER 5: WATER — 90
Randa Harris

CHAPTER 6: CLIMATE CHANGE — 121
Bradley Deline

CHAPTER 7: MATTER AND MINERALS — 140
Randa Harris

CHAPTER 8: IGNEOUS ROCKS — 177
Karen Tefend

CHAPTER 9: VOLCANOES — 205
Karen Tefend

CHAPTER 10: SEDIMENTARY ROCKS — 227
Bradley Deline

CHAPTER 11: METAMORPHIC ROCKS — 261
Karen Tefend

CHAPTER 12: CRUSTAL DEFORMATION — 286
Randa Harris and Bradley Deline

CHAPTER 13: EARTHQUAKES — 313
Randa Harris

CHAPTER 14: PHYSIOGRAPHIC PROVINCES — 334
Bradley Deline

1 Introduction to Physical Geology
Bradley Deline

1.1 INTRODUCTION

The average introductory geology student's perception of geology normally involves the memorization of rocks and discussions of natural disasters, but Geology contains so much more. Geology is the study of our planet, which is vital to our everyday lives from the energy we use, to the growing of the food we eat, to the foundations of the buildings we live in, to the materials that are used to make everyday objects (metals and plastics). The ideal place to start this course is discussing the methods that are used to better understand our planet, the processes that shape it, and its history.

Science is not a set of facts to remember. Instead it is a method to discover the world around us. You are likely already familiar with the **Scientific Method**, but it is worthwhile to review the process. The first step of the scientific method is making an observation or learning the background surrounding the question in which you are interested. This can be done by taking classes on a subject matter as you are doing presently in geology or by simply taking careful notes about your surroundings. Based on your knowledge and observations, you can then make a **hypothesis**, which is a testable prediction on how something works. A hypothesis should be framed in a way that is easy to test and prove wrong. This might sound odd, but science works to rigorously disprove a hypothesis and only those that withstand the tests become accepted. The wonderful aspect of this definition of a hypothesis is that the testing results in a brand new observation that can then be used to formulate a new hypothesis. Therefore, whether the hypothesis is verified or rejected it will lead to new information. The next step is communication to other scientists. This allows other scientists to repeat the experiment as well as alter it in new and unthought-of ways that can then expand on the original idea. These few steps encompass the vast majority of the scientific method and the career of any individual scientist. As hundreds of related observations and tested hypotheses accumulate scientists can formulate a **theory**. The scientific meaning of a theory is an explanation for a natural phenomenon that is supported by a wealth of scientific data. A theory is not yet a law because there still may be some debate on the exact workings of the theory or the reasons why a phenomenon occurs, but there is little debate on the existence of what is being described.

This leads us back to Geology, the scientific study of the Earth. There are aspects in geology that are directly testable, but others are not and geologists must become imaginative in discovering aspects about the earth and its history that we will never be able to directly observe. In this laboratory manual we will discuss the materials that make up the earth (Minerals and Rocks), earth processes both deep inside the earth (Folds and Faults) and on its surface (Rivers and Climate), as well as the theory that helps explain how the earth works (Plate Tectonics). A fundamental aspect of understanding the Earth is a grasp of Geologic Time (the subject of the first chapter), which helps us think about the rate and frequency of geologic events that have formed the planet that we know today.

1.1.2 Learning Outcomes

After completing this chapter, you should be able to:

- Discuss the importance of time in the study of Geology
- Discuss the difference between Relative Time and Absolute Time
- Apply Geologic Laws in the relative dating of geologic events
- Use fossils to date a rock unit
- Use ideas behind radiometric dating to date rock units

1.1.3 Key Terms

- Absolute Dating
- Angular Unconformity
- Carbon-14 Dating
- Daughter Atom
- Disconformity
- Geologic Laws
- Geologic Time Scale
- Half-life
- Index Fossils
- Isotope
- Law of Cross-Cutting

- Law of Faunal Succession
- Law of Original Horizontality
- Law of Superposition
- Nonconformity
- Parent Atom
- Potassium-Argon Dating
- Radiometric Dating
- Relative Dating
- Unconformity
- Uranium Dating

1.2 GEOLOGIC TIME

The amount of time that is involved in the carving of the landscape, the formation of rocks, or the movement of the continents is an important scientific question. Different hypotheses about the age of the earth can essentially change our perspective of the workings of geologic events that molded the Earth. If the geologic time is relatively short then catastrophic events would be required to form the features we see on the surface of the earth, whereas a vast amount of time allows the slow and steady pace that we can easily observe around us today.

Geologists have used many methods attempting to reconstruct geologic time trying to map the major events in earth's history as well as their duration. Scientists studying rocks were able to piece together a progression of rocks through time to construct the **Geologic Time Scale** (Figure 1.1). This time scale was constructed by lining up in order rocks that had particular features such as rock types, environmental indicators, or fossils. Scientists looked at clues within the rocks and deter-

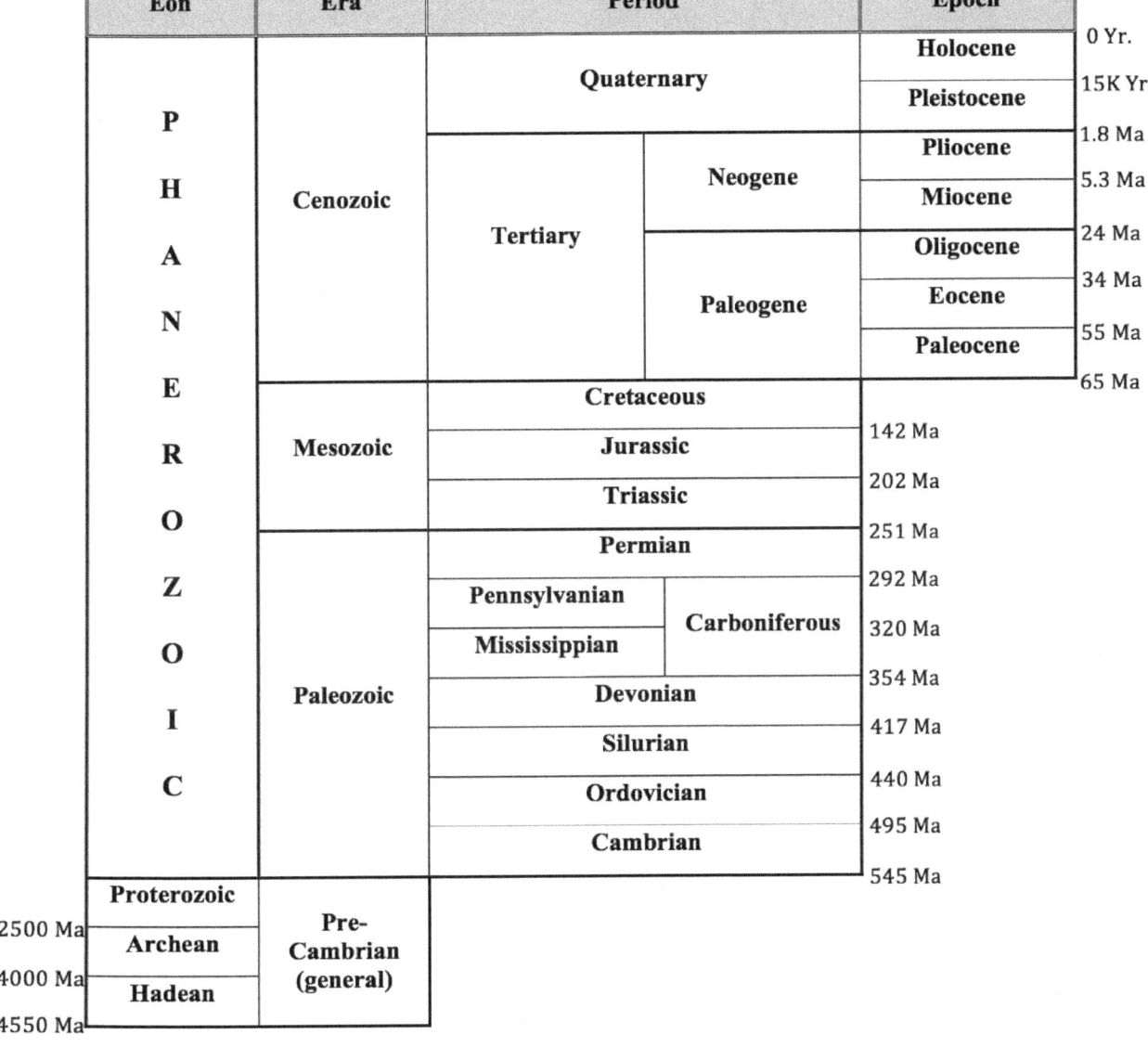

Figure 1.1 | The geologic time scale. Ma, Million years, K. Yr, Thousand years
Author: Bradley Deline
Source: Original Work
License: CC BY-SA 3.0

mined the age of these rocks in a comparative sense. This process is called **Relative Dating**, which is the process of determining the comparative age of two objects or events. For example, you are younger than your parents. It doesn't matter your age or your parents as long as you can establish that one is older than the other. As time progressed, scientists discovered and developed techniques to date certain rocks as well as the Earth itself. They discovered the earth was billions of years old (4.54 billion years old) and put a time frame to the geologic time scale. This process is called **Absolute Dating**, which is the process of determining the exact amount of time that has passed since an object was formed or an event occurred.

Both absolute and relative dating have advantages and are still frequently used by geologists. Dating rocks using relative dating allows a geologist to reconstruct a series of events cheaply, often very quickly, and can be used out in the field on a rocky outcrop. Relative dating also can be used on many different types of rocks, where absolute dating is restricted to certain minerals or materials. However, absolute dating is the only method that allows scientists to place an exact age to a particular rock.

1.2.1 Relative Time and Geologic Laws

The methods that geologists use to establish relative time scales are based on Geologic Laws. A scientific law is something that we understand and is proven. It turns out that, unlike math, it is hard to prove ideas in science and, therefore, Geologic Laws are often easy to understand and fairly simple. Before we discuss the different geologic laws, it would be worthwhile to briefly introduce the different rock types. Sedimentary rocks, like sandstone, are made from broken pieces of other rock that are eroded in the high areas of the earth, transported by wind, ice, and water to lower areas, and deposited. The cooling and crystallizing of molten rock forms igneous rocks. Lastly, the application of heat and pressure to rocks creates metamorphic rocks. This distinction is important because these three different rock types are formed differently and therefore, need to be interpreted differently.

The **Law of Superposition** states that in an undeformed sequence of sedimentary rocks the oldest rocks will be at the bottom of the sequence while the youngest will be on top. Imagine a river carrying sand into an ocean, the sand will spill out onto the ocean floor and come to rest on top of the seafloor. This sand was deposited after the sand of the seafloor was already deposited. We can then create a relative time scale of rock layers from the oldest rocks at the bottom (labeled #1 in Figure 1.2) to the youngest at the top of an outcrop (labeled #7 in Figure 1.2).

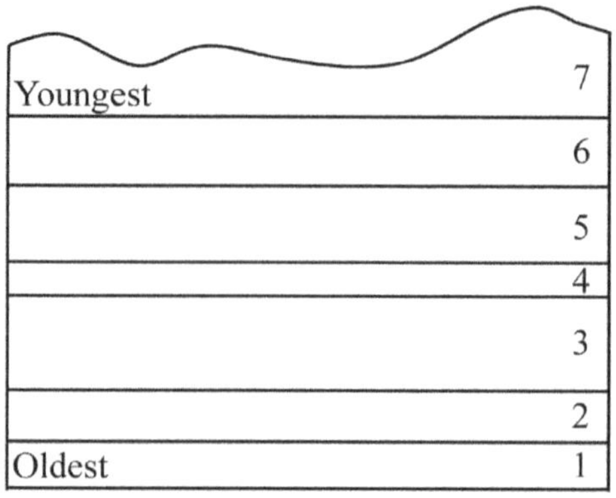

Figure 1.2 | Block diagram showing the relative age of sedimentary layers based on the Law of Superposition.
Author: Bradley Deline
Source: Original Work
License: CC BY-SA 3.0

The **Law of Original Horizontality** states that undeformed sedimentary rock are deposited horizontally. The deposition of sediment is controlled by gravity and will pull it downward. If you have muddy water on a slope, the water will flow down the slope and pool flat at the base rather than depositing on the slope itself. This means that if we see sedimentary rock that is tilted or folded it was first deposited flat, then folded or tilted afterward (Figure 1.3).

The **Law of Cross-Cutting** states that when two geologic features intersect, the one that cuts across the other is younger. In essence, a feature has to be present before something can affect it. For example, if a fault fractures through a series of sedimentary rocks those sedimentary rocks must be older than the fault (Figure 1.4).

One other feature that can be useful in building relative time scales is what is missing in a sequence of rocks. **Unconformities** are surfaces that represent significant weathering and erosion (the breakdown of rock and movement of sediment) which result in missing or erased time. Erosion often occurs in elevated areas like continents or mountains so pushing rocks up (called uplifting) results in erosion and destroying a part of a geologic sequence; much older rocks are then exposed at the earth's surface. If the area sinks (called subsidence), then much younger rocks will be deposited overtop of these newly exposed rocks. The amount of time missing can be relatively short or may represent billions of

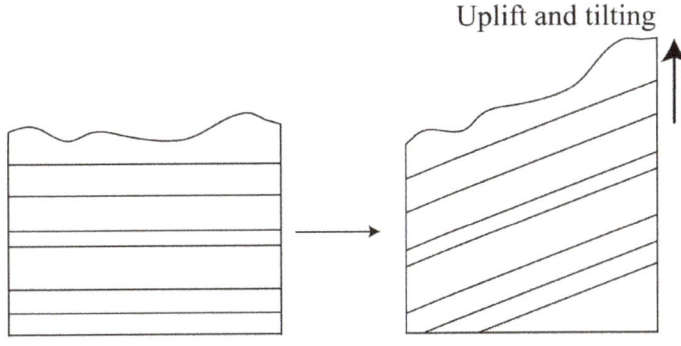

Figure 1.3 | Sedimentary rocks are deposited horizontally such that if the layers are tilted or folded it must have occurred following deposition.
Author: Bradley Deline
Source: Original Work
License: CC BY-SA 3.0

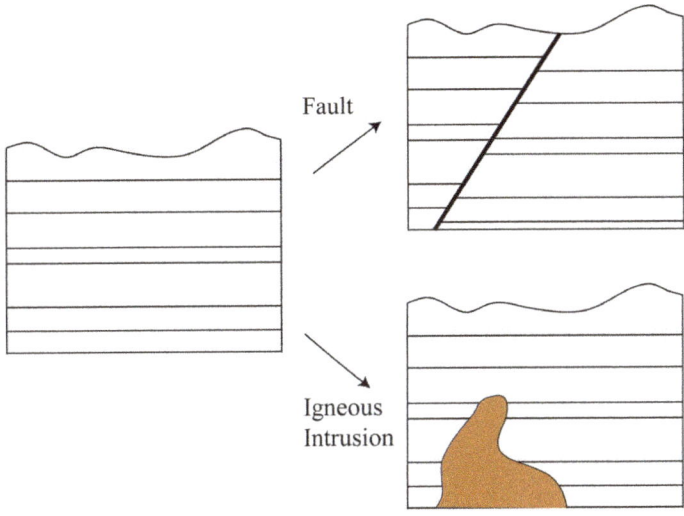

Figure 1.4 | Block diagrams showing the Law of Cross-Cutting. In both instances on the right the geological features (fault or Igneous intrusion) cut across the sedimentary layers and must then be younger.
Author: Bradley Deline
Source: Original Work
License: CC BY-SA 3.0

years. There are three types of unconformities based on the rocks above and below the unconformity (Figure 1.5). If the type of rock is different above and below the unconformity it is called a **Nonconformity.** For example, igneous rock formed deep in the earth is uplifted and exposed at the surface then covered with sedimentary rock. If the rocks above and below the erosion surface are both sedimentary,

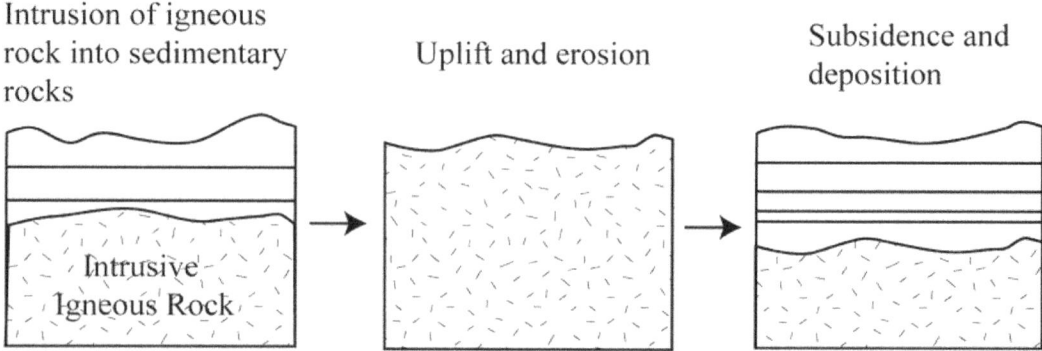

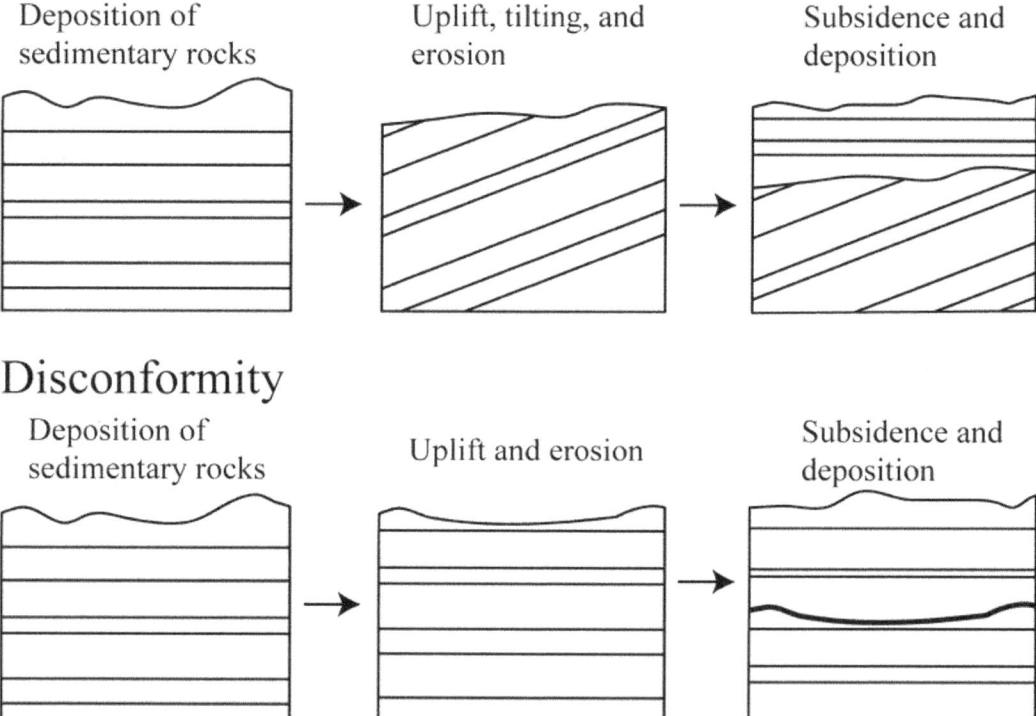

Figure 1.5 | Block diagrams showing the formation of the three types of Unconformities. The three unconformities differ based on the type of rock underneath the erosion surface.
Author: Bradley Deline
Source: Original Work
License: CC BY-SA 3.0

then the orientation of the layers is important. If the rocks below the erosion surface are not parallel with those above, the surface is called an **Angular Unconformity**. This is often the result of the rocks below being tilted or folded prior to the erosion and deposition of the younger rocks. If the rocks above and below the erosion surface are parallel, the surface is called a **Disconformity**. This type of surface is often difficult to detect, but can often be recognized using other information such as the fossils discussed in the next section.

Using these principles we can look at a series of rocks and determine their relative ages and even establish a series of events that must have occurred. Common events that are often recognized can include 1) Deposition of sedimentary layers, 2) Tilting or folding rocks, 3) Uplift and erosion of rocks, 4) Intrusion of liquid magma, and 5) Fracturing of rock (faulting). Figures 1.6 and 1.7 show how to piece together a series of geologic events using relative dating.

Building a Relative Time Sequence

Step 1. Identify and number all of the sedimentary layers.

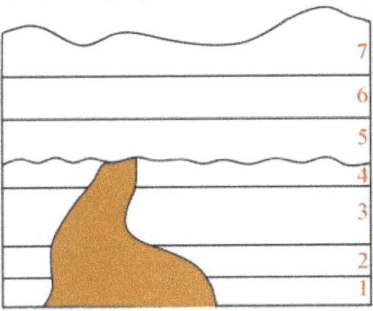

Step 2. Identify any other geologic events

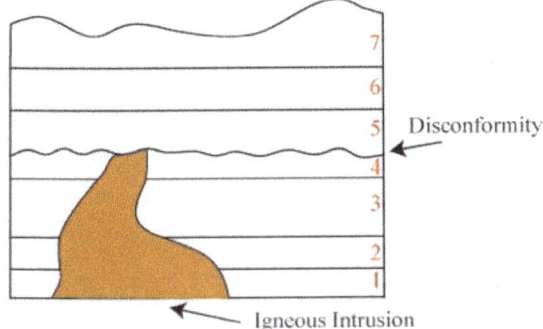

Step 3. Place the sedimentary Layers and geologic events in order based on the Geologic Laws.

1. Deposit sedimentary layers 1-4.
2. Igneous Intrustion.
3. Uplift and Erode - Disconformity.
4. Subsidence and Deposit layers 5-7.
5. Uplift and Erode.

Figure 1.6 | An example showing how to determine a relative dating sequence of events from a block diagram.
Author: Bradley Deline
Source: Original Work
License: CC BY-SA 3.0

Building a Relative Time Sequence

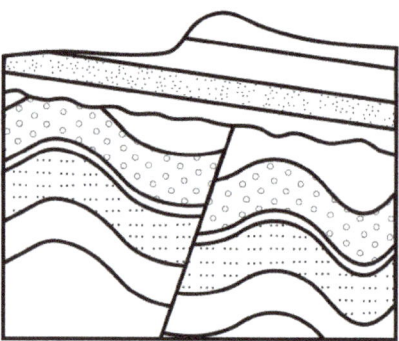

Step 1. Identify and number all of the sedimentary layers.

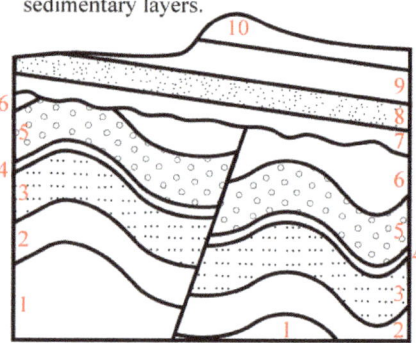

Step 2. Identify any other geologic events

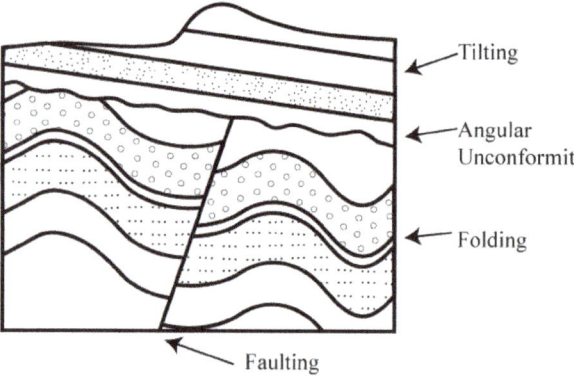

- Tilting
- Angular Unconformity
- Folding
- Faulting

Step 3. Place the sedimentary Layers and geologic events in order based on the Geologic Laws.

1. Deposit sedimentary layers 1-6.
2. Fold and Fault layers 1-6.
3. Uplift and Erode - Angular Unconformity
4. Subsidence and Deposit layers 7-10
5. Tilt Layers 1-10.
6. Uplift and Erode.

Figure 1.7 | An example showing how to determine a relative dating sequence of events from a block diagram.
Author: Bradley Deline
Source: Original Work
License: CC BY-SA 3.0

1.3 LAB EXERCISE

Part A – Relative Time

Relative time is an important tool for geologist to quickly construct series of events, especially in the field. In the following section, apply what you have learned regarding relative time to the questions below.

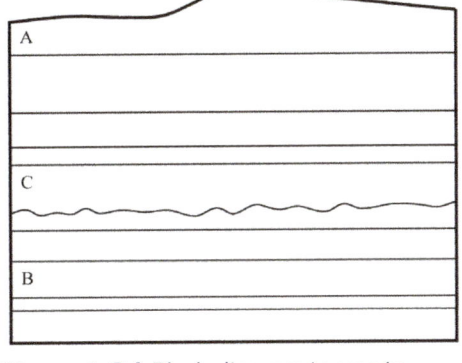

Figure 1.8 | Block diagram to use to answer questions 1 and 2.
Author: Bradley Deline
Source: Original Work
License: CC BY-SA 3.0

1. In Figure 1.8, which of the following rock layers is the **oldest**?

    a. A b. B c. C

2. Which Geologic Law did you use to come to the conclusion you made in the previous question?

    a. The Law of Superposition b. The Law of Cross-Cutting

    c. The Law of Original Horizontality d. Unconformities

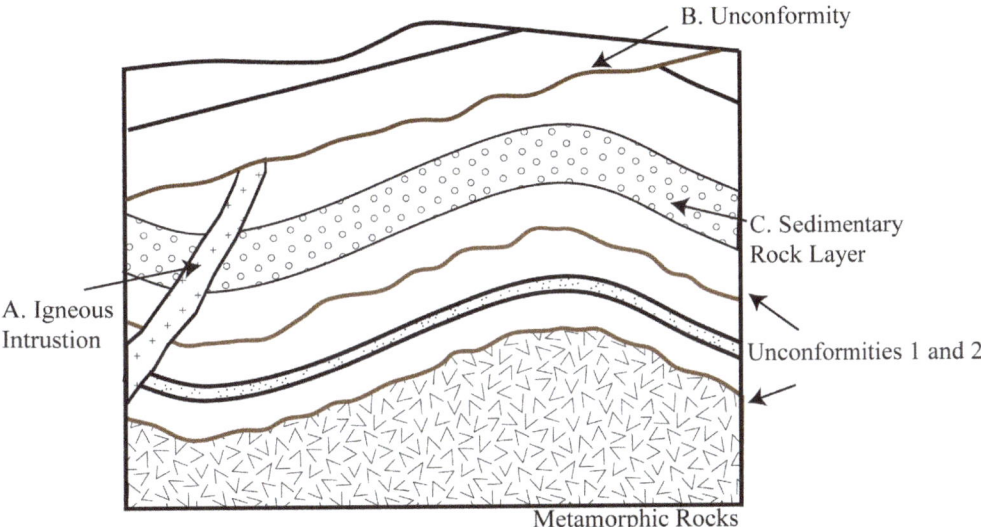

Figure 1.9 | Block diagram to use to answer questions 3, 4, and 5. Unconformities are shown in brown.
Author: Bradley Deline
Source: Original Work
License: CC BY-SA 3.0

3. In Figure 1.9, which of the following geologic structures is the **youngest**?

    a. A b. B c. C

4. Which Geologic Law did you use to come to the conclusion you made in the previous question?

   a. The Law of Superposition
   b. The Law of Cross-Cutting
   c. The Law of Original Horizontality
   d. Unconformities

5. Examine unconformities 1 and 2 indicated in Figure 1.9. Which of the following statements about them is true?

   a. The older unconformity is a Nonconformity, while the younger is an Angular Unconformity.

   b. The older unconformity is a Disconformity, while the younger is a Nonconformity.

   c. The older unconformity is a Nonconformity, while the younger is a Disconformity.

   d. The older unconformity is an Angular Unconformity, while the younger is a Disconformity.

6. Examine the Unconformity shown in Figure 1.10. What type of unconformity is this?

   a. Angular Unconformity

   b. Nonconformity

   c. Disconformity

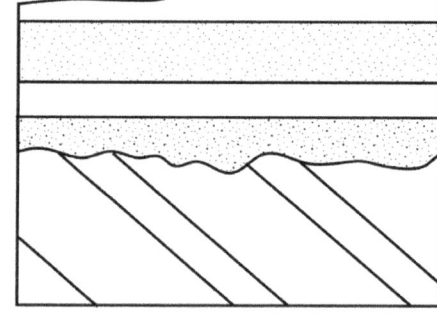

Figure 1.10 | Block diagram to use to answer question 6.
Author: Bradley Deline
Source: Original Work
License: CC BY-SA 3.0

Examine Figure 1.11. Note that all of the layers in this block diagram are composed of sedimentary rock and the unconformities are colored in red. Using the geologic laws discussed earlier and following the examples shown in Figures 1.6 and 1.7, identify the geologic events that occurred in this area. Then place the following geologic events in the correct relative time sequence.

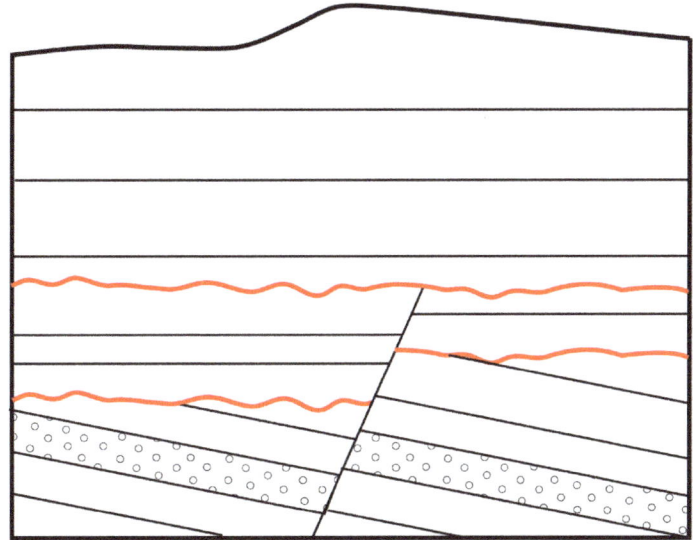

Figure 1.11 | Block diagram to use to answer questions 7, 8, and 9. Unconformities are shown in red.
Author: Bradley Deline
Source: Original Work
License: CC BY-SA 3.0

a. Tilting.

b. Uplift and Erosion (Angular Unconformity).

c. Submergence and deposition of sedimentary layers 10-13.

d. Uplift and Erosion to current position.

e. Submergence and Deposition of sedimentary layers 7-9.

f. Uplift and Erosion (Disconformity)

g. Submergence and deposition of sedimentary layers 1-6.

h. Fault.

```
___   ___   ___   ___   ___   ___   ___   ___
 1     2     3     4     5     6     7     8
Oldest                                  Youngest
```

7. Which of the above geologic events is the second in the sequence?

   a. A b. B c. C d. D

   e. E f. F g. G h. H

8. Which of the above geologic events is the fifth in the sequence?

   a. A b. B c. C d. D

   e. E f. F g. G h. H

9. Which of the above geologic events is the seventh in the sequence?

   a. A b. B c. C d. D

   e. E f. F g. G h. H

1.4 FAUNAL SUCCESSION AND INDEX FOSSILS

Another useful tool in relative dating are fossils. Fossils are the preserved remains of ancient organisms normally found within sedimentary rocks. Organisms appear at varying times in geologic history and go extinct at different times. These organisms also change in appearance through time. This pattern of the appearance, change, and extinction of thousands of fossil organisms creates a recognizable pattern of organisms preserved through geologic time. Therefore, rocks of the same age likely contain similar fossils and we can use these fossils to date sedimentary rocks. This concept is called the **Law of Faunal Succession**.

Some fossils are particularly useful in telling time, these are called **Index Fossils**. These are organisms that we are likely to find because they were abundant when they were alive and were likely to become fossils (for example, having a robust skeleton). These organisms often have a large geographic range so they can be used as an index fossil in many different areas. However, they should also have a short geologic range (the amount of time an organism is alive on Earth), so we can be more precise in the age of the rock if we find the fossil. Index fossils are often the quickest and easiest way to date sedimentary rocks precisely and accurately.

1.5 LAB EXERCISE
Part B – Faunal Succession

The use of animals and their preserved remains (fossils) can help build a highly precise time sequence, often with a higher resolution than absolute dating. In the following section, use this principle to answer the following questions.

A group of geology students stops at a rocky outcrop in Northern Kentucky to examine the rocks and fossils. After looking at geologic maps they conclude that the rocks are Ordovician (~450 Million years old) in age. They make a collection of fossils to better date these rocks. After returning to school they identify the following fossils in order to establish the time frame in which each were alive. They hope to use the Law of Fossil Succession to then plot the ranges of the genera and determine the exact age of the rocks.

   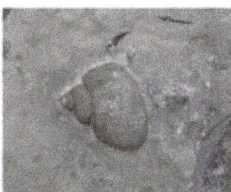

A. The sea lily *Ectenocrinus*. Fulton-Fairmount

B. The trilobite *Isotelus*. Fulton-Arnheim

C. The lampshell *Zygospria*. McMicken- Fairmount

D. The snail *Cyclonema*. McMicken- Fairmount

E. The lampshell *Vinlandostrophia*. Fairmount- Corryville

F. The bryozoan *Parvohallopora*. Southgate- Fairmount

G. The lampshell *Cincinnetina*. Fulton- Fairmount

H. The edrioasteroid *Streptaster*. Southgate- Sunset

I. The clam *Ambonychia*. Fairmount- Bellevue

The geologic range of *Ectenocrinus* (A) is plotted below. Using *Ectenocrinus* as an index fossil you can determine that the rock is from the Late Ordovician and was formed between the Fulton member of the Kope Formation and the Fairmount member of the Maysville Formation. Plot the geologic range of the remaining eight animals to narrow down when this rock was formed.

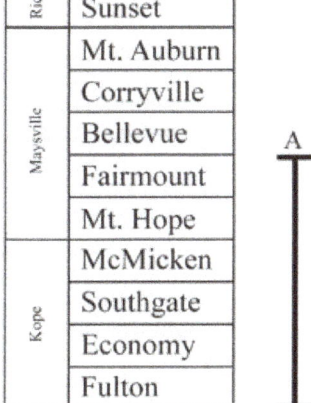

All pictures by B. Deline

Figure 1.12
Author: Bradley Deline
Source: Original Work
License: CC BY-SA 3.0

10. Based on the assemblage of organisms (A-I) in this sample, what is the age of this rock?

    a. Economy   b. Southgate   c. McMicken   d. Mt. Hope

    e. Fairmount   f. Bellevue   g. Corryville   h. Mt. Auburn

11. Which organism was the *most* useful in coming to this conclusion (which is the best index fossil)?

    a. Isotelus   b. Zygospira   c. Cyclonema   d. Vinlandostrophia

    e. Parvohallopora f. Cincinnetina   g. Streptaster h. Ambonychia

12. Which organism was the *least* useful in coming to this conclusion (which is the worst index fossil)?

    a. Isotelus   b. Zygospira   c. Cyclonema   d. Vinlandostrophia

    e. Parvohallopora f. Cincinnetina   g. Streptaster h. Ambonychia

1.6 ABSOLUTE TIME AND RADIOMETRIC DATING

Absolute time is a method for determining the age of a rock or object most often using radiometric isotopes. Atoms are made of three particles, protons, electrons, and neutrons. All three of these particles are important to the study of geology: the number of protons defines a particular element, the number of electrons control how that element bonds to make compounds, and the number of neutrons changes the atomic weight of an element. **Isotopes** are atoms of an element that differ in the number of neutrons in their nucleus and, therefore, their atomic weight. If an element has too many or too few neutrons in its nucleus then the atom becomes unstable and breaks down over time, which is called **radioactive decay**. The process of radioactive decay involves the emitting of a particle from a radioactive atom, called the **parent atom**, which changes it to another element, called the **daughter atom**. We can study and measure the radioactivity of different elements in the lab and calculate the rate of decay. Though the rate of decay varies between isotopes from milliseconds to billions of years, all radiometric isotopes decay in a similar way. Radiometric decay follows a curve that is defined by a radiometric isotope's **half-life**. The half-life is defined as the amount of time it takes for half of the atoms of the radiometric parent isotope to decay to the daughter. The half-life is independent of the amount of atoms at a given time so it takes the same amount of time to go from 100% of the parent isotope remaining to 50% as it does to go from 50% of the parent isotope remaining to 25%. If we know the length of the half-life for a particular radiometric isotope and we measure the amount of parent and daughter isotope in a rock, we can then calculate the age of the rock, which is called **Radiometric Dating**. Given the shape of the decay curve, a material never runs out of the parent isotope, but we can only effectively measure the parent up to 10-15 half-lives.

1.7 LAB EXERCISE

Part C – Radiometric Dating

Complete the following chart by calculating the amount of parent isotope remaining for all of the given half-lives, then plot your findings on the graph (Figure 1.13). Make sure you connect the data points on the graph by drawing in the decay curve.

Use the completed chart and graph to answer the questions below.

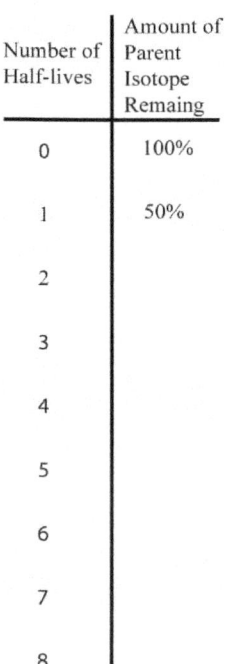
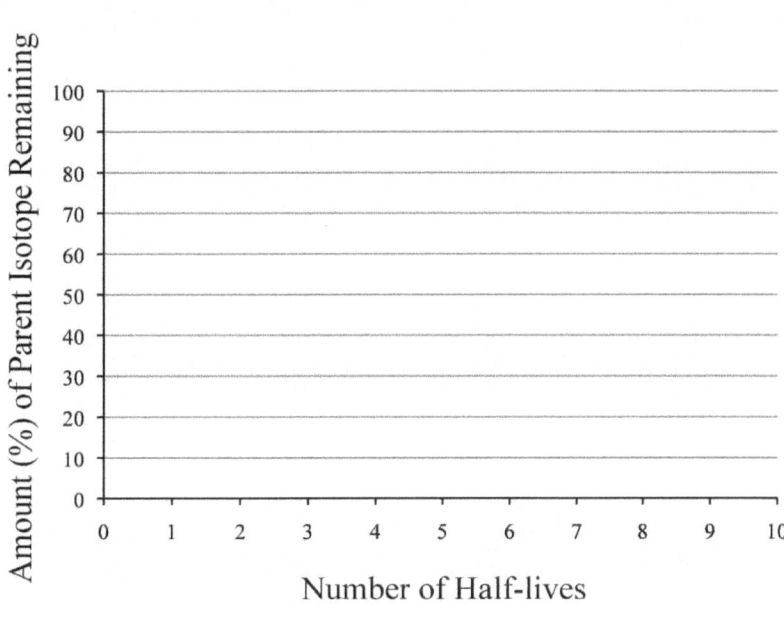

Figure 1.13
Author: Bradley Deline
Source: Original Work
License: CC BY-SA 3.0

13. How much of the parent isotope would be remaining after 7 half-lives have passed?

    a. 6.25% b. 1.56% c. 0.78% d. 0.39%

14. If a radiometric element has a half-life of 425 years, how old would a rock be that only had 3.125% of the parent isotope remaining?

    a. 2125 years b. 1700 years c. 2550 years d. 3400 years

15. Based on your graph above, approximately how much of the parent isotope would be remaining after 3.5 half-lives?

    a. 16% b. 12% c. 4% d. 8%

16. Based on your graph above, approximately how many half-lives have passed when only 35% of the parent isotope is remaining?

    a. 0.75 b. 1.5 c. 2.1 d. 2.5

1.8 DATING SYSTEMS

There are several different radiometric isotopes that are commonly used in absolute dating. Each of these systems have different uses within geology in that they require different materials and can date objects within specific time frames.

Carbon-14 dating is of limited use within geology, but is still the system that is familiar to most people. Carbon-14 (parent isotope) is found in organic material including bone, tissue, plants, and fiber. This isotope is found naturally in small amounts in the atmosphere within CO_2 and is incorporated into plants during photosynthesis and then filters throughout the food chain. You currently have Carbon-14 in your body that is decaying to Nitrogen-14 (daughter isotope), but you replace it whenever you eat. When an animal stops eating or a plant stops photosynthesizing, the radioactive carbon starts to decay without being replaced, which can be easily measured. Carbon-14 has a very short half-life of 5,730 years and can only be used to date materials up to approximately 70,000 years. Given the age of the Earth is 4.54 billion years, carbon-14 can only be used to date very recent materials.

Uranium dating involves a complex system of multiple isotopes that decay through a chain reaction until it reaches non-radiogenic lead. Surprising to most students, uranium can be found in many places, but it is normally in very miniscule amounts. Another issue with this system is that the daughter isotope, lead, is also found naturally in many different places, which makes it difficult to differentiate between lead formed from radiometric decay and lead found naturally in the environment. The mineral zircon solves both of these issues, by concentrating uranium and excluding lead from its mineral structure. Therefore, we use Uranium dating on zircons found within igneous rocks (such as volcanic ash or rocks formed deep in the earth). Uranium has a very long half-life of 4.5 billion years, which is more than long enough to date most rocks on Earth. It takes about one million years for the complex system to normalize such that Uranium dates of less than that are unreliable.

Potassium-Argon dating is also a useful method of dating rocks. Potassium decays into two separate daughter isotopes, Argon and Calcium. We measure the amount of Argon in the rocks because unlike calcium it is rare within minerals since it is a Noble Gas and doesn't normally bond with other elements. Therefore, any argon within a mineral is from the decay of potassium. The use of Argon also has its drawbacks, for instance a gas can easily escape from a rock and, therefore, special care needs to be taken in the lab to prevent this. This system works well when there are multiple materials to examine that contain abundant potassium, like the rock granite that is full of potassium-rich pink minerals called feldspars. The half-life of Potassium is 702 million years, so it is similar to Uranium in that it is most useful dating older rocks.

With all of these methods there is still the chance for error such that it is best to think of any particular radiometric date as a scientific hypothesis that needs to be further tested. Error can come from the addition or subtraction of either parent or daughter isotopes in the rock following its formation. This can be done in several ways, most commonly through the adding of heat and pressure (metamorphism). There are ways to correct for these issues that allows the scientist to date both the rock and the metamorphic event as long as the geologic history is known.

As you may have guessed from the previous exercise, it is rare to find a rock that contains an amount of the parent remaining that falls exactly on one of the half-lives. In most cases we need to use a simple formula to calculate the age of a rock using the length of the half-life and the amount of parent remaining.

The formula is:

$$Age = -(\frac{t_{1/2}}{0.693})\ln(P)$$

$t_{1/2}$ = The length of the half-life in years

P = The amount of the parent remaining in decimal form. For example, if there is 50% of the parent remaining it would equal 0.5.

Let's work an example using the equation that we already know the answer to in advance. You have a sample of bone that has 25% of the Carbon-14 (Half-life= 5730 years) remaining, how old is the sample? We can answer this question in two ways:

1. We know that if there is 25% remaining, two half-lives have passed and with each half-life being 5730 the bone would be 11,460 years old.
2. We could use the above equation and insert both the length of the half-life and the amount of the parent remaining:

$$Age = -(\frac{5730}{0.693})\ln(0.25)$$

To solve the equation, take the Natural Log (ln) of 0.25 and multiply by the term in the parentheses (make sure to include the negative sign). If you do this you will get 11,460 as well.

1.9 LAB EXERCISE

Part D – Isotopic Systems

Using what you learned in the previous section regarding absolute dating, determine the most appropriate methods and the ages of the materials in the following questions.

17. An Archeologist finds some cotton cloth at a burial site and wants to determine the age of the remains. Which isotopic system should they use?

    a. Carbon-14   b. Uranium   c. Potassium-Argon

18. The Archeologist determines that there is 16.7% of the parent isotope remaining in the cloth sample. How old is the burial site? Hint: you can find the length of the half-life in the reading above.

    a. 13,559 years   b. 14,798 years   c. 16,743 years

    d. 1.66 billion years   e. 1.81 billion years   f. 2.05 billion

19. A geologist is trying to date a sequence of sedimentary rocks with abundant fossils and sandstones. Within the sequence is a distinctive clay layer that under closer inspection is fine-grained volcanic ash. Which of the following is the best way to obtain an absolute date for the sequence of rocks?

    a. Carbon date the fossils   b. Potassium-Argon date the sands

    c. Uranium date the Zircons in the ash   d. Identify the index fossils

20. The geologist determines there is 78.3% of the parent remaining in the sample that they examine. How old is the sequence of rocks? Hint: you can find the length of the half-life in the reading above.

    a. 187.5 million years   b. 247.8 million years   c. 390.7 million years

    d. 2.504 billion years   e. 1.588 billion years   f. 1.202 billion years

1.10 STUDENT RESPONSES

The following is a summary of the questions in this lab for ease in submitting answers online.

1. In Figure 1.8, which of the following rock layers is **oldest**?

   a. A b. B c. C

2. Which Geologic Law did you use to come to the conclusion you made in the previous question?

   a. The Law of Superposition b. The Law of Cross-Cutting

   c. The Law of Original Horizontality d. Unconformities

3. In Figure 1.9, which of the following geologic structures is **youngest**?

   a. A b. B c. C

4. Which Geologic Law did you use to come to the conclusion you made in the previous question?

   a. The Law of Superposition b. The Law of Cross-Cutting

   c. The Law of Original Horizontality d. Unconformities

5. Examine unconformities 1 and 2 indicated in Figure 1.9. Which of the following statements about them is true?

   a. The older unconformity is a Nonconformity, while the younger is an Angular Unconformity.

   b. The older unconformity is a Disconformity, while the younger is a Nonconformity.

   c. The older unconformity is a Nonconformity, while the younger is a Disconformity.

   d. The older unconformity is an Angular Unconformity, while the younger is a Disconformity.

6. Examine the Unconformity shown in Figure 1.10. What type of unconformity is this?

   a. Angular Unconformity b. Nonconformity c. Disconformity

7. Which of the above geologic events is the second in the sequence?

    a. A   b. B   c. C   d. D

    e. E   f. F   g. G   h. H

8. Which of the above geologic events is the fifth in the sequence?

    a. A   b. B   c. C   d. D

    e. E   f. F   g. G   h. H

9. Which of the above geologic events is the seventh in the sequence?

    a. A   b. B   c. C   d. D

    e. E   f. F   g. G   h. H

10. Based on the assemblage of organisms (A-I) in this sample, what is the age of this rock?

    a. Economy   b. Southgate   c. McMicken   d. Mt. Hope

    e. Fairmount f. Bellevue   g. Corryville h. Mt. Auburn

11. Which organism was the *most* useful in coming to this conclusion (which is the best index fossil)?

    a. Isotelus   b. Zygospira   c. Cyclonema   d. Vinlandostrophia

    e. Parvohallopora f. Cincinnetina   g. Streptaster   h. Ambonychia

12. Which organism was the *least* useful in coming to this conclusion (which is the worst index fossil)?

    a. Isotelus   b. Zygospira   c. Cyclonema   d. Vinlandostrophia

    e. Parvohallopora f. Cincinnetina   g. Streptaster   h. Ambonychia

13. How much of the parent isotope would be remaining after 7 half-lives have passed?

    a. 6.25%   b. 1.56%   c. 0.78%   d. 0.39%

14. If a radiometric element has a half-life of 425 years, how old would a rock be that only had 3.125% of the parent isotope remaining?

    a. 2125 years b. 1700 years c. 2550 years d. 3400 years

15. Based on your graph above, approximately how much of the parent isotope would be remaining after 3.5 half-lives?

    a. 16% b. 12% c. 4% d. 8%

16. Based on your graph above, approximately how many half-lives have passed when only 35% of the parent isotope is remaining?

    a. 0.75 b. 1.5 c. 2.1 d. 2.5

17. An Archeologist finds some cotton cloth at a burial site and wants to determine the age of the remains. Which isotopic system should they use?

    a. Carbon-14 b. Uranium c. Potassium-Argon

18. The Archeologist determines that there is 16.7% of the parent isotope remaining in the cloth sample. How old is the burial site? Hint: you can find the length of the half-life in the reading above.

    a. 13,559 years b. 14,798 years c. 16,743 years

    d. 1.66 billion years e. 1.81 billion years f. 2.05 billion

19. A geologist is trying to date a sequence of sedimentary rocks with abundant fossils and sandstones. Within the sequence is a distinctive clay layer that under closer inspection is fine-grained volcanic ash. Which of the following is the best way to obtain an absolute date for the sequence of rocks?

    a. Carbon date the fossils b. Potassium-Argon date the sands

    c. Uranium date the Zircons in the ash d. Identify the index fossils

20. The geologist determines there is 78.3% of the parent remaining in the sample that they examine. How old is the sequence of rocks? Hint: you can find the length of the half-life in the reading above.

    a. 187.5 million years b. 247.8 million years c. 390.7 million years

    d. 2.504 billion years e. 1.588 billion years f. 1.202 billion years

2 Earth's Interior

Randa Harris and Bradley Deline

2.1 INTRODUCTION

Studying the Earth's interior poses a significant challenge due to the lack of direct access. Many processes observed at the Earth's surface are driven by the heat generated within the Earth, however, making an understanding of the interior essential. Volcanism, earthquakes, and many of the Earth's surface features are a result of processes happening within the Earth.

Much of what we know regarding the Earth's interior is through indirect means, such as using seismic data to determine Earth's internal structure. Scientists discovered in the early 1900's that **seismic waves** generated by earthquakes could be used to help distinguish the properties of the Earth's internal layers. The velocity of these waves (called primary and secondary waves, or P and S waves) changes based on the density of the materials they travel through. As a result, seismic waves do not travel through the Earth in straight lines, but rather get reflected and refracted, which indicates that the Earth is not homogeneous throughout.

The Earth's interior consists of an inner and outer core, the mantle, and the crust. Located in the center of the Earth is the **inner core**, which is very dense and under incredible pressure, and is thought to be composed of an iron and nickel alloy. It is solid, and surrounded by a region of liquid iron and nickel called the **outer core**. The outer core is thought to be responsible for the generation of the Earth's **magnetic field**. A very large portion of the Earth's volume is in the **mantle**, which surrounds the core. This layer is less dense than the core, and consists of a solid that can behave in a plastic (deformable) manner. The thin outer layer of the Earth is the **crust**. The two types, continental and oceanic crust, vary from each other in thickness, composition, and density.

2.1.1 Learning Outcomes

After completing this chapter, you should be able to:
- Determine the different layers of the Earth and the distinguishing properties of each layer

- Understand how seismic waves behave within the different layers of the Earth
- Understand how seismic tomography has been used to gain a better understanding of the Earth's interior
- Understand the Earth's magnetic field and how it changes over time
- Learn how to use the program Google Earth for geological applications

2.1.2 Key Terms

- Crust
- Inner Core
- Magnetic Field
- Mantle
- Outer Core
- Polar Wandering Curves
- Seismic Tomography
- Seismic Waves

2.2 INTERIOR OF THE EARTH

The study of seismic waves and how they travel through the Earth has been very useful in helping to determine the changes in density and composition within the Earth and in locating the boundaries between the inner core, outer core, mantle, and crust. **Seismic waves** are energy waves generated during earthquakes; two types known as P and S waves propagate through the Earth as wave fronts from their place of origin. P-waves are compressional waves that move back and forth like an accordion, while S-waves are shear waves that move material in a direction perpendicular to the direction of travel, much like snapping a rope. The velocity of both of these waves increases as the density of the materials they are traveling through increases. Because most liquids are less dense than their solid counterparts, and seismic velocity is dependent on density, then seismic waves will be affected by the presence of any liquid phase in the Earth's interior. In fact, S waves are not able to travel through liquids at all, as the side to side motion of S waves can't be maintained in fluids; because of this, we know that the outer core is liquid.

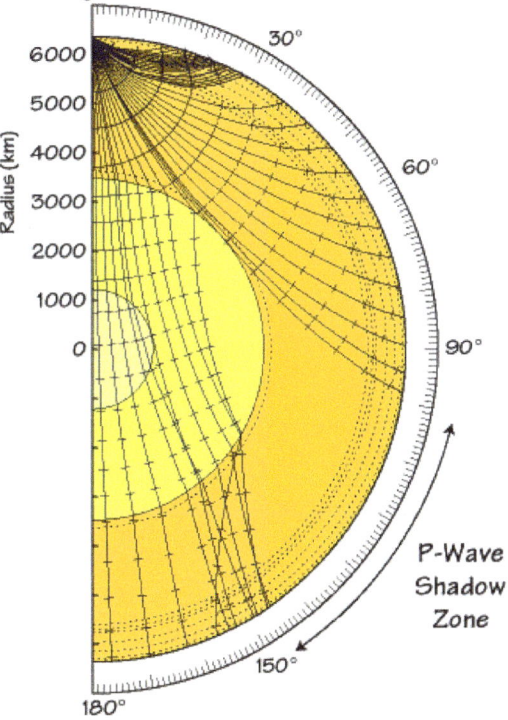

Figure 2.1 | A depiction of the P-wave shadow zone.
Author: USGS
Source: Wikimedia Commons
License: Public Domain

If the Earth was completely homogeneous, the P and S waves would flow in straight lines. They don't behave this way, however. As the waves travel through materials of different densities, they are refracted, or bent, as their direction and velocity alter. Sometimes these refractions can result in shadow zones, which are areas along the Earth where no seismic waves are detected. Due to the presence of a liquid outer core, a P-wave shadow zone exists from 103°-143° (see Figure 2.1) from the earthquake origination point (focus), and a larger S-wave shadow zone exists in areas greater than 103° from the earthquake focus.

Based on the way that the Earth travels through space, we know that the average density of the Earth is 5.52 g/cm³. When rocks at the Earth's surface are analyzed, we find that most crustal rocks have densities in the range of 2.5-3 g/cm³, which is lower than the Earth's average. This means that there must be denser material inside the Earth to arrive at that higher average density; in fact, the core region of the Earth is estimated to have a density of 9-13 g/cm³. The composition of the Earth's layers also changes with depth. The bulk Earth composition is mostly made up of iron (~32%), oxygen (~30%), silicon (~16%), and magnesium (15%). If you examine rocks at the Earth's surface, however, you will find that oxygen is the most abundant element by far (~47%), followed by silicon (~28%) and aluminum (~8%), and lesser amounts of iron, calcium, sodium, potassium, and magnesium. Minerals made from silicon and oxygen are very important and are called silicates. So, if iron is present in lower numbers in the crustal rocks, where has that iron gone? Much of it can be found in the core of the Earth, which accounts for the major increase in density there. Review Table 2.1 below for general information about each layer of the Earth, and note how much thicker continental crust is compared to oceanic crust. Examine Figure 2.2 for a depiction of the layers of the Earth.

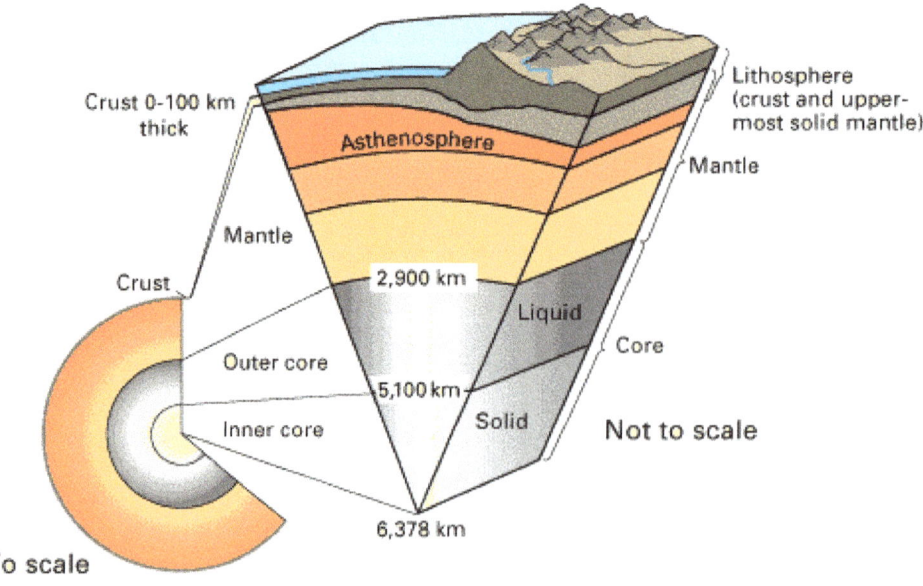

Figure 2.2 | A depiction of the inner layers of the Earth. Note that the image to the bottom left is to scale, while the image to the right is not.
Author: USGS
Source: USGS
License: Public Domain

Note that this figure includes the terms lithosphere and asthenosphere. The lithosphere is the outer, rigid part of the Earth made up of the upper mantle, oceanic crust, and continental crust. The asthenosphere is just beneath the lithosphere and, rather than being rigid, behaves plastically and flows.

Table 2.1

Earth's Layer	Density (g/cm3)	Thickness (km)	Composition
Continental Crust	~2.7-2.9	~20-70	Felsic rocks
Oceanic Crust	~3.0	~8-10	Mafic rocks
Mantle	~3.4-5.6	~2,885	Ultramafic rocks
Outer Core	~9.9-12.2	~2,200	Iron, some sulfur, nickel, oxygen, silicon
Inner Core	~12.8-13.1	~1,220	Iron, some sulfur and nickel

Relatively recent advances in imaging technology have been used to better understand the Earth's interior. **Seismic tomography** has been used to give a more detailed model of the Earth's interior. In CAT scans, x-rays are aimed at a person and rapidly rotated, generating cross-sectional images of the body. In a similar fashion, repeated scans of seismic waves are stacked to produce a three-dimensional image in seismic tomography. This technique has been used in many ways, from searching for petroleum near the Earth's surface to imaging the planet as a whole. Figure 2.3 depicts an image of the mantle created from seismic tomography.

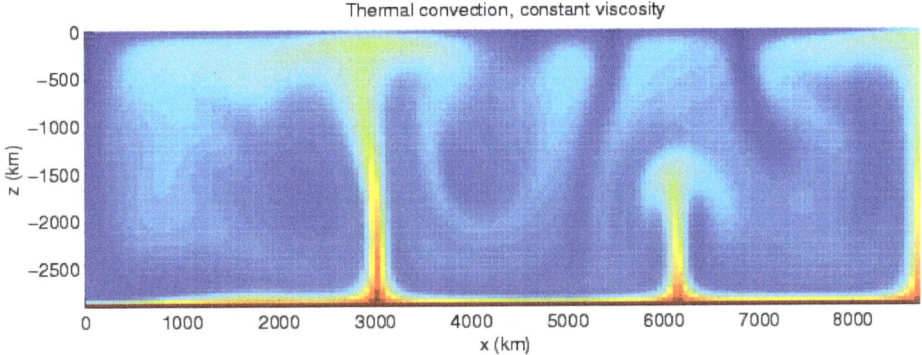

Figure 2.3 | A model of thermal convection in the mantle, created using seismic tomography. This model depicts areas of cool mantle material in blue and areas of warm mantle material in red. The thin red areas represent rising plumes.
Author: User "Harroschmeling"
Source: Wikimedia Commons
License: CC BY-SA 3.0

2.3 LAB EXERCISE

Part A – Interior of the Earth

The following graph (Figure 2.4) displays seismic velocities (in kilometers per second) of P and S waves with depth (measured in kilometers) inside the Earth. Examine the graph closely and answer the following questions.

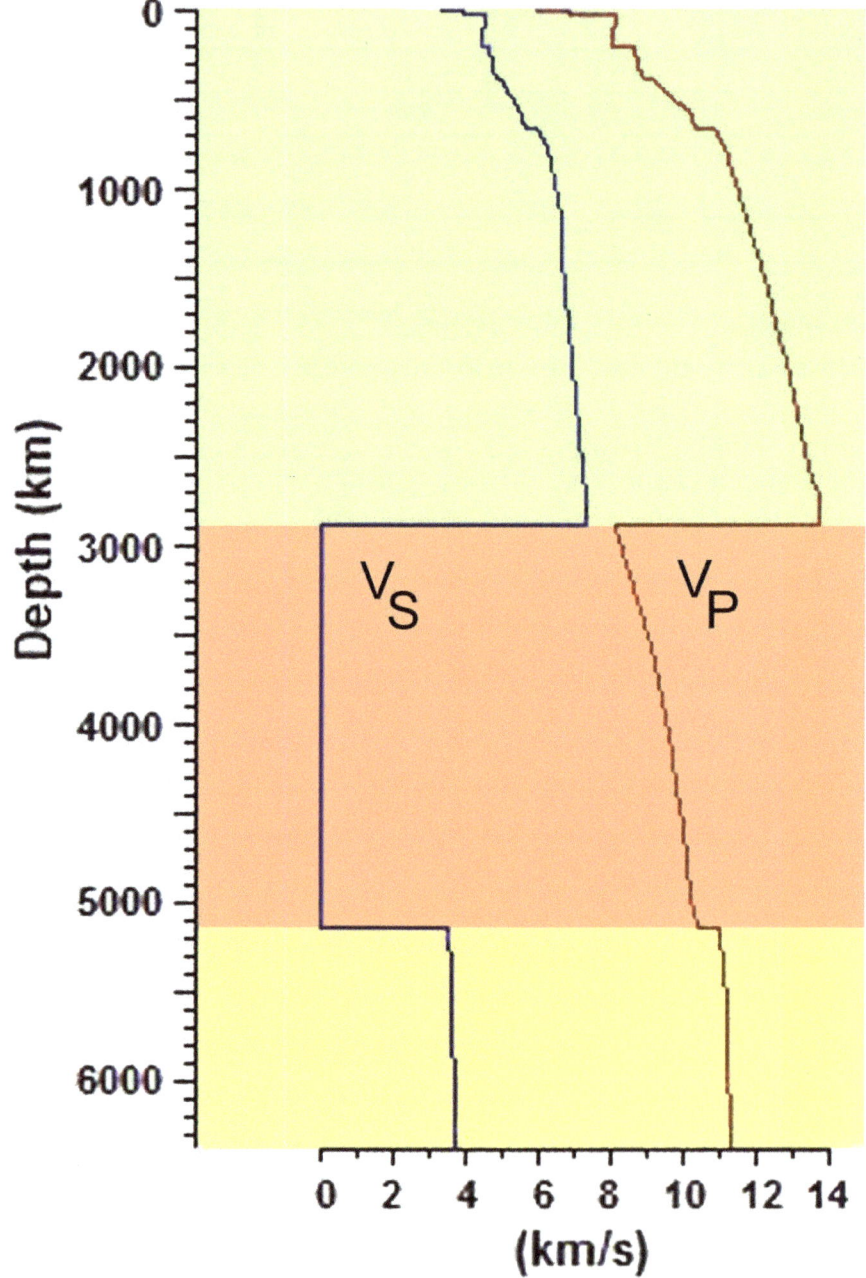

Figure 2.4 | Graph of seismic velocity with depth within the Earth's Interior.
Author: User "Actualist"
Source: Wikimedia Commons
License: CC BY-SA 3.0

1. Observe the velocities of the waves on the graph. Which one travels faster?

    a. P waves	b. S waves

2. Inspect the P wave velocities. Where do the P wave velocities abruptly change?

    a. ~20 km	b. ~2,900 km	c. ~5,100 km	d. All of the above

3. In which zones do the P wave velocities appear to be steadily increasing?

    a. ~20–2,900 km	b. ~2,900-5,100 km	c. ~5,100-6,400 km	d. Both a & b

4. Observe the S wave velocities. Where do the S wave velocities abruptly change?

    a. ~20 km	b. ~2,900 km	c. ~5,100 km	d. All of the above

5. At ~2,900 km, the S wave velocity falls to 0. Why?

    a. S waves can't travel through solids, and this depth is where the solid inner core exists.

    b. S waves can't travel through liquids, and this depth is where the liquid outer core exists.

    c. S waves can't travel through solids, and this depth is where the solid mantle exists.

    d. S waves entered the shadow zone.

Observe closely the changes in seismic wave velocity. You may add lines to your graph to denote the abrupt changes. Label each zone with the internal layers of the Earth and answer the following questions.

6. The zone from ~0-20 km represents the Earth's:

    a. crust	b. mantle	c. inner core	d. outer core

7. The zone from ~20-2,900 km represents the Earth's:

    a. crust	b. mantle	c. inner core	d. outer core

8. The zone from ~2,900-5,100 km represents the Earth's:

    a. crust	b. mantle	c. inner core	d. outer core

9. The zone from ~5,100-6,400 km represents the Earth's:

    a. crust	b. mantle	c. inner core	d. outer core

2.4 EARTH'S MAGNETIC FIELD

The thermal and compositional currents moving within the liquid outer core, coupled with the Earth's rotation, produce electrical currents that are responsible for the Earth's magnetic field. The shape of the magnetic field is similar to that of a large bar magnet. The ends of the magnet are close to, but not exactly at, the geographic poles on Earth. The north arrow on a compass, therefore, does not point to geographic north, but rather to magnetic north. The magnetic field plays a role in making the Earth hospitable to humans. Solar wind sends hot gases called plasma to Earth, and the magnetic field deflects most of this plasma. Without the work of the magnetic field, these damaging rays would harm life on the planet. As the solar wind approaches the Earth, the side of the Earth's magnetic field closest to the Sun gets pushed in, while the magnetic field on the opposite side away from the sun stretches out (Figure 2.5). You may have heard of the Aurora Borealis or "Northern Lights." Solar storms can create disturbances within the magnetic field, producing these magnificent light displays (Figure 2.6).

Figure 2.5 | Solar wind interacting with the Earth's magnetic field.
Author: NASA
Source: Wikimedia Commons
License: Public Domain

The magnetic field changes constantly and has experienced numerous reversals of polarity within the past, although these reversals are not well understood. Study of past reversals relies on paleomagnetism, the record of remnant magnetism preserved within certain rock types. Iron-bearing minerals that form from lava can align with the Earth's magnetic field and thus provide a record of the magnetic field in the Earth's past. However, this preserved magnetism could be lost if the mineral in the rocks has not been heated above a temperature known as the Curie point (a temperature above which minerals lose their magnetism). Essentially, the iron atoms "lock" into position, pointing to the magnetic pole. This records the alignment of the magnetic field at that time (we currently are in a normal polarity, in which north on a compass arrow aligns closely with geographic north, or the North Pole). If the magnetic field was stationary, all of the magnetic

Figure 2.6 | An example of the beautiful Aurora Borealis, light displays created by solar storm interaction with the Earth's magnetic field.
Author: User "Soerfm"
Source: Wikimedia Commons
License: CC BY-SA 3.0

minerals would point in the same direction. This is not the case, however. Reversals occur rather frequently on the geologic time scale.

Not only do magnetic poles reverse over geologic time, they also wander. Paleomagnetic data show that the magnetic poles move systematically, wandering across the globe. **Polar wandering curves** have been created to display the migration of the poles across the Earth's surface over time. Apparent polar wander refers to the perceived movement of the Earth's paleomagnetic poles relative to a continent (the continent remains fixed) (Figure 2.7). As you will learn in the Plate Tectonics chapter, polar wandering curves provide excellent evidence of the theory that the plates move, as curves for different continents do not agree on the magnetic pole locations. They all converge on the current pole location at present day, however.

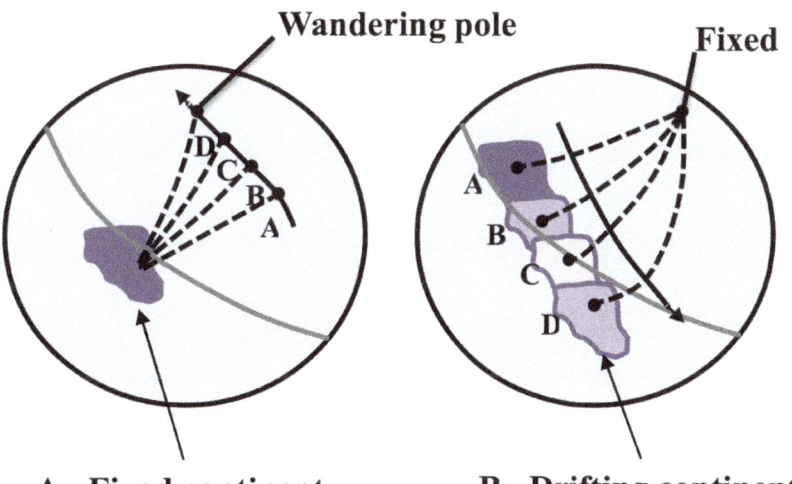

A. Fixed continent   **B. Drifting continent**

Figure 2.7 | If continents are fixed, as in A in the figure, the pole must be wandering. However, the pole is relatively fixed around the pole (with some movement), so the drifting continent (B) is the correct model.
Author: Randa Harris
Source: Original Work
License: CC BY-SA 3.0

2.5 LAB EXERCISE

Part B – Earth's Magnetic Field

You will use the polar map given (Figure 2.8) to plot the changing locations of the magnetic pole over time. To view the polar map, imagine that you are above the North Pole looking down on it. 90° N latitude is directly in the center of the map, and the lines of latitude, measured in 2 degree increments, spread out in circles from the center. Values of longitude are also given, and are represented as lines that radiate out from the center in increments of 30 degrees. A scale bar, in kilometers, is provided. To familiarize yourself with the map, first practice plotting some of the locations given. The first two, A and B, have been done for you. You will be plotting actual locations of the magnetic North Pole over time ranging from 1400-1900 AD onto Figure 2.8. The table below includes the year and latitude and longitude (as degrees from 0 to 360) for each location. Once you have completed this, answer the questions that follow. Remember that 1,000 meters = 1 kilometer.

Point	Year	Latitude	Longitude
A	—	87.7	160.3
B	—	85.6	6.6
C	1400	84.8	228.3
D	1500	86.3	301.5
E	1600	85.6	316.7
F	1700	81.1	307.1
G	1800	81.1	297.1
H	1900	82.3	288.2

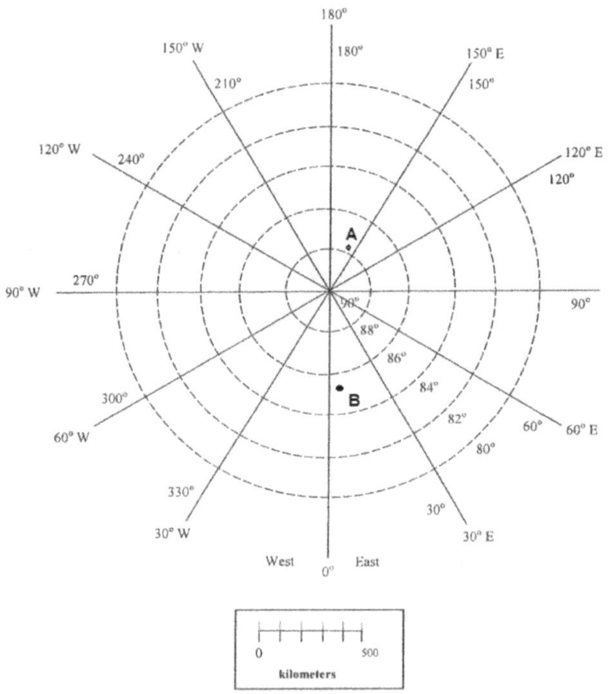

10. Measure the distance, using the bar scale, between the pole in 1400 and 1500 (locations C and D). How far did the pole move?

    a. ~50 km   b. ~150 km

    c. ~600 km   d. ~1,000 km

11. How far (in km) did the pole move in one year during this time period?

    a. 0.5 km   b. 1.5 km   c. 6 km   d. 10 km

12. How far did the pole move in meters in one year during this time period?

    a. 1,500 m   b. 500 m   c. 10,000 m   d. 6,000 m

Figure 2.8 | Add the magnetic pole locations to this polar map and answer the questions below.
Author: NASA
Source: NASA
License: Public Domain

13. Approximately how far did the pole move per day?

    a. <1 m   b. ~16 m   c. ~20 m   d. ~12 m

14. Measure the distance, using the bar scale, between the pole in 1400 and 1700 (locations C and F). How far did the pole move?

    a. ~50 km   b. ~150 km   c. ~600 km   d. ~1,000 km

15. Do pole movements tend to be steady every 100 years or variable?

    a. movements are steady   b. movements are variable

2.6 GOOGLE EARTH INTRODUCTION

Google Earth is a great tool to visualize and explore many of the geologic features that we will discuss in this class. This program is free and easy to use. This first Google Earth assignment will focus on familiarizing you with the program and some of the tools that we will use in later labs. Note that the optimal way to view geology is to go outside. Since that is not an option for an online class, the next best thing is using Google Earth. This is a practical and useful program that has many applications.

NOTICE: Google Earth updates versions periodically. If this occurs the instructions in the labs for this chapter may refer to the older version of the program. If you think this has occurred please let your instructor know.

If Google Earth is not already installed on the computer you are using, **then** please do the following:

1. Go to http://earth.google.com
2. Click on the Download Google Earth tab at the top of the page, **review** the Privacy Policy, and click Agree and Download to download the latest version.
3. Save the file to **your** desktop, open it and follow the instructions to install.
4. Open Google Earth.

Before we begin the assignment, let's first familiarize ourselves with Google Earth. Read each step and spend a few minutes trying things out, which will make things easier later. Also note that the Mac and PC versions of Google Earth are a little different.

You will use this program extensively throughout the course. Take the time to learn how to navigate it now.

Step 1 – Navigation

Watch each of the tutorial videos at: http://www.google.com/earth/learn/beginner.html#navigation. It is important that you dedicate time to review these virtual resources to help you better **understand Google Earth and its capabilities**. These will be key to mastering the tool. Navigating in Google Earth can be done in two ways:

First, you can use the Search panel in the upper left **hand side of the screen**. Just type in a location, address, or coordinate and it will zoom into the position (give it a try **now**).

The second approach is that you can also navigate manually:

- To move position you can left click with the mouse and drag the map or click on the hand icon in the upper right corner.
- You can zoom in and out using the mouse wheel, by right clicking and dragging the mouse up or down, or by sliding the lowest bar in the upper right corner.
- Click and hold the mouse wheel in order to rotate the map (left and right) or tilt the scenery (up and down). This can also be done using the arrows surrounding the eye icon in the upper right corner.

Along the bottom of the image it gives several important pieces of information:

1. Latitude and Longitude
2. Elevation in reference to sea level
3. Eye altitude, which indicates how zoomed in or out you are.

For example, let's check out Niagara Falls. In the Search panel in the upper left, type in Niagara Falls, NY. To better tell where the Falls are, you want to zoom out a little bit. Notice your eye altitude along the bottom right. Use the minus button for the zoom until you are at ~10,000' eye altitude. Check your latitude (~43°04'39"N – read as 43 degrees, 4 minutes, and 57 seconds North) and longitude (~79°04'28"W). It will move as you move your cursor across the screen, as will your elevation. If you want to see a picture of the Falls, just click on one of the many photo icons to see one. Note that the river is headed in a general northerly route – you can zoom in close to see the actual Falls.

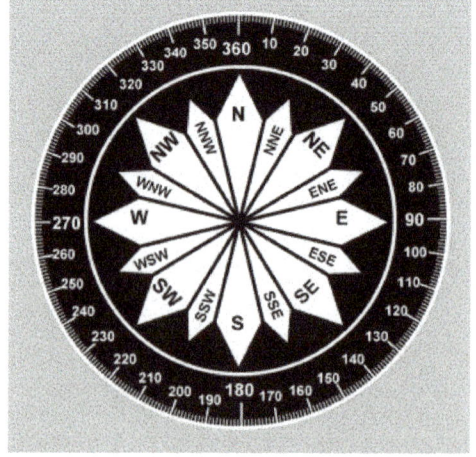

Figure 2.9 | A compass displaying the degrees.
Author: FAA
Source: Wikimedia Commons
License: Public Domain

Also important to understand with navigation is the concept of bearing. A bearing is the compass direction as measured between two points. It can be expressed as an azimuth bearing in degrees between 0 and 360, as along a circle. 0 and 360 degrees would be north, 90 degrees would be east, south would be 180 degrees, and west would be 270 degrees (Figure 2.9).

Bearing can be useful in determining how to get from point A to point B. To measure a bearing in Google Earth, follow this procedure. You want the bearing from point A to point B. Locate point A. Select the Ruler tool. Measure from A to B. In the last line in the Ruler tool box, read the Heading in degrees – that is measuring bearing (Tip: when point A is very far from point B, it can be helpful to add a placemark. Once at point A, choose the item in the menu bar right above the image that looks like a yellow pushpin. That is a placemark. Place it at point A. Also add a placemark at point B. Zoom out far enough that you can see both placemarks, then use the ruler tool to measure the heading between them. This works well for measuring headings across the entire United States). Let's practice getting a bearing again with Niagara Falls. For this example, point A will be the Scotiabank Convention Center located at 43°04'34.08"N and 79°05'05.98"W. Point B will be the name label for Goat Island (located across the river, just before the Falls begin). The bearing from point A to point B is ~74 degrees.

Step 2 – Measuring

In order to examine features, we will need to be able to measure them, which is easily managed in Google Earth. Measuring is done using the Ruler Tool, which can be accessed either by clicking the ruler icon in the toolbar above the image or by selecting from the menu across the top Tools, then Ruler.

There are two options with the ruler tool, line and path. The line option (which is the default option) gives the distance and direction between two points; notice the pull down menu that gives 11 different options for units of measurements. To make a measurement, after you have selected the ruler tool, you simply click on two different points. The path option gives the distance for a set of two or more points giving the ability to measure a distance that isn't a straight line.

NOTE: When measuring features you want to use the Map Length – using the ground length can lead to an incorrect answer.

Gradient will often be measured for this and future labs. Gradient is similar to slope which indicates how steep or flat an area is. It is calculated as the difference in elevation divided by the horizontal distance. When calculating gradient, maintain the same units in the numerator and denominator.

Gradient = change in elevation/horizontal distance

Let's practice again at Niagara Falls. First move the image slightly higher to fully see the start of the white water, just before the Falls begin (do this by left clicking the mouse and using the hand to move the image). With the eye altitude still at ~10,000', let's measure the distance across the river right at the start of the whitewater before the Falls (where the whitewater stops). First click your ruler icon, then select a point on one side of the river, then move your mouse straight across to the other side. In feet, this should measure ~4,800 feet (don't stop at the island – measure all the way to the other river bank). Using the pull-down menu, you can change the feet to miles, and the result should be ~0.9 miles. Now let's practice gradient across the actual Falls. Position your cursor over the actual Falls, and zoom in to an eye altitude of ~1,000 feet. Hold your cursor over the top of the Falls and record the elevation (remember, this is located along the bottom bar). Now move your cursor to the bottom of the Falls and record the elevation. The change in elevation (highest-lowest) will be your numerator. Use the Ruler tool to measure the distance between the two places – this will be your horizontal distance (the denominator). There will be variation in this answer depending on your exact spot along the Falls, but results should be similar to this:

$$\text{Gradient} = (500' - 325') / 75' = 175'/75' = 2.3$$

Step 3 – Changing the Options

For a few tasks it will be important to change some of the default settings on Google Earth in order to see a feature better or make your work easier. These changes can all be made by going to Tools in the menu bar across the top, then Options in the PC version (for the MAC, go to Google Earth, then Preferences).

1. Changing the unit for Elevation – From the 3D view tab, in the middle of the box there is a section entitled "Units of Measurement" that you can change between metric and English units.

2. Exaggerating Features – Since differences in elevations are much smaller than geographic distances it is sometimes hard to see features. To exaggerate features (that is, make a mountain look taller than it actually is in order to see it better), click on the 3D view tab, in the lower left side at the "Terrain" section of the box, look for "Elevation Exaggeration (also scales 3D buildings and trees)". If you want to exaggerate a feature increase this value up to 3. To view the area without any exaggeration, return the value to the default of 1.

2.7 LAB EXERCISE

Part D – Google Earth

Answer the following questions using the skills discussed above. In order to prepare you for examining geologic features in Google Earth, let's first examine a more familiar area, Washington, D.C.

16. Search for the "Washington Monument" and zoom into ~3,000 feet eye altitude. What is the latitude of the monument?

   a. 38° 53'N   b. 38° 53'S   c. 77° 02'E   d. 77° 02'W

17. What is the longitude of the monument?

   a. 38° 53'N   b. 38° 53'S   c. 77° 02'E   d. 77° 02'W

18. Zoom into an eye altitude of ~500 feet. Locate the base of the monument (the base is square and is located in the center of the circle). What is the elevation of the base of the monument in feet?

   a. 15 feet   b. 40 feet   c. 65 feet   d. 80 feet

19. What is the elevation of the base of the monument in meters?

   a. 40 meters   b. 25 meters   c. 12 meters   d. 7 meters

20. How big is the base (area) of the Washington Monument in square feet? (Hint – make sure the 3D Buildings is selected in the Layers box, located in the lower left of the screen. Also, area is measured in square feet, so make sure you are multiplying two measurements.)

   a. ~1040 feet^2 b. ~3100 feet^2 c. ~4700 feet^2 d. ~6030 feet^2

21. If you were standing at the Washington Monument, what direction would you need to walk to go to the United States Capitol Building?

   a. North   b. East   c. South   d. West

22. The direction system is useful, but imprecise. It is better to use a bearing. If you were standing at the Washington Monument, what bearing would you need to walk to go to the Smithsonian National Museum of Natural History, which houses many important geological specimens (Hint: Many of the buildings on the National Mall are part of the Smithsonian – make sure you get the correct building!)?

   a. 30°   b. 75°   c. 90°   d. 260°

23. If you decided instead to walk from the Washington Monument to the White House, how far would you have to walk, in miles (assume you could walk right to the entrance of the building)?

   a. 0.56 miles   b. 0.78 miles   c. 1.20 miles   d. 1.87 miles

24. How far is this distance in kilometers?

    a. 0.5 kilometers   b. 0.9 kilometers   c. 1.4 kilometers   d. 1.9 kilometers

25. Overall, would you be walking uphill or downhill?

    a. No change in elevation   b. Slightly uphill   c. Slightly downhill

26. How much does the elevation change in feet?

    a. 0 feet   b. 10 feet   c. 20 feet   d. 30 feet

27. What is the gradient of your walk?

    a. ~0.003   b. ~0.05   c. ~0.5   d. ~1

28. You decide to start walking from the Washington Monument to the White House, but as soon as you start a Park Ranger yells at you for walking on the grass. How far would the walk be (in miles) if you stayed on the sidewalks? There are many possible routes – try to take one of the shortest routes possible.

    a. 0.5 miles   b. 0.8 miles   c. 2 miles   d. 5 miles

29. Now, let's look at a geologic feature. Put 36 05 35.38 N 113 14 43.70 W into the search bar and zoom out to an eye altitude of ~25,000 feet. This is the Grand Canyon. If you started out in Washington, D.C., what bearing would you need to travel in to go to the Grand Canyon?

    a. 210°   b. 245°   c. 275°   d. 315°

30. The Grand Canyon is an extraordinarily steep area. If you go from the river at the bottom of the Canyon (36 05 35.38 N 113 14 43.70 W) straight north up the canyon rim (the area that flattens out above the red layers of rock – 36 06 10.12 N 113 14 48.88 W), what is the gradient of the Grand Canyon? (**Hint:** You can tilt the image, which will make the flat rim easier to see, by either pressing down on the middle mouse wheel and moving the mouse forward and backward or by clicking on the arrows above and below the eye in the upper right hand portion of the window.)

    a. 0.214   b. 0.673   c. 0.976   d. 1.245

INTRODUCTORY GEOLOGY   EARTH'S INTERIOR

2.8 STUDENT RESPONSES

1. Observe the velocities of the waves on the graph. Which one travels faster?

    a. P waves   b. S waves

2. Inspect the P wave velocities. Where do the P wave velocities abruptly change?

    a. ~20 km   b. ~2,900 km   c. ~5,100 km   d. All of the above

3. In which zones do the P wave velocities appear to be steadily increasing?

    a. ~20–2,900 km   b. ~2,900-5,100 km   c. ~5,100-6,400 km   d. Both a & b

4. Observe the S wave velocities. Where do the S wave velocities abruptly change?

    a. ~20 km   b. ~2,900 km   c. ~5,100 km   d. All of the above

5. At ~2,900 km, the S wave velocity falls to 0. Why?

    a. S waves can't travel through solids, and this depth is where the solid inner core exists.

    b. S waves can't travel through liquids, and this depth is where the liquid outer core exists.

    c. S waves can't travel through solids, and this depth is where the solid mantle exists.

    d. S waves entered the shadow zone.

6. The zone from ~0-20 km represents the Earth's:

    a. crust   b. mantle   c. inner core   d. outer core

7. The zone from ~20-2,900 km represents the Earth's:

    a. crust   b. mantle   c. inner core   d. outer core

8. The zone from ~2,900-5,100 km represents the Earth's:

    a. crust   b. mantle   c. inner core   d. outer core

9. The zone from ~5,100-6,400 km represents the Earth's:

    a. crust   b. mantle   c. inner core   d. outer core

INTRODUCTORY GEOLOGY   EARTH'S INTERIOR

10. Measure the distance, using the bar scale, between the pole in 1400 and 1500 (locations C and D). How far did the pole move?

    a. ~50 km   b. ~150 km   c. ~600 km   d. ~1,000 km

11. How far (in km) did the pole move in one year during this time period?

    a. 0.5 km   b. 1.5 km   c. 6 km   d. 10 km

12. How far did the pole move in meters in one year during this time period?

    a. 1,500 m   b. 500 m   c. 10,000 m   d. 6,000 m

13. Approximately how far did the pole move per day?

    a. <1 m   b. ~16 m   c. ~20 m   d. ~12 m

14. Measure the distance, using the bar scale, between the pole in 1400 and 1700 (locations C and F). How far did the pole move?

    a. ~50 km   b. ~150 km   c. ~600 km   d. ~1,000 km

15. Do pole movements tend to be steady every 100 years or variable?

    a. movements are steady   b. movements are variable

16. Search for the "Washington Monument" and zoom into ~3,000 feet eye altitude. What is the latitude of the monument?

    a. 38 53'N   b. 38 53'S   c. 77 02'E   d. 77 02'W

17. What is the longitude of the monument?

    a. 38 53'N   b. 38 53'S   c. 77 02'E   d. 77 02'W

18. Zoom into an eye altitude of ~500 feet. Locate the base of the monument (the base is square and is located in the center of the circle). What is the elevation of the base of the monument in feet?

    a. 15 feet   b. 40 feet   c. 65 feet   d. 80 feet

19. What is the elevation of the base of the monument in meters?

    a. 40 meters   b. 25 meters   c. 12 meters   d. 7 meters

INTRODUCTORY GEOLOGY   EARTH'S INTERIOR

20. How big is the base (area) of the Washington Monument in square feet? (Hint – make sure the 3D Buildings is selected in the Layers box, located in the lower left of the screen. Also, area is measured in square feet, so make sure you are multiplying two measurements.)

    a. ~1040 feet   b. ~3100 feet   c. ~4700 feet   d. ~6030 feet

21. If you were standing at the Washington Monument, what direction would you need to walk to go to the United States Capitol Building?

    a. North   b. East   c. South   d. West

22. The direction system is useful, but imprecise. It is better to use a bearing. If you were standing at the Washington Monument, what bearing would you need to walk to go to the Smithsonian National Museum of Natural History, which houses many important geological specimens (Hint: Many of the buildings on the National Mall are part of the Smithsonian – make sure you get the correct building!)?

    a. 30   b. 75   c. 90   d. 260

23. If you decided instead to walk from the Washington Monument to the White House, how far would you have to walk, in miles (assume you could walk right to the entrance of the building)?

    a. 0.56 miles   b. 0.78 miles   c. 1.20 miles   d. 1.87 miles

24. How far is this distance in kilometers?

    a. 0.5 kilometers   b. 0.9 kilometers   c. 1.4 kilometers   d. 1.9 kilometers

25. Overall, would you be walking uphill or downhill?

    a. No change in elevation   b. Slightly uphill   c. Slightly downhill

26. How much does the elevation change in feet?

    a. 0 feet   b. 10 feet   c. 20 feet   d. 30 feet

27. What is the gradient of your walk?

    a. ~0.003   b. ~0.05   c. ~0.5   d. ~1

28. You decide to start walking from the Washington Monument to the White House, but as soon as you start a Park Ranger yells at you for walking on the grass. How far would the walk be (in miles) if you stayed on the sidewalks? There are many possible routes – try to take one of the shortest routes possible.

    a. 0.5 miles   b. 0.8 miles   c. 2 miles   d. 5 miles

29. Now, let's look at a geologic feature. Put 36 05 35.38 N 113 14 43.70 W into the search bar and zoom out to an eye altitude of ~25,000 feet. This is the Grand Canyon. If you started out in Washington, D.C., what bearing would you need to travel in to go to the Grand Canyon?

    a. 210   b. 245   c. 275   d. 315

30. The Grand Canyon is an extraordinarily steep area. If you go from the river at the bottom of the Canyon (36 05 35.38 N 113 14 43.70 W) straight north up the canyon rim (the area that flattens out above the red layers of rock – 36 06 10.12 N 113 14 48.88 W), what is the gradient of the Grand Canyon? (**Hint:** You can tilt the image, which will make the flat rim easier to see, by either pressing down on the middle mouse wheel and moving the mouse forward and backward or by clicking on the arrows above and below the eye in the upper right hand portion of the window.)

    a. 0.214   b. 0.673   c. 0.976   d. 1.245

3 Topographic Maps

Karen Tefend and Bradley Deline

3.1 INTRODUCTION

A topographic map is an extremely useful type of map that adds a third dimension (vertical) to an otherwise two dimensional map defined by the north, south, east, and west compass directions. This third dimension on a topographic map is represented by contour lines, which are imaginary lines drawn on a map that represent an elevation above average sea level (a.s.l.) or mean sea level (m.s.l). A map with such elevation lines will provide the map reader with detailed information regarding the shape of the Earth's surface. Knowledge of how to interpret a topographic map will allow a person to locate and identify features on the Earth's surface such as hills, valleys, depressions, steep cliffs and gentle slopes. In addition, the map reader will be able to identify areas that may be prone to geologic hazards such as landslides and flooding. Any person interested in purchasing property, landscaping, planning a hike or camping trip, or who needs to survey an area for construction of a road, dam, or building will want to first consult a topographic map.

3.1.1 Learning Outcomes

After completing this chapter, you should be able to:
- Recognize topographic patterns and geologic patterns
- Read and construct contour lines
- Determine gradients
- Read map scales and convert fractional scales
- Construct a topographic profile

3.1.2 Key Terms

- Bar Scale
- Benchmarks
- Contour Interval
- Contour Line
- Equator
- Fractional (Ratio) Scale

- Gradient
- Hachure Marks
- Index Contour
- Latitude
- Longitude
- Prime Meridian
- Relief
- Topographic Profile
- Verbal Scale
- Vertical Exaggeration

3.2 MAP ORIENTATION AND SCALE

All topographic maps produced by the U.S. Geological Survey (U.S.G.S) are oriented with north at the top of the map. Therefore if you locate a position on the map, and move towards the top of the map you are moving in a northerly direction, and if you are moving to the bottom of the map, you are moving towards the south. Any movement to the right will be towards the east, and a movement towards the left will be towards the west. These maps are oriented with their sides oriented parallel to lines of longitude, which are imaginary lines that circle the globe and are oriented so that they pass through the north and south geographic poles . Starting with the 0° longitude line (known as the Prime Meridian) that passes through the town of Greenwich, England, these lines increase up to 180° in both directions east and west of the Prime Meridian (Figure 3.1). It may help to visualize longitude lines if you think of an orange, which when peeled will show the sections of orange oriented like longitude lines that section the Earth. All longitude lines converge at the navel of the orange (or the geographic north and south poles of the Earth).

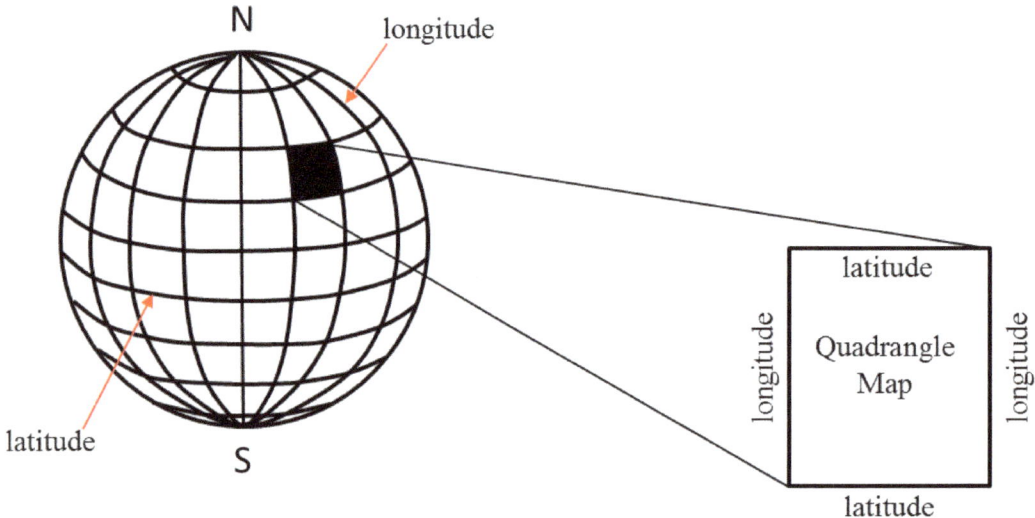

Figure 3.1 | Latitude and Longitude grid system of the Earth.
Author: Karen Tefend
Source: Original Work
License: CC BY-SA 3.0

The top edge and bottom edges of a topographic map are oriented so that they are parallel to lines of **latitude**, which are imaginary lines that circle the globe and are oriented at right angles to the Earth's axis. The 0° latitude line is the Earth's **Equator**; latitude lines increase up to 90° north or 90° south of the Equator, so that the North Pole has a latitude of 90°N, and the South Pole has a latitude of 90°S.

This grid system of latitude and longitude allows a position on the Earth to be uniquely defined, provided that the values for latitude are always identified by their position N or S of the Equator, and longitude is identified as E or W of the Prime Meridian. A degree of latitude or longitude represents a large distance on the Earth, therefore degrees have been further subdivided into minutes (a minute of distance is not the same as a minute of time!), and these minutes of distance are further subdivided into seconds. There are 60 minutes (60') of distance in 1° of latitude or longitude, and there are 60 seconds (60") of distance in one minute. An example of a precise location on the Earth's surface would be 33°34'22"N, 85°05'46"W (the author's location at the time of writing this chapter!), which is read as "33 degrees, 34 minutes, 22 seconds north latitude, and 85 degrees, 5 minutes, 46 seconds west longitude."

The latitude and longitude coordinates of each topographic map are found at the corners of the map. Often these maps represent an area of the Earth that is smaller than one degree of distance. For example a common topographic map will show only 7.5 minutes of distance (7' 30") for both latitude and longitude; in this case, the top and bottom edges of a map will represent a distance of 7.5 minutes of latitude, and the left and right edges of the map will represent a distance of 7.5 minutes of longitude. Because the degree of distance is unchanging on the map, a

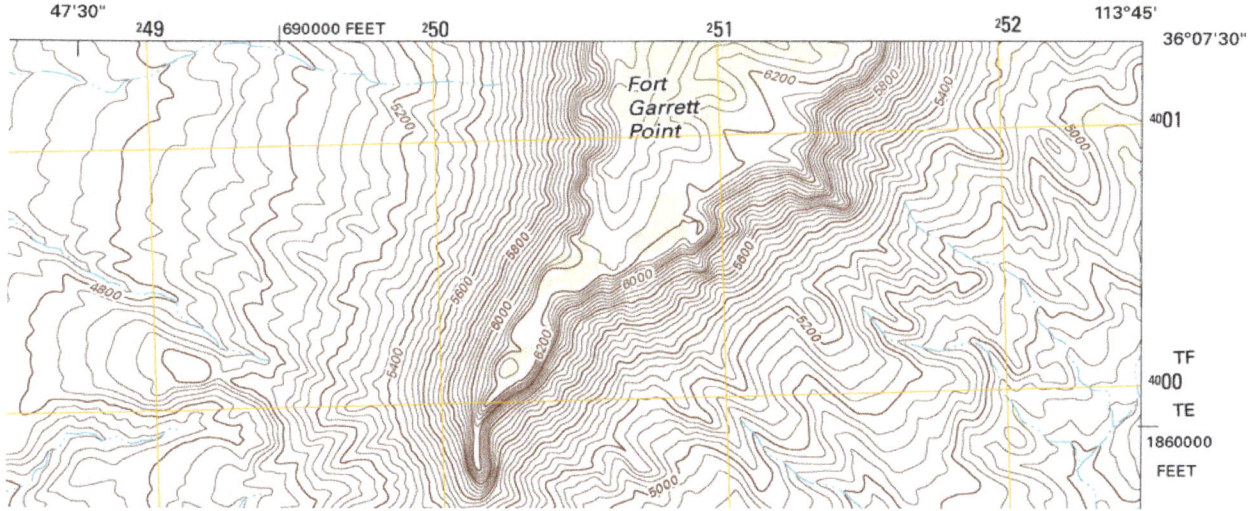

Figure 3.2 | This is the top right corner of a topographic map covering an area of 7' and 30" of latitude and longitude. Note that the coordinates are written out completely in the corner of the map, but shorthand notation is used for the longitude coordinate on the top edge of the map (47'30"). Since the number increased towards the left (by a distance of 2'30"), we know that these are west longitude numbers (meaning the longitude at that tick mark is 113°47'30" W), since only lines of longitude located west of the Prime Meridian increase with distances to the left.
Author: USGS
Source: USGS
License: Public Domain

shorthand notation of just the change in minutes and seconds is labelled at certain positions on the map's edge. Only at the corners of the map will the degrees be included for the latitude and longitude coordinates. See the example above, which shows the top right corner of a 7.5 minute map or 7.5 minute quadrangle (the shape of most topographic maps). Additional numbers other than latitude and longitude are also shown on the edges of the map; these are a different grid system and will not be explained here.

Remember that the top and bottom edges of a topographic map are oriented parallel to lines of latitude. In Figure 3.2, the top edge of the map is a latitude line, so the coordinates that are changing as you move along the top of the map (in an East-West direction), must be longitude coordinates, since the latitude should not change. The bottom edge of this map is not shown, but you can predict what the latitude on the bottom edge should be: this map is of a region in Arizona, and since Arizona is located north of the Equator the latitude lines should increase from the bottom of the map, towards the top of the map (as all latitude lines increase as you move north of the Equator). For a 7.5-minute map, the increase in latitude from the bottom corner to the top corner should be exactly 0°7'30"N. This gives the bottom corner of the map a latitude of 36°00'00"N.

All maps are a scaled down version of the region of the world that they depict; if this were not the case, then the map that a person must carry would be the exact same size as a city (if it is a city map) or the size of a state (if it is a state map). Imagine trying to carry around with you a map of the entire country! The word "scale" refers to the amount of reduction, and all maps provide a map scale to indicate how much the area on the map has been reduced. Map scales are provided so that a map reader can determine exactly how much distance is actually represented on their map, or to measure the distance between two points on a map, or even to calculate the gradient of a hill or river. The two commonly used map scales on a topographic map are the **bar scale** (or **graphical scale**) and the **fractional scale** (also known as a **ratio scale**).

In Figure 3.3 there are three bar scales; each bar is a graphical representation of distance on the map, and it is up to the map reader to decide if they want to measure distances in kilometers, meters, miles, or feet. To find the distance between

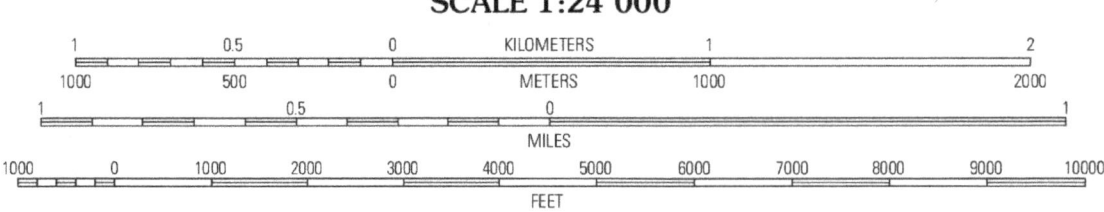

CONTOUR INTERVAL 40 FEET
NORTH AMERICAN VERTICAL DATUM OF 1988

Figure 3.3. Map scales typically located on the bottom of a topographic map. Note that the bar scales start at zero in the interior of each bar for kilometers, miles, and feet. The bar to the left of zero is further subdivided for more accurate distance determinations.
Author: USGS
Source: USGS
License: Public Domain

any two points on a map, a person could use a piece of paper to transfer the two points down to the bar scale and read the distance directly from the bar scale. Notice that each bar scale has the starting point (zero) within the interior of the scale, and not on the end of each scale (Figure 3.3).

The other type of map scale is the fractional scale; in the Figure 3.3 the fractional scale is 1:24,000. No units are reported as this ratio of 1 to 24,000 is valid for any unit of measure, provided that it is the same unit. For example, if using inches, then this map scale indicates that 1 inch on the map is actually covering 24,000 inches of ground (the distance between two locations in the real world). Or if using centimeters, then 1 centimeter on the map is actually covering 24,000 centimeters of ground. If our map was the same size as the area that it is representing (say for example, a map of the room you are currently sitting in), then the fractional scale of your map would be 1:1, and your map would be the exact same size as your room! This brings us again to the definition of a map, which is a scaled down version of a region that it is made to represent; maps that are greatly scaled down (greatly reduced) are called small scale maps even though they represent large sections of the Earth. For example, a 1:500,000 map will show a large section of the Earth, but small details are lost (such as building locations or small streets), whereas a 1:12,000 map is a large scale map even though it shows a much smaller region of the Earth's surface, but details can be seen (such as buildings, roads, and other landmarks). Placing your fingertip on the surface of a small scale map may cover an area of several miles, but placing your fingertip on a large scale map (such as 1:12,000) may cover only 1/10 of a mile.

One advantage to using a fractional or ratio scale is that any unit of measure can be used, and conversions are easy to make when needed. For example, for a 1:12,000 map 1 inch on the map is equal to 12,000 inches on the Earth's surface, and since there are 12 inches in 1 foot, we can also say that 1 inch on the map is equal to 1000 feet on the Earth's surface (convert the 12,000 inches to feet by multiplying by the conversion factor 1ft/12in). Simply verbalizing this scale by saying "on this map, 1 inch represents 1000 feet" is a third type of map scale, which for obvious reasons is called a **verbal scale**. Writing the phrase "1 inch equals 1000 feet" is a way of adding a verbal scale to your map.

Table 3.1 | Some useful conversion factors:

1 foot = 12 inches	1 meter = 3.28 feet
1 mile = 5280 feet	1 mile = 63,360 inches
1 kilometer = 1000 meters	1 kilometer = 0.62 miles

Remember that since conversion factors are equalities, such as 1 foot = 12 inches, then dividing one by the other (1ft/12in) gives you 1, and since multiplying anything by 1 does not change any value, all we really are doing is changing the units. Therefore the calculation 5.5 ft x (12in/1foot) will allow 5.5 ft to be expressed as inches, which in this case would be 66 inches.

3.3 LAB EXERCISE

Part A – Practice Questions (non-graded)

The following problems are for practice; answers to these questions are provided at the end of the chapter.

1. A 15 minute quadrangle map of a region within the United States with a longitude of 76°00'00" in the right corner of the map, will read what longitude in the left corner?

2. A 15 minute quadrangle map of a region within the United States with a latitude of 43°15'00" in the top corner of the map, will read what latitude in the bottom corner?

3. A fractional scale of 1:24,000 means that 1 inch = _____ feet.

4. A fractional scale of 1:24,000 means that 1 foot = _____ kilometers.

3.4 CONTOUR LINES

Contour lines allow a vertical dimension to be added to a map and represent elevations above sea level. Since each individual contour line connects points of equal elevation, then following that line in the real world means that you are staying at the same elevation while walking along that imaginary line. If you were to move off that line, you are either walking up or down in elevation. Imagine if you are on a small circular island in the ocean, and you walk from the shore up to 10 feet above the shoreline. If you were to walk around the island and stay exactly 10 feet above shore, you would be walking a contour line that represents 10ft of elevation above sea level. If you move off that line, you are either moving uphill or downhill. If you could walk uphill another 10ft and again stay at that elevation (now 20ft above sea level) while circling the island, then you are now walking the 20ft contour line. The vertical change in elevation between these two adjacent contour lines is called the **contour interval**, which in this case is 10 feet. If you were to transfer these imaginary lines onto a map, you would see three lines forming concentric circles that represent 0 ft (the seashore or sea level), 10ft and 20ft, and your map would look like a bull's eye pattern. Congratulations, you've made your first topographic map!

A topographic map will have contour lines shown as brown lines, and all maps will have a contour interval that is specific for that map. However, the elevations represented by the contour lines are not always labeled on each line (see Figure 3.2). Instead, every 5th contour line is labelled with an elevation, and is darkened; such a contour line is called an **index contour**. The use of index contours allows a map to be visually more appealing, especially when the contour lines are numerous and closely spaced to one another.

To determine the elevation of each contour line you must first know the contour interval for the map. By using the values of two adjacent index contours, one can easily calculate the contour interval between each line. For example, there are 4 contour lines between the 5200ft and 5400ft index contours (see Figure 3.2), which means that there are 4 contour lines separating the 200ft of elevation between the index contours into 5 sections. Dividing this 200ft elevation change between the index contours by 5 gives a contour interval of 40 ft (just as cutting a ruler in half creates two 6 inch pieces, or dividing the ruler into 3 evenly spaced cuts yields four 3 inch pieces). To verify this, locate the 5200ft index contour on the western side of the map in Figure 3.2, and increase the elevation by 40ft each time you cross a contour line while traveling east (to the right) towards the 5400ft contour line. Luckily there is no need to do this calculation to find the contour interval on a complete topographic map, as all topographic maps give the contour interval at the bottom of the map near the bar and fractional scales (see Figure 3.3). The contour interval must be obeyed for each contour line on a map; for example if the contour interval is 50 ft, then an example of possible contour lines on such a map could be 50ft, 100ft, 150ft, 200ft, etc.

You may be wondering why some contour lines are closely spaced in some areas of a map (such as the central portion of the map in Figure 3.2) and why they are farther apart in other areas of a map (such as the western part of the map in Figure 3.2). Imagine yourself again on the circular island in the ocean, and you are standing 10ft above sea level (on the 10ft contour line). If you want to walk up the hill to reach the 20ft elevation, how far did you have to walk? It depends on how steep the hill is; if it is a gentle slope you may have to walk a long time before you reach a higher elevation of 20ft. On a topographic map, the contour lines for this hill would be spaced far apart. However, if the hill's slope is very steep, you do not need to walk as far up the hill to reach a 20ft elevation, and the contour lines representing such a steep slope will be closely spaced on a topographic map. Recall that a slope (**gradient**) is the change in elevation divided by the distance; you can easily calculate the slope of your hill or any region on a topographic map if you know the change in elevation between two points, and if you know the distance between those same two points. Gradients are usually reported in feet per mile (ft/mile), but other units are also used. Remember to use the contour lines to determine the elevations, and the bar scale on your map to measure the distance.

In addition to contour lines, topographic maps will also have **benchmarks** (actual surveyed points) in various locations on your map. These surveyed points are exact elevations above sea level and are commonly used to mark the elevations of mountains, hilltops, road intersections and airport runways. These benchmarks are rarely located on a contour line and instead are usually identified by a black "x" or identified with the letters "BM" and with the elevations included in black numbers (as opposed to the brown numbers on index contours). Benchmark locations will normally be found in the area between contour lines; for example a benchmark of 236ft will be found somewhere between the 230ft and 240ft contour line (if the contour interval is 10ft), or between the 235ft and 240ft contour line (if the contour interval is 5ft).

In addition to obeying the set contour interval for a map, contour lines should never branch (split) or simply end inside of the mapped region. Instead these lines are continuous, although they can continue off the edge of the map. Contour lines also never touch or overlap, unless certain rare instances occur, such as if there is a vertical or overhanging cliff. In the case of a vertical cliff, the contour lines will appear to merge.

The entire third dimension (elevation) represented by the contour lines on a topographic map is called the **relief**, and is easily determined if you can find the highest and lowest contour line elevations and subtract the two values to determine the vertical relief represented in the map. The hardest part is finding these highest and lowest elevations on the map. Start by finding the highest index contour line and continue counting lines until you reach the lowest contour line. In Figure 3.2, the highest contour line is the line that runs through the letter "r" in Fort (of Fort Garrett Point). This same contour line circles back and goes through the letter "o" in Fort. The elevation of this line is 6360ft (based on the contour interval of 40ft). Recall that this is only a small portion of a 7.5 minute map (or quadrangle), and because of this, some of the index contours appear to be missing the identifying elevation numbers, but it is still easy to identify the index contours because all index contours are in bold (darkened lines). To find the lowest elevation on the map, find the lowest index contour line and continue counting lines in the downhill direction. An easy way to determine which way is downhill is to find a water feature on the map; water is colored blue on topographic maps, and flowing water such as a river or stream is a blue line. A dashed blue line such as in Figure 3.2 implies that the stream is dry part of the year (this is called an intermittent stream). Since water collects in low spots, such as a basin (where ponds, lakes, or oceans are found) or a valley (such as a stream or river valley), then the contour lines should represent decreasing elevation as you move towards a water feature on a map. Referring back to Figure 3.2, it is apparent that the highest portion of the map is the central portion where Fort Garrett Point is located, and that any point west, south and east of this is a downhill direction. Note all of the streams are flowing away from this Fort Garrett Point region. The lowest elevation will be a contour line that is crossing the stream just before leaving the map area. Close examination of the contour lines reveals that the lowest contour line is in the lower right corner of the map; the contour line that is crossing the stream in this portion of the map represents an elevation of 4560ft.

So for this small portion of the 7.5 minute map shown in Figure 3.2, the relief of the map region is 6360ft (highest contour) − 4560ft (lowest contour) = 1800ft.

An interesting feature regarding flowing water such as streams and rivers is that they erode the landscape and as a result the topography of the land is affected; we see this as a deflection of the contour lines on a map as they cross flowing water. Notice in Figure 3.2 that the contour lines form a "v" shape as they cross the water, and that the pointed end of this "v" is pointing in the upstream direction. We can use this to easily determine which way water is flowing without even paying attention to the elevation of the contour lines. Notice in Figure 3.2 that the contour lines that cross the streams are pointing toward the central hill (Fort Garrett Point), which means that the streams are all flowing away from the central portion of this map and towards the edges of the map region.

3.5 LAB EXERCISES

Part B - Practice Questions

For Questions 5 through 9, refer to Figure 3.4 below, which shows a hill, an intermittent stream, and two index contours (darkened contour lines). Assume the contour interval for this map is 5ft, and the index contour that is crossing the stream has an elevation of 70ft.

5. Which way is the stream flowing, to the North to the South?

6. What is the elevation of the highest contour on this portion of the map?

Figure 3.4 | Portion of the 7.5 minute Quadrangle of Bat Cave, Arizona
Author: USGS
Source: USGS
License: Public Domain

7. Calculate the relief of this map (Hint: Review the "Contour Lines" section in this chapter for assistance calculating relief).

8. Calculate the gradient of the stream between the highest and lowest contour lines that you can see cross the stream. These two contour lines are 2 miles apart.

9. The hill in the above diagram has a slightly steeper side on which side of the hill, the west or east side?

You have learned that the spacing between contour lines indicates the slope of the Earth's surface, and that the shape of the contour line as it crosses flowing water can indicate the slope direction. You have also learned that enclosed (circular) contour lines indicate a hill or mountain. However, sometimes there are circular depressions (for example, a sinkhole) found on the Earth's surface and these depressions may appear as hilltops on a topographic map unless a new convention is used. Contour lines with small perpendicular lines (called **hachure marks**) are used for such depressions on a topographic map. The contour interval for the map is still obeyed when contouring a depression. The only difference is that the hachure marks on the contour lines indicate that you should count down in elevation, not up, as you move towards the center of the hachured contour circles. However, if there is a depression at the top of a hill or mountain (for example, a volcanic crater), then the first contour line that is hachured must be the same elevation of the closest contour line that is not hachured. The reason for the repeat is that a person climbing the hill will reach the highest contour line, and walk a little higher still, before descending into the depression (crater), and will therefore encounter the same elevation line while descending (see Figure 3.5).

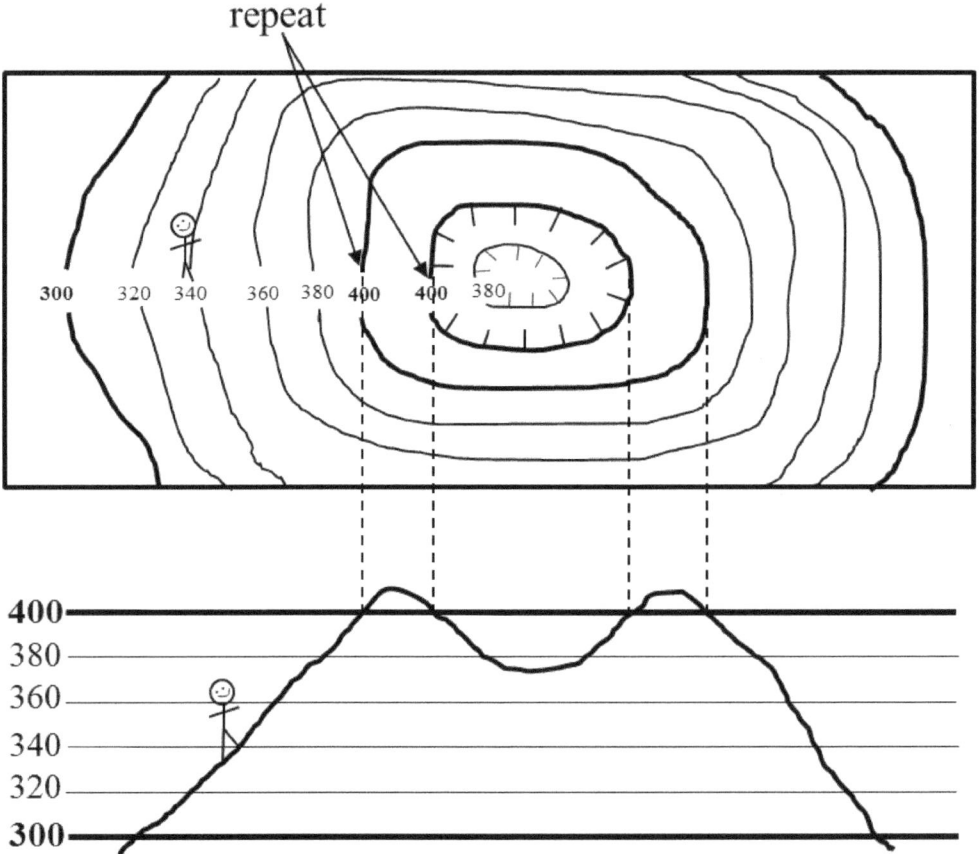

Figure 3.5 | Contours and hachured contours for a depression at the top of a hill. Notice that the first hachured depression is a repeat of the closest non-hachured contour line.
Author: Karen Tefend
Source: Original Work
License: CC BY-SA 3.0

3.6 DRAWING CONTOUR LINES AND TOPOGRAPHIC PROFILES

Constructing a topographic map by drawing in contours can be easily done if a person remembers the following rules regarding contour lines: 1) contour lines represent lines connecting points of equal elevation above sea level; 2) contour lines never cross, split or die off; 3) contour intervals must be obeyed, therefore the contour line elevations can only be multiples of the contour interval; and 4) contour lines make a "v" pattern as they cross streams and rivers, and the "v" always points towards the upstream direction.

As you draw a contour line on a map you will notice that the elevations on one side of your line will be lower elevations, and elevations on the other side of your line will be higher elevations. Once your contour lines are drawn, you will notice that you had to draw some lines closer together in some areas and wider apart in other areas, and that you may have even enclosed an area by drawing a contour line in a circular pattern. These circular patterns indicate hilltops, like in the diagram below (Figure 3.6). To illustrate what these hills look like in profile (or, how they would look if you saw them while standing on the ground and looking at them from a distance), you can draw what is known as a topographic profile. Essentially a topographic profile is a side image of a topographic map, but the image is only a representation of the area shown on the line on the topographic map (line A-B on Figure 3.6). To construct a profile, you need graph paper, a ruler and a pencil. You want to have the y-axis of the graph paper represent the elevations of the contour lines that intersect your drawn line (line A-B in this case). By using a ruler, you can transfer these elevation points from your topographic map straight down onto your graph paper such as shown in Figure 3.6. Be sure to only plot those elevations that are at the intersection of the contour line with line A-B. Once your points are plotted on the graph paper, you simply connect the dots. As a rule, hill tops will be slightly rounded to show a slight increase in elevation to represent the crest of the hill, but be careful not to draw the hill top

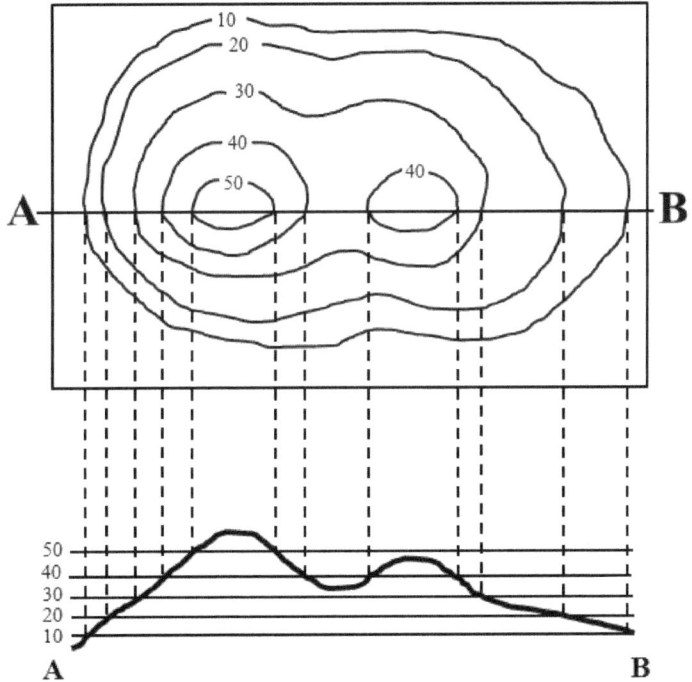

Figure 3.6 | Contour map and topographic profile of two hills and a valley between them.
Author: Karen Tefend
Source: Original Work
License: CC BY-SA 3.0

too high on your graph paper. For example, the first hill on the left has a top contour line of 50ft. Because there isn't a 60ft contour line on this hill top, we know that the hill's highest point (the crest) is some elevation between 50 and 60ft. When connecting the points on your graph paper in the area between the two hills in the Figure 3.6, you again want to round out the area to represent the base of your valley between the hills, but be careful not to make the valley floor too deep, as according to the topographic map the elevation is below 40ft, but not as low as 30ft.

If you examine the graph showing the topographic profile in Figure 3.6, can you image what would happen to your profile if we changed the spacing for elevations on the y-axis? When the vertical dimension of your graph is different from the horizontal dimension on your map, you may end up with a graph that shows a **vertical exaggeration**, and the features of the Earth represented by your topographic profile may be deformed such as in Figure 3.7.

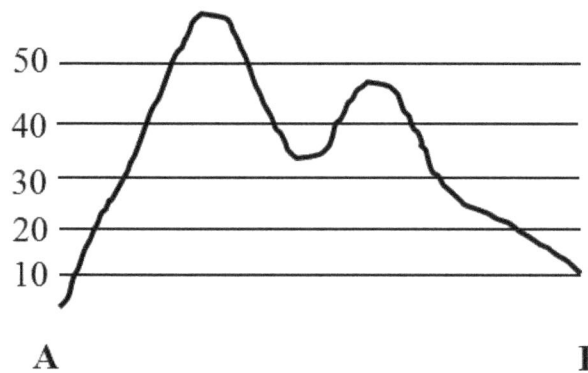

Figure 3.7 | A stretched profile (from Figure 3.6) to demonstrate what vertical exaggeration can do to an image. Notice how much steeper the slopes are in this image.
Author: Karen Tefend
Source: Original Work
License: CC BY-SA 3.0

Sometimes vertical exaggeration is desired, but in some cases you may not want it. To avoid having your profile distorted so that it accurately conveys what the surface of the Earth really looks like in profile, you will want both the vertical and horizontal scales to match. For example, if your map scale is 1 inch = 50 ft, then one inch on your graph's y-axis should only represent 50ft of elevation. If the topographic map in Figure 3.6 has a fractional scale of 1:12,000 then 1 inch is equal to 12,000 inches or 1000ft; this 1inch = 1000ft equivalency is for the horizontal scale. When we hold a ruler to the y-axis of the topographic profile in Figure 3.6, we see that 0.5 inches = 50ft, which means 1 inch = 100ft on the vertical scale. To calculate the vertical exaggeration in the topographic profile shown in Figure 3.6 we divide the horizontal scale by the vertical scale: (1000ft/1inch)/(100ft/inch) = 10. Therefore the topographic profile in Figure 3.6 represents a profile of the map surface (along the A-B line) that has been vertically exaggerated by 10 times (10X).

Answers to Practice Lab Exercises, Parts A and B

1. 76°15'00" W longitude
2. 43°00'00" N latitude
3. 2000 ft
4. 1 ft = 24,000 ft, and 24,000 ft x (1km/3280ft) = 7.32 km
5. South

6. 95 ft (this is the index contour at the top of the hill)
7. 95 ft – 65 ft = 30 ft
8. (80ft – 65 ft)/2miles = 7.5 ft/mile
9. East

3.7 LAB EXERCISE
Part C - Topographic Maps

This is a graded activity. The following pages must be printed and completed by the student, and <u>mailed</u> to the instructor in order to be scored. Alternatively, you may scan the activity and send it electronically to the instructor; unreadable scans will not be accepted, so be sure to send in legible work. Please remember to include your name in the provided blank.

INTRODUCTORY GEOLOGY   TOPOGRAPHIC MAPS

Name_____

3.8 TOPOGRAPHIC MAPS LAB ASSIGNMENT

Note: This lab is in color. Therefore, if you print it out in black and white please refer back to the electronic copy to avoid confusion.

This Lab Assignment is to be mailed to your Instructor at the contact address recorded in the Syllabus. Make sure that you use additional postage if needed. There is no online assessment for the Topographic Maps Lab.
<u>Complete the entire assignment and mail to your instructor postmarked by the assignment deadline</u>. You should make an extra copy to practice on and mail in a clean and neat version for grading. Make sure to include your name on every page and staple all of the pages together.

Please take advantage of all of the resources available to you. Be sure to read the corresponding lecture which contains directions to work out the solutions to the problems below. You should also review the instructional videos located in the unit content area within the course for additional assistance. Finally, check the Topographic Map Unit Discussion forum and the tutor talk area for additional resources and hints.

3.8.1 Topographic Maps Lab

NOTE: For all of the following figures, assume North is up.

1. (10 pts) The following topographic map (Map 3.1) is from a coastal area and features an interesting geological hazard in addition to the Ocean. Using a contour interval of 40 meters, label the elevation of every contour line on the map below. (Note: elevation is meters above sea level, which makes sea level = ____m).

Map 3.1
Author: Brad Deline
Source: Original Work
License: CC BY-SA 3.0

2. (10 pts) Imagine you are a geologist for the United States Geological Survey. You are tasked with creating your own coastal Topographic map, so you hike around the area with a GPS receiver (Global Positioning System) and every so often you record your position along with the elevation in meters at that point, which results in the following map (Map 3.2). Complete Map 3.2 by adding in the contour lines using a <u>contour interval of 100 meters</u>. Draw the contour lines so that they are continuous (do not die off), and either continue off the map or form an enclosed circle (look at the topographic map in the problem 1 for an example). More often than not, your contour lines will fall between the GPS points on your map, so do your best to determine the contour line positions.

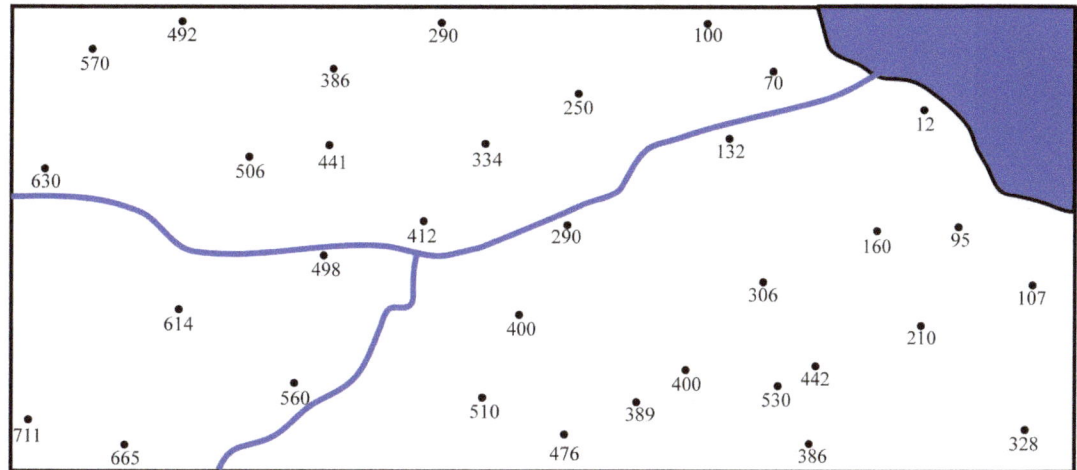

Map 3.2
Author: Brad Deline
Source: Original Work
License: CC BY-SA 3.0

For questions 3-7 refer to the Map 3.3. The following topographic map shows an interesting and informative geological feature called a drumlin, which is a pile of sediment left behind by a retreating glacier.

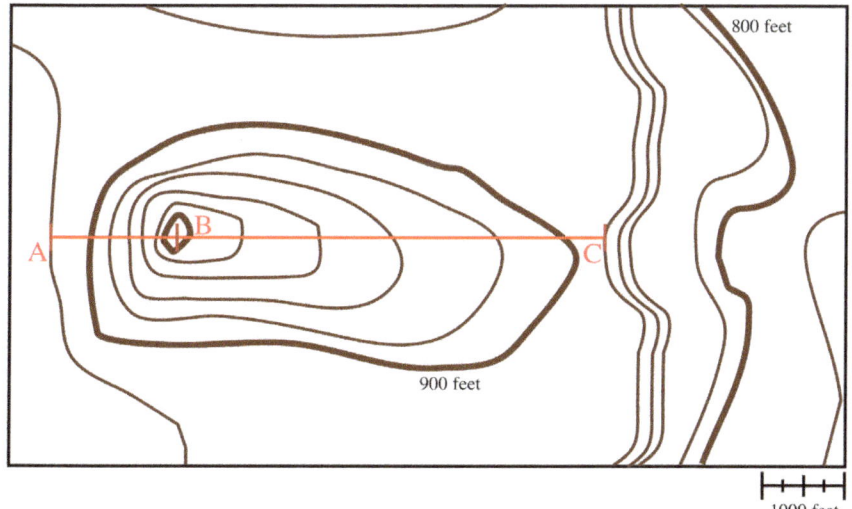

Map 3.3
Author: Brad Deline
Source: Original Work
License: CC BY-SA 3.0

3. (2 pts) What is the contour interval on Map 3.3?

4. (2 pts) What is the regional relief on Map 3.3?

5. (5 pts) Using the contour lines on Map 3.3, which area along the red line is steeper A to B or B to C? Explain how you came to this conclusion.

6. (5 pts) What is the gradient from A to B and B to C on Map 3.3? Show your work.

7. (2 pts) Drumlins can be used to determine the direction of movement in the glacier with the glacier moving toward the shallower side of the structure. Using your previous answers for Map 3.3, what direction was the glacier traveling? Note: unless indicated otherwise, assume that North is up (towards the top of the map).

8. (20 pts) Construct a topographic profile from A to A' on the graph paper below.

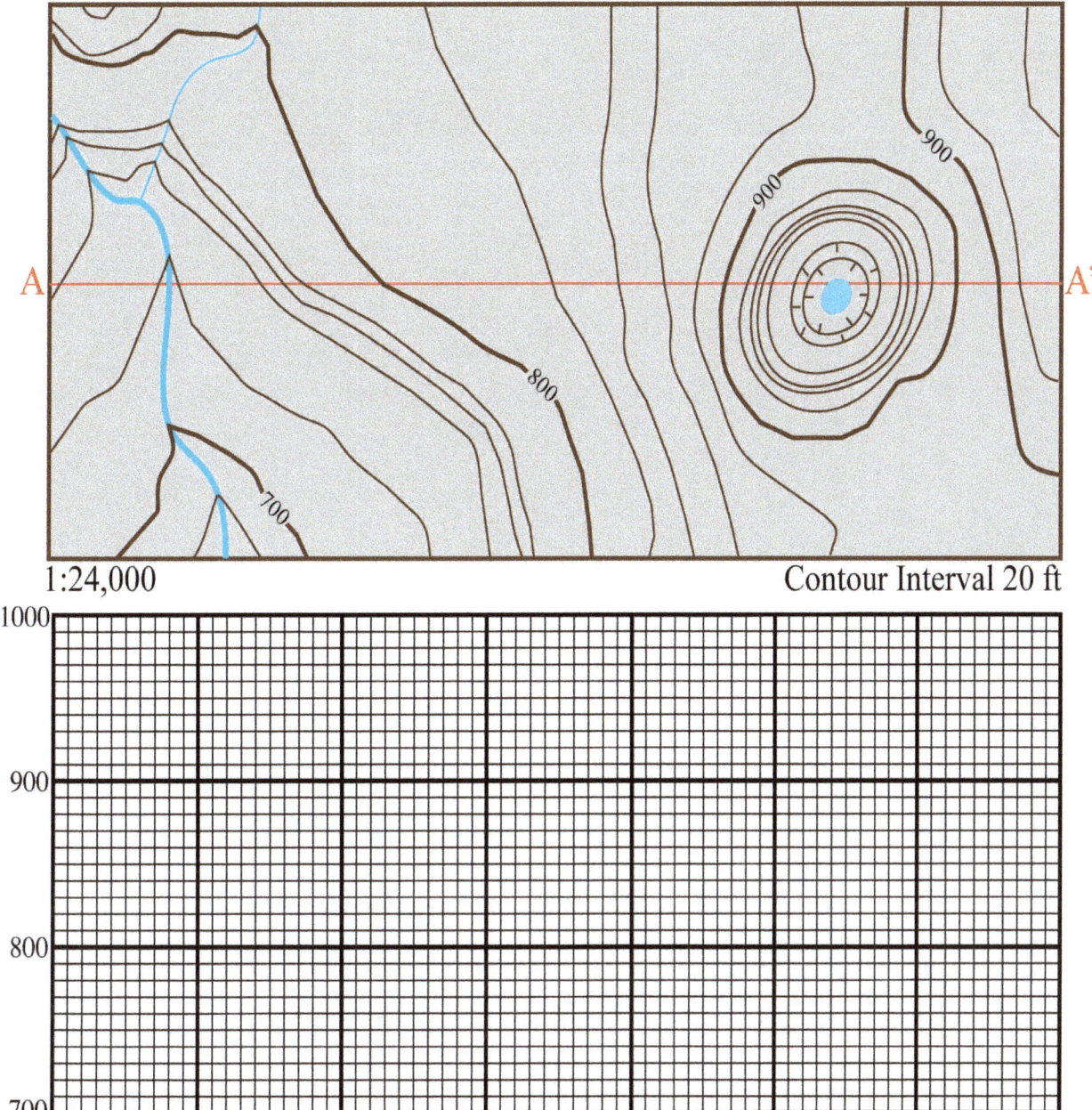

Map 3.4
Author: Brad Deline
Source: Original Work
License: CC BY-SA 3.0

9. (5 pts) Based on the scale you choose for the topography (vertical axis) in question 8, calculate the amount of vertical exaggeration on the topographic profile you constructed above. Show your work.

INTRODUCTORY GEOLOGY TOPOGRAPHIC MAPS

For this part of the lab you will need to use Maps 3.5 and 3.6 (appearing at the end of this chapter). Following Maps 3.5 and 3.6 is a Map Key that you can use to identify the various symbols found on topographic maps. Also, note that the maps are in color and the colors have significance in terms of the symbols.

Questions 10-18: Rome North Quadrangle (27 pts)

10. (2 pts) What is the ratio scale of this map?

11. (2 pts) Explain in a sentence how this type of scale works.

12. (2 pts) What is the latitude on the north edge of the map?

13. (2 pts) What is the longitude on the east edge of the map?

14. (6 pts) Find Big Dry Creek, which is north of Rome. What direction does that river flow? Explain two reasons why you came to this conclusion.

15. (2 pts) Examine the large Ridge in the Northwestern portion of the map. What is the tallest point in this ridge? How tall is it?

16. (2 pts) How much higher is that point from Lake Conasauga?

17. (5 pts) What is the gradient between Lake Conasauga and the tallest point in the ridge? Show your work (Hint: zooming out will let you see both features on the map at the same time and may make it easier to measure).

18. (4 pts) How would the gradient change if you measured from Swan Lake to the tallest point in the ridge rather than Lake Conasauga? Explain why.

Questions 19-22: Grand Tetons (12 pts)

19. (3 pts) Explain why this map is referred to as a 7.5 minute map?

20. (3 pts) What is the relief on this map?

21. (2 pts) Does Taggert Creek flow into Taggert Lake or Lake Taminah? What direction does the creek flow?

22. (4 pts) Garnet Canyon (a little to the west of the word Garnet) is a common camping location for hikers and mountain climbers at the Grand Teton's National Park. Examine the topography surrounding Garnet Canyon and the Middle Teton. What would be the easiest and safest route from Garnet Canyon to the top of the Middle Teton? Explain why. (Drawing a simple map will help).

Maps 3.5, 3.6, and Map Key (appearing on following pages)
Author: USGS
Source: USGS
License: Public Domain

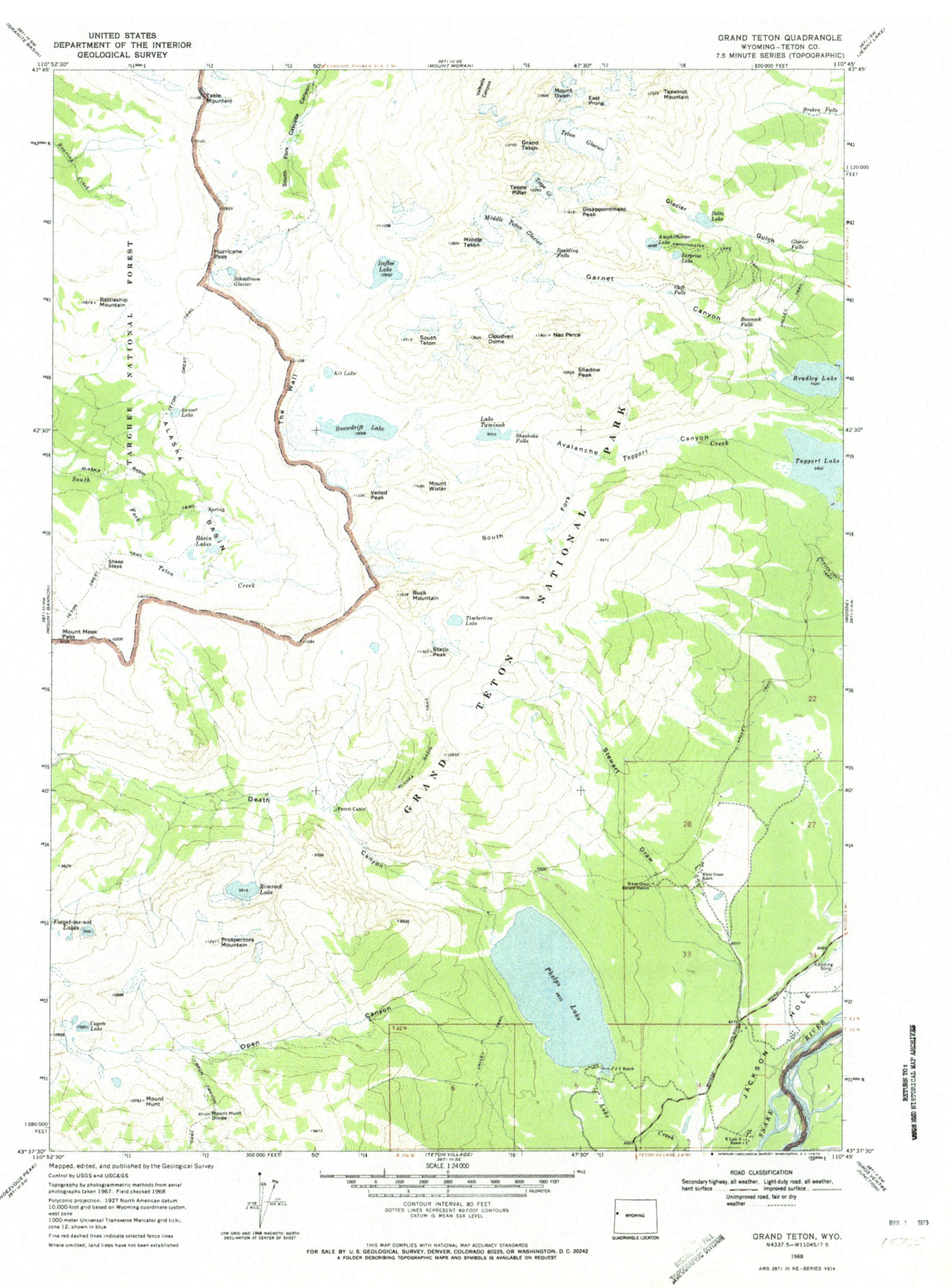

BATHYMETRIC FEATURES

Area exposed at mean low tide; sounding datum line***	
Channel***	=====
Sunken rock***	+

BOUNDARIES

National	
State or territorial	
County or equivalent	
Civil township or equivalent	
Incorporated city or equivalent	
Federally administered park, reservation, or monument (external)	
Federally administered park, reservation, or monument (internal)	
State forest, park, reservation, or monument and large county park	
Forest Service administrative area*	
Forest Service ranger district*	
National Forest System land status, Forest Service lands*	
National Forest System land status, non-Forest Service lands*	
Small park (county or city)	

BUILDINGS AND RELATED FEATURES

Building	
School; house of worship	
Athletic field	
Built-up area	
Forest headquarters*	
Ranger district office*	
Guard station or work center*	
Racetrack or raceway	
Airport, paved landing strip, runway, taxiway, or apron	
Unpaved landing strip	
Well (other than water), windmill or wind generator	
Tanks	
Covered reservoir	
Gaging station	
Located or landmark object (feature as labeled)	
Boat ramp or boat access*	
Roadside park or rest area	
Picnic area	
Campground	
Winter recreation area*	
Cemetery	

COASTAL FEATURES

Foreshore flat	
Coral or rock reef	
Rock, bare or awash; dangerous to navigation	
Group of rocks, bare or awash	
Exposed wreck	
Depth curve; sounding	
Breakwater, pier, jetty, or wharf	
Seawall	
Oil or gas well; platform	

CONTOURS

Topographic

Index	6000
Approximate or indefinite	
Intermediate	
Approximate or indefinite	
Supplementary	
Depression	
Cut	
Fill	
Continental divide	

Bathymetric

Index***	
Intermediate***	
Index primary***	
Primary***	
Supplementary***	

CONTROL DATA AND MONUMENTS

Principal point**	⊕ 3-20
U.S. mineral or location monument	▲ USMM 438
River mileage marker	+ Mile 69

Boundary monument

Third-order or better elevation, with tablet	BM □ 9134 BM ⊕ 277
Third-order or better elevation, recoverable mark, no tablet	□ 5628
With number and elevation	67 □ 4567

Horizontal control

Third-order or better, permanent mark	△ Neace ⊕ Neace
With third-order or better elevation	BM △ 52 ⊕ Pike BM393
With checked spot elevation	△ 1012
Coincident with found section corner	△ Cactus ⊕ Cactus
Unmonumented**	+

CONTROL DATA AND MONUMENTS – continued

Vertical control

Third-order or better elevation, with tablet	BM × 5280
Third-order or better elevation, recoverable mark, no tablet	× 528
Bench mark coincident with found section corner	BM + 5280
Spot elevation	× 7523

GLACIERS AND PERMANENT SNOWFIELDS

Contours and limits	
Formlines	
Glacial advance	
Glacial retreat	

LAND SURVEYS

Public land survey system

Range or Township line	
Location approximate	
Location doubtful	
Protracted	
Protracted (AK 1:63,360-scale)	
Range or Township labels	R1E T2N R4W T4S
Section line	
Location approximate	
Location doubtful	
Protracted	
Protracted (AK 1:63,360-scale)	
Section numbers	1 - 36 1 - 36
Found section corner	+
Found closing corner	+
Witness corner	WC +
Meander corner	MC
Weak corner*	+

Other land surveys

Range or Township line	
Section line	
Land grant, mining claim, donation land claim, or tract	
Land grant, homestead, mineral, or other special survey monument	□
Fence or field lines	

MARINE SHORELINES

Shoreline	
Apparent (edge of vegetation)***	
Indefinite or unsurveyed	

MINES AND CAVES

Quarry or open pit mine	✕
Gravel, sand, clay, or borrow pit	✕
Mine tunnel or cave entrance	⊰
Mine shaft	■
Prospect	X
Tailings	Tailings
Mine dump	
Former disposal site or mine	

PROJECTION AND GRIDS

Neatline	39°15′ 90°37′30″
Graticule tick	55′
Graticule intersection	+
Datum shift tick	+

State plane coordinate systems

Primary zone tick	640 000 FEET
Secondary zone tick	247 500 METERS
Tertiary zone tick	260 000 FEET
Quaternary zone tick	98 500 METERS
Quintary zone tick	320 000 FEET

Universal transverse mercator grid

UTM grid (full grid)	273
UTM grid ticks*	269

RAILROADS AND RELATED FEATURES

Standard guage railroad, single track	
Standard guage railroad, multiple track	
Narrow guage railroad, single track	
Narrow guage railroad, multiple track	
Railroad siding	
Railroad in highway	
Railroad in road	
Railroad in light duty road*	
Railroad underpass; overpass	
Railroad bridge; drawbridge	
Railroad tunnel	
Railroad yard	
Railroad turntable; roundhouse	

RIVERS, LAKES, AND CANALS

Perennial stream	
Perennial river	
Intermittent stream	
Intermittent river	
Disappearing stream	
Falls, small	
Falls, large	
Rapids, small	
Rapids, large	
Masonry dam	
Dam with lock	
Dam carrying road	

RIVERS, LAKES, AND CANALS – *continued*

Perennial lake/pond	
Intermittent lake/pond	
Dry lake/pond	
Narrow wash	
Wide wash	
Canal, flume, or aqueduct with lock	
Elevated aqueduct, flume, or conduit	
Aqueduct tunnel	
Water well, geyser, fumarole, or mud pot	
Spring or seep	

ROADS AND RELATED FEATURES

Please note: Roads on Provisional-edition maps are not classified as primary, secondary, or light duty. These roads are all classified as improved roads and are symbolized the same as light duty roads.

Primary highway	
Secondary highway	
Light duty road	
Light duty road, paved*	
Light duty road, gravel*	
Light duty road, dirt*	
Light duty road, unspecified*	
Unimproved road	
Unimproved road*	
4WD road	
4WD road*	
Trail	
Highway or road with median strip	
Highway or road under construction	
Highway or road underpass; overpass	
Highway or road bridge; drawbridge	
Highway or road tunnel	
Road block, berm, or barrier*	
Gate on road*	
Trailhead*	

SUBMERGED AREAS AND BOGS

Marsh or swamp	
Submerged marsh or swamp	
Wooded marsh or swamp	
Submerged wooded marsh or swamp	
Land subject to inundation	

SURFACE FEATURES

Levee	
Sand or mud	
Disturbed surface	
Gravel beach or glacial moraine	
Tailings pond	

TRANSMISSION LINES AND PIPELINES

Power transmission line; pole; tower	
Telephone line	
Aboveground pipeline	
Underground pipeline	

VEGETATION

Woodland	
Shrubland	
Orchard	
Vineyard	
Mangrove	

* USGS-USDA Forest Service Single-Edition Quadrangle maps only.
In August 1993, the U.S. Geological Survey and the U.S. Department of Agriculture's Forest Service signed an Interagency Agreement to begin a single-edition joint mapping program. This agreement established the coordination for producing and maintaining single-edition primary series topographic maps for quadrangles containing National Forest System lands. The joint mapping program eliminates duplication of effort by the agencies and results in a more frequent revision cycle for quadrangles containing National Forests. Maps are revised on the basis of jointly developed standards and contain normal features mapped by the USGS, as well as additional features required for efficient management of National Forest System lands. Single-edition maps look slightly different but meet the content, accuracy, and quality criteria of other USGS products.

** Provisional-Edition maps only.
Provisional-edition maps were established to expedite completion of the remaining large-scale topographic quadrangles of the conterminous United States. They contain essentially the same level of information as the standard series maps. This series can be easily recognized by the title "Provisional Edition" in the lower right-hand corner.

*** Topographic Bathymetric maps only.

Topographic Map Information

For more information about topographic maps produced by the USGS, please call:
1-888-ASK-USGS or visit us at http://ask.usgs.gov/

ISBN 0-607-96942-3

Printed on recycled paper

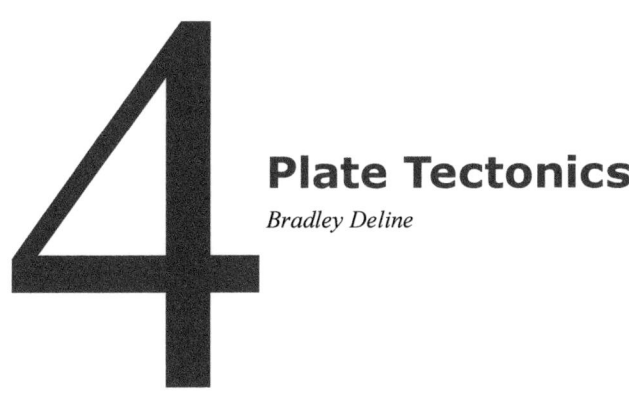

Plate Tectonics
Bradley Deline

4.1 INTRODUCTION

In chapter one, we reviewed the scientific method and the exact meaning of a theory, which is a well-supported explanation for a natural phenomenon that still cannot be completely proven. A **Grand Unifying Theory** is a set of ideas that is central and essential to a field of study such as the theory of gravity in physics or the theory of evolution in biology. The Grand Unifying Theory of geology is the **theory of Plate Tectonics**, which defines the outer portion of the earth as a brittle outer layer that is broken into moving pieces called **tectonic plates**. This theory is supported by many lines of evidence including the shape of the continents, the distribution of fossils and rocks, the distribution of environmental indicators, as well as the location of mountains, volcanoes, trenches, and earthquakes. The movement of plates can be observed on human timescales and easily measured using GPS satellites.

Plate tectonics is integral to the study of geology because it aids in reconstructing earth's history. This theory helps to explain how the first continents were built, how oceans formed, and even helps inform hypotheses for the origin of life. The theory also helps explain the geographic distribution of geologic features such as mountains, volcanoes, rift valleys, and trenches. Finally, it helps us assess the potential risks of geologic catastrophes such as earthquakes and volcanoes across the earth. The power of this theory lies in its ability to create testable hypotheses regarding Earth's history as well as predictions regarding its future.

4.1.1 Learning Outcomes

After completing this chapter, you should be able to:
- Explain several lines of evidence supporting the movement of tectonic plates
- Accurately describe the movement of tectonic plates through time
- Describe the progression of a Hawaiian Island and how it relates to the Theory of Plate Tectonics

- Describe the properties of tectonics plates and how that relates to the proposed mechanisms driving plate tectonics
- Be able to describe and identify the features that occur at different plate boundaries

4.1.2 Key Terms

- Continental Crust
- Convergent Boundary
- Divergent Boundary
- Grand Unifying Theory
- Hot Spot
- Oceanic Crust
- Ridge Push
- Slab Pull
- Slab Suction
- Subduction
- Tectonic plates
- Theory of Plate Tectonics
- Transform Boundary
- Wadati-Benioff Zone

4.2 EVIDENCE OF THE MOVEMENT OF CONTINENTS

The idea that the continents appear to have been joined based on their shapes is not new, in fact this idea first appeared in the writings of Sir Francis Bacon in 1620. The resulting hypothesis from this observation is rather straightforward: the shapes of the continents fit together because they were once connected and have since broken apart and moved. This hypothesis is discussing a historical event in the past and cannot be directly tested without a time machine. Therefore, geoscientists reframed the hypothesis by assuming the continents used to be connected and asking what other patterns we would expect to find. This is exactly how turn of the century earth scientists (such as Alfred Wegener) addressed this important scientific question.

Wegener compiled rock types, fossil occurrences, and environmental indicators within the rock record on different continents (focusing on Africa and South America) that appear to have been joined in the past and found remarkable similarities. Other scientists followed suit and the scientific community was able to compile an extensive dataset that indicated that the continents were linked in the past in a supercontinent called Pangaea (coined by Alfred Wegener) and have shifted to their current position over time. Dating these rocks using the methods discussed in chapter one allowed the scientists to better understand the rate of motion, which has assisted in trying to determine the mechanisms that drive plate tectonics.

4.3 LAB EXERCISE

This lab will use two different ways to input your answers. Most of the questions will be multiple choice and submitted online as you have in previous labs. Other questions will give you a blank box to input your answer as text. Your professor will manually grade this text, such that the format is not as important as your answer. This format allows you the opportunity to show your work using simple symbols and allows the instructor to better see your thought process. Also note, that for many of these questions there is not a single correct answer and seeing your thought process and understanding the material is more important than your answer. Therefore, it is important to show your work.

In addition, several of the exercises that follow use Google Earth. For each question (or set of questions) paste the location that is given into the "search" box. Examine each location at multiple eye altitudes and differing amounts of tilt. For any measurements use the ruler tool, this can be accessed by clicking on the ruler icon above the image.

Part A – Plate Motion and Evidence

As was mentioned above one of the most striking things about the geography of the continents today is how they appear to fit together like puzzle pieces. The reason for this is clear: they once were connected in the past and have since separated shifting into their current positions. Open Google Earth and zoom out to an eye altitude of ~8000 miles. Examine the coastlines of eastern South America and Western Africa and notice how well they match in shape.

There are scientifically important rock deposits in southern Brazil, South America and Angola, Africa that show the northernmost glacial deposits on the ancient continent of Pangaea, which indicates these two areas were once connected. Based on the shape of the two coastlines, give the present day latitude and longitude of two sites along the coast of these countries that used to be connected when the two continents were joined as a part of Pangaea (note: there are multiple correct answers):

1. Brazil (Latitude and Longitude)

2. Angola (Latitude and Longitude)

3. Measure in centimeters the distance (Map Length) between the two points you recorded in the previous question. Given that this portion of Pangaea broke apart 200,000,000 years ago, calculate how fast South America and Africa are separating in cm/year? (Hint: Speed= Distance/Time)

4. When will the next supercontinent form? Examine the Western Coast of South America, the Eastern Coast of Asia, and the Pacific Ocean. If South America and Africa are separating and the Atlantic Ocean is growing, then the opposite must be occurring on the other side of the earth (the Americas are getting closer to Asia and the Pacific Ocean is shrinking). How far apart are North America and Mainland Asia in cm? (measure the distance across the Pacific at 40 degrees north latitude- basically measure between Northern California and North Korea)? Take that distance and divide it by the speed you calculated in question 3 to estimate when the next supercontinent will form. Show your work!

Use the Figures 4.1 and 4.2 to answer questions 5-7.

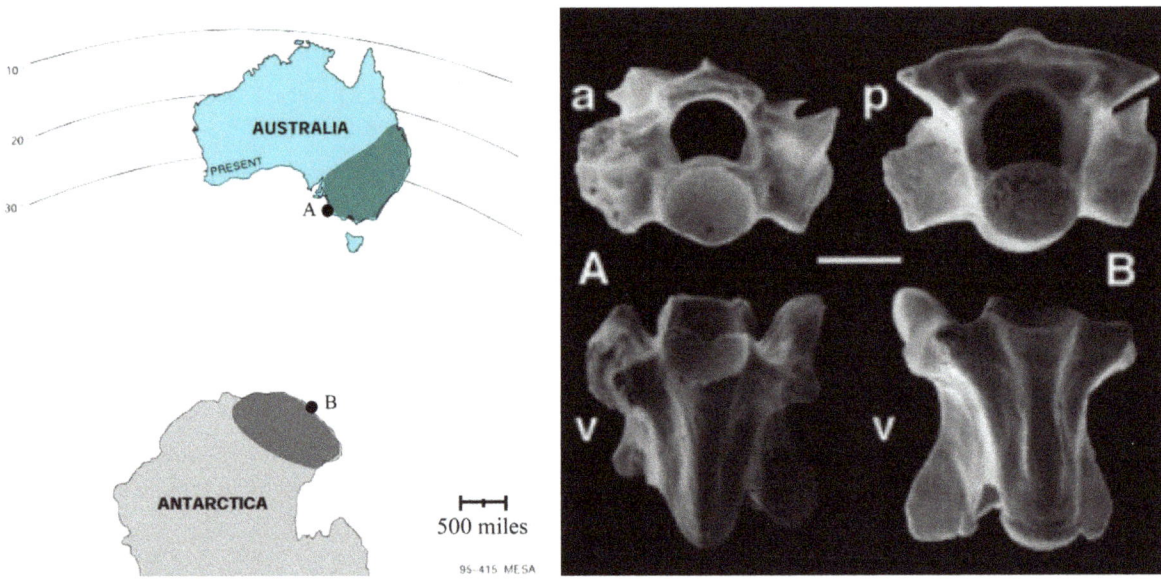

Figure 4.1

Figure 4.2

Figures 4.1 and 4.2 | The distribution across Australia and Antartica (Figure 4.1) of the fossil snake *Patagoniophis* (Figure 4.2). Obviously, this small snake was unable to swim the immense distance between the contients and, therefore, lived while Australia and Antarctica were still joined together. Figure modified from the Australia Department of Natural Resources and Scanlon (2005), Memoirs of the Queensland Museum.

Figure 4.1
Author: Bradley Deline
Source: Original Work
License: CC BY-SA 3.0

Figure 4.2
Author: The Queensland Museum
Source: The Queensland Museum
License: CC BY-NC-ND 3.0

5. How far have the snake fossils moved apart since they were originally deposited?

   a. 1250 miles   b. 1700 miles   c. 2150 miles   d. 2700 miles

6. Given that this portion of the Australian plate moves at a speed of 2.2 inches per year, how old are the snake fossils?

   a. 310 million years old

   b. 217 million years old

   c. 98 million years old

   d. 62 million years old

   e. 34 million years old

7. There are fossils such as *Glossopteris* and *Lystrosaurus* that are found in rocks in South America and Africa that indicate they were part of Pangaea approximately 200 million years ago. These same fossils can be found in Australia, which indicates it, along with Antarctica, was also part of Pangaea at that time. Based on your answer to question 6 which of the following statements about the break-up of Pangaea is TRUE?

   a. Australia and Antarctica separated before the break-up of Pangaea.

   b. Australia and Antarctica separated during the break-up of Pangaea.

   c. Australia and Antarctica separated after the break-up of Pangaea.

4.4 HOT SPOTS

Another line of evidence that can be used to track plate motion is the location of **hot spots**. Hot spots are volcanically active areas on the Earth's surface that are caused by anomalously hot mantle rocks underneath. This heat is the result of a mantle plume that rises from deep in the mantle toward the surface resulting in melted rocks and volcanoes. These mantle plumes occur deep in the Earth such that they are unaffected by the movement of the continents or the crust under the ocean. Mantle plumes appear to be stationary through time, but as the tectonic plate moves over the hot spot a series of volcanoes are produced. This gives geologists a wonderful view of the movement of a plate through time with the distribution of volcanoes indicating the direction of motion and their ages revealing the rate at which the plate was moving.

One of the most striking examples of a hot spot is underneath Hawaii. The mantle plume generates heat that results in an active volcano on the surface of the crust. Each eruption causes the volcano to grow until it eventually breaks the surface of the ocean and forms an island. As the crust shifts the volcano off the hot spot the volcano loses its heat and become inactive. The volcano then cools down, contracts, erodes, sinks slowly beneath the ocean surface, and is carried by the tectonic plate as it moves through time. As each island moves away from the mantle plume a new island will then be formed at the hot spot in a continual conveyor belt of islands. Therefore, the scars of ancient islands near Hawaii give a wonderful view of the movement of the tectonic plate beneath the Pacific Ocean.

INTRODUCTORY GEOLOGY   PLATE TECTONICS

4.5 LAB EXERCISE
Materials

Type "Hawaii" into the search bar of Google Earth and examine the chain of Hawaiian Islands. On a separate sheet of paper please draw yourself a map of the islands and label the following on your map (making sure to include the names), which will be used to answer the following questions.

Islands to include:

Big Island, Maui, Kauai, Nihoa (23 03 32.79N 161 55 11.94W)

a. Put a North arrow on your map.

b. Label the ages of each of the islands determined by radiometric dating of the lava.

Big Island- 0 (active), Maui – 1.1 million, Kauai- 4.7 million, Nihoa- 7.2 million years

a. Place a dot at the center of each island and measure on Google earth the distance between the center of an island and its adjacent island in centimeters (for example measure the distance between the center of the Big Island and Maui).

b. Look closely at each island in Google Earth and record their maximum elevation in centimeters. Remember elevation can be found by placing your cursor over a point and reading the elevation on the lower right of the image by the latitude and longitude. The elevation will be given in meters, but can be converted to centimeters by multiplying by 100. (Hint: tilting the image of the island will help to find the highest point.)

Part B - Hawaii

8. Consider the ages and positions of the islands listed above along with what you know about plate tectonics and hotspots. In what general direction is the Pacific Plate moving?

    a. Northwest   b. Southeast   c. Northeast   d. Southwest

9. How fast was the Pacific plate moving during the last 1.1 million years between the formation of the Big Island and Maui in cm/year? To calculate this divide the distance (in centimeters) between the two islands by the difference in their ages.

10. How fast was the Pacific plate moving from 7.2 million years ago to 4.7 million years ago between the formation of Kauai and Nihoa in cm/year? To calculate this divide the distance (in centimeters) between the two islands by the difference in their ages.

11. Examine the headings of the measurements that you took for the previous two questions. The headings indicate the direction the Pacific Plate is moving over the hot spot. How does the direction of motion of the Pacific Plate during the last 1.1 million years differ from direction of movement between 4.7 and 7.2 million years ago? The direction of plate movement in the last 1.1 million years_____.

    a. shows no change   b. has become more southerly   c. has become more northerly

12. Zoom out and examine the dozens of sunken volcanoes out past Nihoa, named the Emperor Seamounts. As one of these volcanic islands on the Pacific Plate moves off the hotspot it becomes inactive, or extinct, and the island begins to sink as it and the surrounding tectonic plate cool down. The speed the islands are sinking can be estimated by measuring the difference in elevation between two islands and dividing by the difference in their ages (this method assumes the islands were a similar size when they were active). Calculate how fast the Hawaiian Islands are sinking, by using the ages and elevations of Maui and Nihoa.

13. Using the speed you calculated in the previous question (and ignoring possible changes in sea level), when will the Big Island of Hawaii sink below the surface of the ocean? Divide the current maximum elevation of the Big Island by the rate you calculated in the previous question.

14. Now zoom out to ~4000 miles eye altitude and look at the chain of Hawaiian Islands again. Notice the chain continues for thousands of miles up to Aleutian Islands (between Alaska and Siberia). Examine the northernmost sunken volcano (50 49 16.99N 167 16 36.12E) in this chain. Where was that volcano located when it was still active, erupting, and above the surface of the ocean?

    a. 50 49 16.99N 167 16 36.12E   b 52 31 48.72N 166 25 43.14W

    c. 27 45 49.27N 177 10 08.75W   d. 19 28 15.23N 155 19 14.43W

4.6 PLATE MATERIALS

By now you can see many different lines of evidence that the tectonic plates are moving (there are many additional lines of evidence as well). To build a theory we need an explanation or a mechanism that explains the patterns that we see. The **theory of plate tectonics** states that the outer rigid layer of the earth (the lithosphere) is broken into pieces called tectonic plates (Figure 4.3) and that these plates move independently above the flowing plastic-like portion of the mantle (Asthenosphere).

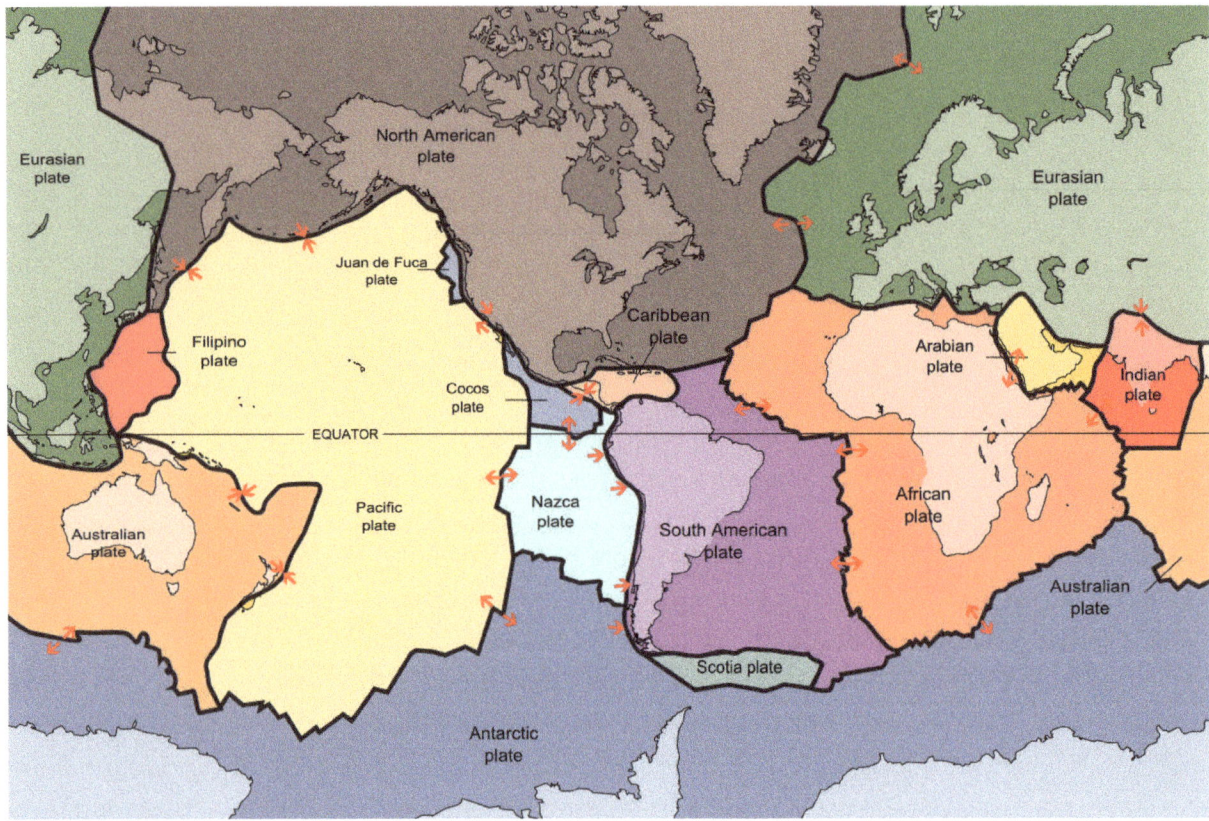

Figure 4.3 | Tectonic plates on Earth.
Author: USGS
Source: USGS
License: Public Domain

Tectonic plates are composed of the crust and the upper most mantle that functions as a brittle solid. These plates can be composed of oceanic crust, continental crust or a mixture of both. The **Oceanic Crust** is thinner and normally underlies the world's oceans, while the **Continental Crust** is thicker and like its name consists of the continents. The interaction of these tectonic plates is at the root of many geologic events and features, such that we need to understand the structure of the plates to better understand how they interact. The interaction of these plates is controlled by the relative motion of two plates (moving together, apart, or sliding past) as well as the composition of the crustal portion of the plate (continental or ocean crust). Continental crust has an overall composition similar to the igneous rock granite, which is a solid, silica-rich crystalline rock typically consisting of a mixture of pink (feldspar), milky white (feldspar), clear (quartz), and black (biotite) minerals. Oceanic crust is primarily composed of the igneous rock gabbro, which is a solid, iron and magnesium-rich crystalline rock consisting of a mixture of black and dark gray minerals (pyroxene and feldspar). The difference in rock composition results in distinctive physical properties that you will determine in the next set of questions.

4.7 LAB EXERCISE
Part C – Plate Densities

An important property of geological plates is their density (mass/volume). Remember the asthenosphere has fluid-like properties, such that tectonic plates 'float' relative to their density. This property is called isostasy and is similar to buoyancy in water. For example, if a cargo ship has a full load of goods it will appear lower than if it were empty because the density of the ship is on average higher. Therefore, the relative density of two plates can control how they interact at a boundary and the types of geological features found along the border between the two plates. Measuring the density of rocks is fairly easy and can be done by first weighing the rocks and then calculating their volume. The latter is best done by a method called fluid displacement using a graduated cylinder. Water is added to the cylinder and the level is recorded, a rock is then added to the cylinder and the difference in water levels equals the volume of the rock. Density is then calculated as the mass divided by the volume (Figure 4.4).

The information needed to calculate density was collected for four rocks and can be used to answer the following questions including the weight (in grams) as well as

Figure 4.4 | Method to find the density of a rock. First the weight is measured on a digital scale and then the fluid displacement method is used to determine the volume.
Author: Bradley Deline
Source: Original Work
License: CC BY-SA 3.0

the volume of water recorded by a graduated cylinder (in milliliters) before and after the rock was added. Note: each line on the graduated cylinder represents 10 ml. When measuring the volume please round to the nearest 10 milliliter line on the graduated cylinder. **Hint**: Surface tension will often cause the water level to curve up near the edges of the graduated cylinder creating a feature called a meniscus. To accurately measure the volume, use the lowest point the water looks to occupy.

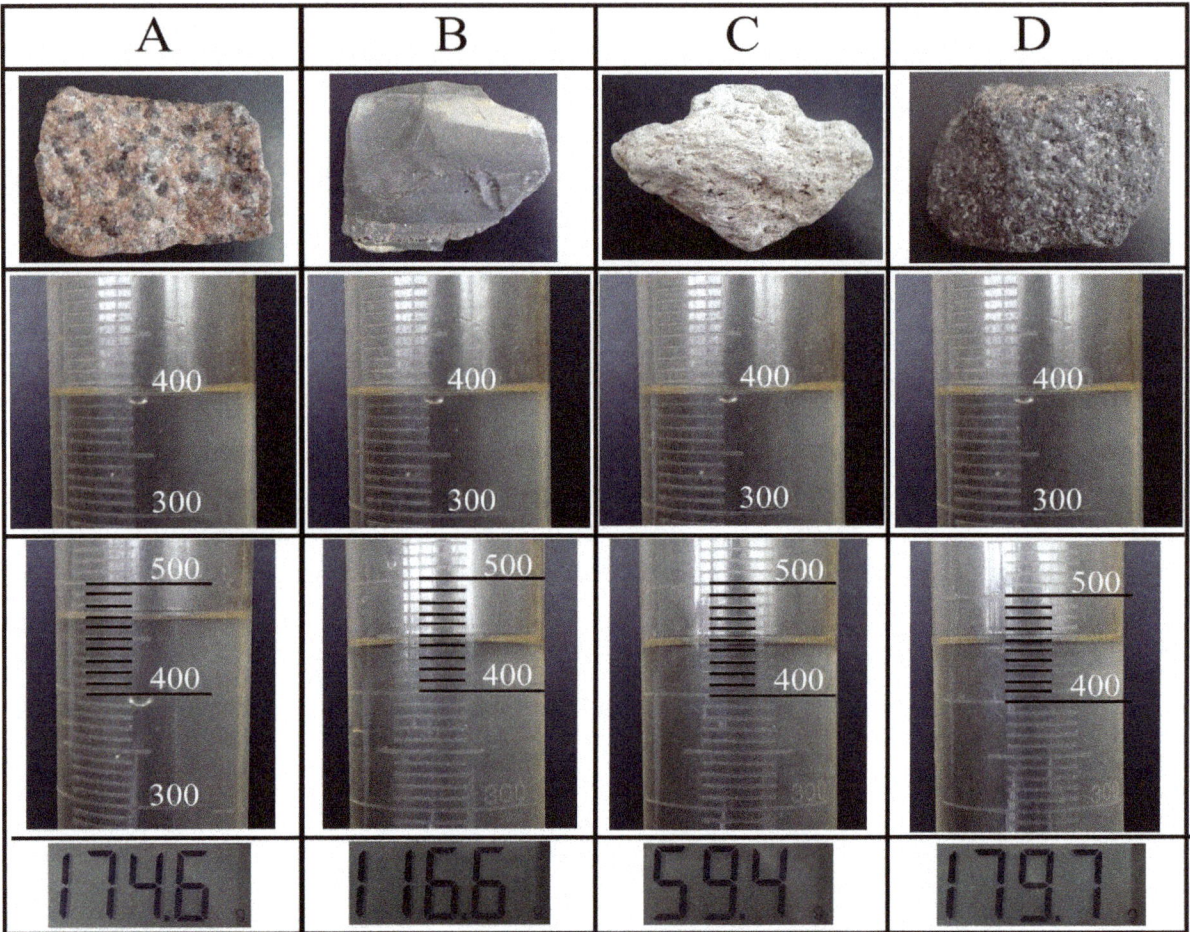

Figure 4.5 | Figure to use to answer questions 15-19. The first row shows images of the four rocks. The second and third rows show the volume (in milliliters) of material in the graduated cylinder before and after the rock was added. The last row shows the mass (in grams) of the four rocks.
Author: Bradley Deline
Source: Original Work
License: CC BY-SA 3.0

15. The rock that most closely resembles the composition of continental crust based on the description in the previous section is:

    a. A   b. B   c. C   d. D

16. Based on the choice you made for question 15, what is the density of the rocks that make up continental crust? Please give your answer in grams/milliliter.

17. The rock that most closely resembles the composition of oceanic crust based on the description in the previous section is:

   a. A   b. B   c. C   d. D

18. Based on the choice you made for question 17, what is the density of the rocks that make up oceanic crust? Please give your answer in grams/milliliter.

19. Remember, because of isostasy the denser plate will be lower than the less dense plate. If oceanic and continental crust collided, based on their densities the _____ crust would sink below the _____ crust.

   a. continental; oceanic   b. oceanic; continental

4.8 PLATE BOUNDARIES

Tectonic plates can interact in three different ways they can come together, they can pull apart, or they can slide by each other (Figure 4.6). The other factor that can be important is the composition of the plates (oceanic or continental crust) that are interacting as was explored in the previous section. These three types of motions along with the type of plates on each side of the boundary can produce vastly different structures and geologic events (Table 4.1).

Two plates that are moving apart from each other are called Divergent. **Divergent boundaries** are important because they are the way that continents split apart and break into separate plates as well as where new ocean crust is formed. If a divergent boundary forms within a continent that area stretches apart. This results in the area becoming thinner creating a topographic low or a valley. This extension is not a smooth process so the area is prone to earthquakes as well as volcanic activity. Eventually, the crust gets so thin it will rupture forming a gap between the plates, which will be filled with molten rock forming new oceanic crust. A thin and dense plate will be topographically low and will be covered in water forming a long and narrow sea. As the plates persist in pulling apart new crust is continually being formed at the plate boundary along an elevated crest known as a mid-oceanic ridge.

Two plates that are moving together are called Convergent. **Convergent boundaries** are important because they are the way distinctive plates can join (suture) together to form larger plates as well as where ocean crust is destroyed. The resulting structures we see at convergent boundaries depend on the types of tectonic plates. If two thick and lower density continental plates converge we get a large collision which results in mountains. This is a violent process resulting in many earthquakes, deformation (folds and faults) of rock, and the uplift of mountains. The rocks are also under immense pressure and heat and will eventually become stuck together as a single plate. If a continental plate and an ocean plate converge (continent-ocean convergent plate boundary) there will be **subduction**, where the oceanic plate sinks downward underneath the continental plate. This will result in several features including a deep trench near the subducting plate, abundant earthquakes, and the formation of magma which results in a line of volcanoes along the coast. Associated with this type of plate boundary is the **Wadati-Benioff zone**, a zone where earthquakes are produced; this zone ranges in depth from shallow (at the trench) to deep (~600km), indicating that the oceanic plate is sinking into the mantle. If two oceanic plates converge it will also result in subduction with similar features as were just discussed. The only exception will be that the volcanoes will appear on an oceanic plate and will eventually form islands along the tectonic boundary.

When the two plates slide past each other it is called a **Transform boundary**. This type of boundary differs from the previous two in that no new crust is being formed and no old crust is being destroyed. Therefore, there won't be as many striking geologic features. Transform boundaries are often marked by abundant earthquakes that can be close to the surface as well as distinctive patterns of rivers that become offset as the land is moving underneath them. Transform boundaries are also often associated with mid-oceanic ridges. If a the ridge has a jagged or stair-stepped edge the pulling apart of the two tectonic plates will also result in transform motion as you can see in Figure 4.6.

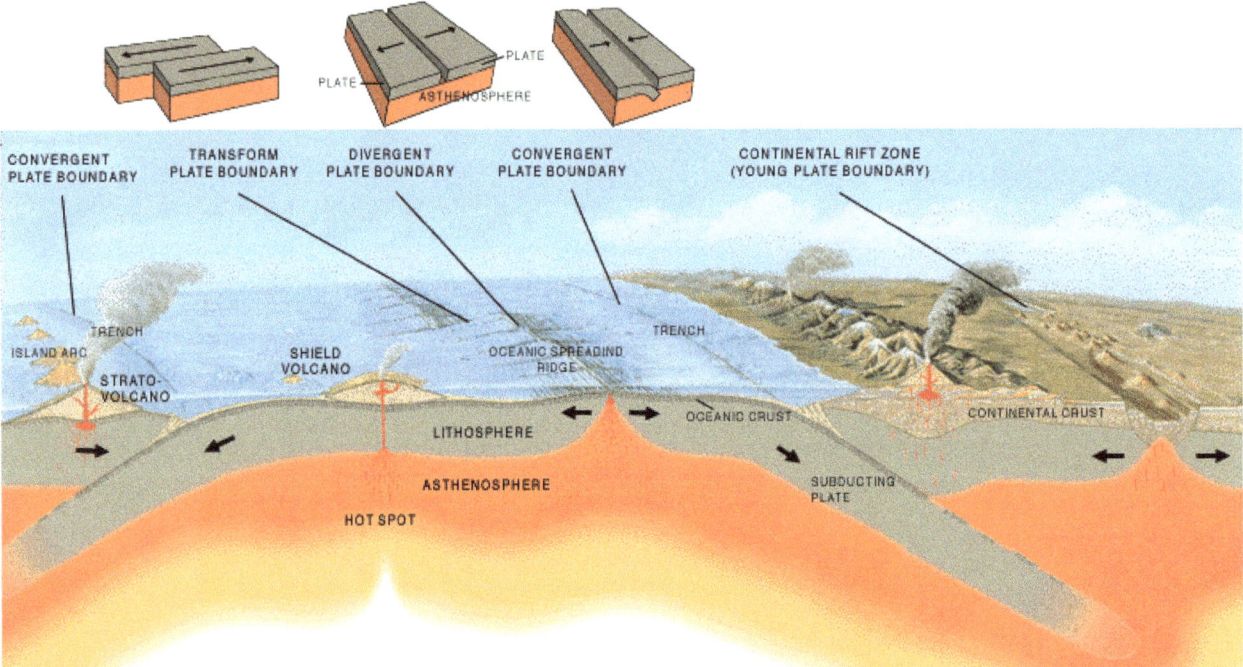

Figure 4.6 | Figure showing multiple plate boundaries and the features associated with them.
Author: José F. Vigil
Source: USGS
License: Public

Table 4.1. Characteristics of the different Plate Boundaries

Boundary Type	Plate Compositions	Earthquake Depth	Change in Crust	Identifying Features
Divergent				
	Continent-Continent	Shallow	No change	Rift Valley and Volcanoes
	Ocean-Ocean	Shallow	Formation of New Crust	Ocean Ridges
Convergent				
	Continent-Continent	Shallow to Intermediate	Metamorphism of Crust	Mountains
	Continent-Ocean	Shallow to Deep	Melting of Crust	Trench and Coastal Volcanoes
	Ocean-Ocean	Shallow To Deep	Melting of Crust	Trench and Volcanic Islands
Transform				
	Continent-Continent	Shallow	No change	Offset rivers
	Ocean-Ocean	Shallow	No change	Often associated with Ocean ridges.

4.9 LAB EXERCISE

Part D – Origin of Magma

Magma is formed from the melting of rock at both convergent and divergent boundaries. However, the processes that occur to melt the rock are quite different. Three different processes are involved in the melting of rocks as we will explore in the following exercise. In Figure 4.7 you can see a graph depicting a variety of temperature and pressure conditions. The increasing temperature with pressure on rocks as you go deeper within the earth through the crust and mantle lithosphere is called the geothermal gradient (shown in black). This gradient shows the actual temperature conditions that exist in the lithosphere. Obviously, the addition or subtraction of heat or pressure can move rocks off that gradient and cause potential change. The orange line represents the temperature and pressure required for a dry mantle rock to start to melt and any point to the right of this line is where melting of lithospheric rock can occur. The blue line represents the temperature and pressure required for a lithospheric rock to melt if water is present.

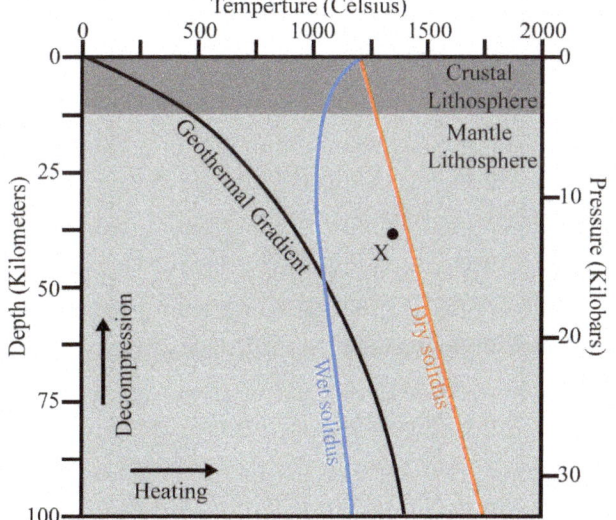

Figure 4.7 | Melting diagram for mantle rock.
Author: Bradley Deline
Source: Original Work
License: CC BY-SA 3.0

20. According to the geothermal gradient, rocks buried 75 km beneath the surface would normally be at what temperature?

    At 75 km depth, rocks will be heated to about _____ degrees Celsius.

    a. 1500   b. 1250   c. 1000   d. 750

21. According to the geothermal gradient, rocks at 500 degrees Celsius will be buried how deep?

    At 500 degrees Celsius, rocks will be buried to about _____ km depth.

    a. 8   b. 12.5   c. 20   d. 27

22. What is the physical state of a dry mantle rock at point X?

    a. Completely melted   b. Starting to melt   c. Completely solid

23. What happens when the lithosphere at point X is heated to 1500 °C?

    a. No change   b. Starts to crystallize   c. Starts to melt

24. At what depth will the dry mantle rock at point X begin to melt if it is uplifted closer to Earth's surface and its temperature remains the same?

    a. 35 km   b. 25 km   c. 18 km   d. 12 km

25. What would happen to the mantle rock at point X if water is added to it?

    a. No change   b. Starts to crystallize   c. Starts to melt

Part E – Boundaries

Earthquakes are great indicators of plate boundaries and are associated with all three boundary types. One type of boundary is unique in having a Wadati-Benioff zone. Answer the following questions using Figure 4.8.

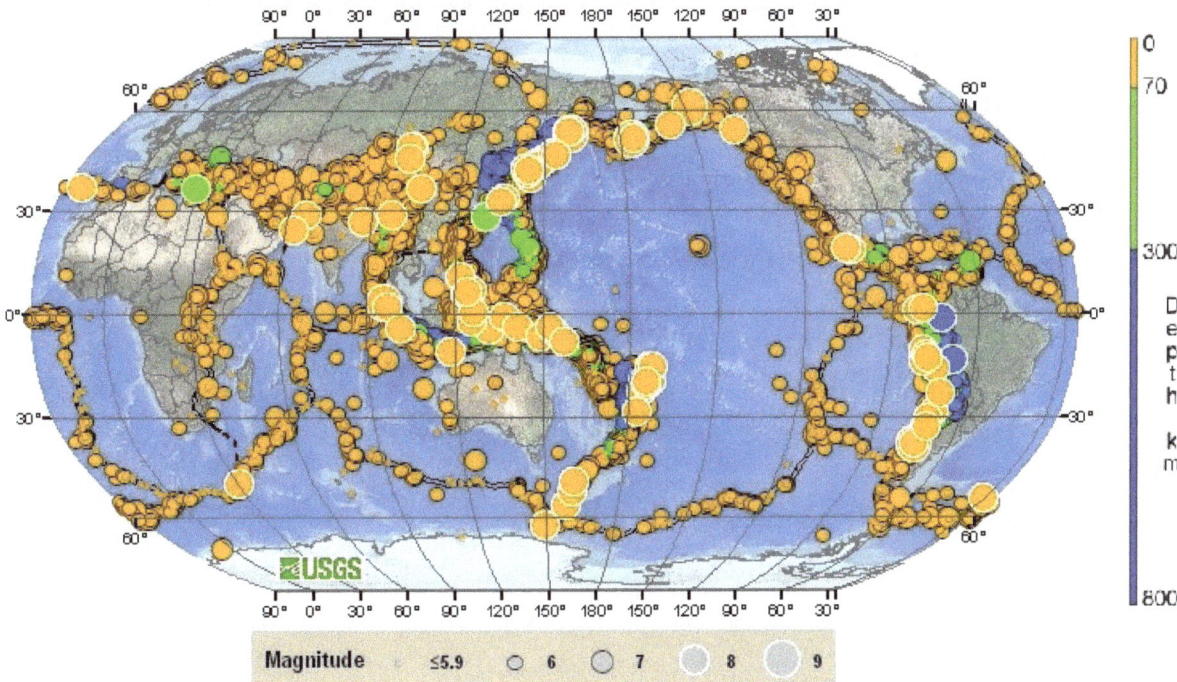

Figure 4.8 | Distribution and depth of Earthquakes.
Author: USGS
Source: USGS
License: Public Domain

26. Which of the following places represent a Wadati-Benioff zone?

    a. 10°S, 110°W   b. 0°, 0°   c. 15°S, 180°   d. 30°N, 75°E

27. The Wadati-Benioff zone is associated with which type of plate boundary?

    a. Divergent   b. Convergent (Continent-Continent)

    c. Convergent (Continent-Ocean or Ocean-Ocean)   d. Transform

Type 34 46 16.16 N 118 44 58.19 W into the search bar in Google Earth and zoom out to an eye altitude of 10 miles. Quail Lake is a dammed river that is sitting directly over top of the San Andres Fault, which is a well-known transform boundary with the North American Plate on the northern side and the Pacific Plate on the southern side. This boundary is running East-West in this area and you may be able to see the boundary better by zooming out.

28. Examine the path of the river that feeds into and flows out of Quail Lake. What direction is the North American plate moving in comparison to the Pacific Plate at this location?

    a. East   b. West

29. Given that San Francisco is located on the North American Plate and Los Angles is located on the Pacific Plate, are these two cities getting closer together or farther apart over time?

    a. Closer　　　　　　　　　　b. Farther

Google Earth: Identifying Plate Boundaries

Re-read the section on Plate Boundaries before answering the following questions.

30. Type "15 19 48.78 S 75 12 03.41 W" into the Google Earth Search bar. What type of tectonic plates are present?

    a. Ocean- Ocean　　　b. Ocean- Continent　　　c. Continent- Continent

31. What type of plate tectonic boundary is present?

    a. Transform　　　b. Convergent　　　c. Divergent

32. Type "6 21 49.68 S 29 35 37.87 E" into the Google Earth Search bar. What type of process is going on at this location?

    a. Seafloor spreading　　　b. Continental rifting　　　c. Subduction

33. What features would you expect to occur at this type of boundary?

    a. Earthquakes and a trench　　　b. Volcanoes and a valley

    c. Mountains and landslides　　　d. Earthquakes and offset rivers

34. Type "28 04 27.04N 86 55 26.84E" into the Google Earth Search bar. What type of tectonic plates are present?

    a. Ocean- Ocean　　　b. Ocean- Continent　　　c. Continent- Continent

35. What type of plate tectonic boundary is present?

    a. Transform　　　b. Convergent　　　c. Divergent

36. Type "46 55 25.66 N 152 01 25.17 E" into the Google Earth Search bar. What type of tectonic plates are present? Make sure to zoom out to get a good view of the relevant features.

    a. Ocean- Ocean　　　b. Ocean- Continent　　　c. Continent- Continent

37. What features would you expect to occur at this type of boundary?

   a. Volcanos, earthquakes and a trench   b. Volcanoes and a linear valley

   c. Mountains and landslides   d. Earthquakes and offset rivers

38. Type "43 41 07.81 N 128 16 56.29 W" into the Google Earth Search bar. What type of tectonic plates are present? Hint- make sure to re-read the section on plate boundaries before answering!

   a. Ocean- Ocean   b. Ocean- Continent   c. Continent- Continent

39. What type of plate tectonic boundary is at this exact location?

   a. Transform   b. Convergent   c. Divergent

40. This plate boundary isn't as simple as the previous examples, meaning another nearby plate boundary directly influences it. Zoom out and examine the area, what other type of boundary is nearby?

   a. Transform   b. Convergent   c. Divergent

4.10 PLATE TECTONIC MECHANISMS

The question still remains, why do tectonic plates move? The answer comes down to gravity and mantle convection. You have already studied in chapter two how the mantle flows through time creating convection currents. These convection currents flow underneath the plates and through friction pull them along at the surface as well as when they are subducted which is a force called **slab suction**. Related to this force is **slab pull,** which is a gravitational force pulling the cold subducting plate down into the mantle at a subduction zone. In addition, there is a force from potential energy at ocean ridges called **ridge push**. This is a gravitational force pushing down on the elevated ridge and because of the plates curvature it results in a horizontal force pushing the plate along the earth's surface. These forces all occur deep inside the Earth and operate on very large geographic scales making them difficult to measure. There are several competing models for the mechanisms behind plate motion, such that there are still some areas of debate surrounding the mechanics of plate tectonics which is why Plate Tectonics is a scientific theory. Documenting an event is much easier and more straightforward than explaining why it occurred.

4.11 LAB EXERCISE

Part F – Mechanisms

41. Go back to the location in Google Earth that you examined for question 36 (46 55 25.66 N 152 01 25.17 E). Which of the three proposed plate tectonic mechanisms would NOT occur at this location?

    a. Slab pull	b. Ridge push	c. Slab suction

Reference

Scalon, J. D. 2005. Australia's oldest known snakes: *Patagoniophis*, *Alamitophis* and *c.f. Madtsoia* (Squamata: Madtsoiidae) from the Eocene of Queensland. Memoirs of the Queensland Museum 51: 215-235.

4.11 STUDENT RESPONSES

The following is a summary of the questions in this lab for ease in submitting answers online.

1. Brazil (Latitude and Longitude)

2. Angola (Latitude and Longitude)

3. Measure in centimeters the distance (Map Length) between the two points you recorded in the previous question. Given that this portion of Pangaea broke apart 200,000,000 years ago, calculate how fast South America and Africa are separating in cm/year? (Hint: Speed= Distance/Time)

4. When will the next supercontinent form? Examine the Western Coast of South America, the Eastern Coast of Asia, and the Pacific Ocean. If South America and Africa are separating and the Atlantic Ocean is growing, then the opposite must be occurring on the other side of the earth (the Americas are getting closer to Asia and the Pacific Ocean is shrinking). How far apart are North America and Mainland Asia in cm? (measure the distance across the Pacific at 40 degrees north latitude- basically measure between Northern California and North Korea)? Take that distance and divide it by the speed you calculated in question 3 to estimate when the next supercontinent will form. Show your work!

5. How far have the snake fossils moved apart since they were originally deposited?

    a. 1250 miles   b. 1700 miles   c. 2150 miles   d. 2700 miles

6. Given that this portion of the Australian plate moves at a speed of 2.2 inches per year, how old are the snake fossils?

   a. 310 million years old

   b. 217 million years old

   c. 98 million years old

   d. 62 million years old

   e. 34 million years old

7. There are fossils such as *Glossopteris* and *Lystrosaurus* that are found in rocks in South America and Africa that indicate they were part of Pangaea approximately 200 million years ago. These same fossils can be found in Australia, which indicates it, along with Antarctica, was also part of Pangaea at that time. Based on your answer to question 6 which of the following statements about the break-up of Pangaea is TRUE?

   a. Australia and Antarctica separated before the break-up of Pangaea.

   b. Australia and Antarctica separated during the break-up of Pangaea.

   c. Australia and Antarctica separated after the break-up of Pangaea.

8. Consider the ages and positions of the islands listed above along with what you know about plate tectonics and hotspots. In what general direction is the Pacific Plate moving?

   a. Northwest   b. Southeast   c. Northeast   d. Southwest

9. How fast was the Pacific plate moving during the last 1.1 million years between the formation of the Big Island and Maui in cm/year? To calculate this divide the distance (in centimeters) between the two islands by the difference in their ages.

10. How fast was the Pacific plate moving from 7.2 million years ago to 4.7 million years ago between the formation of Kauai and Nihoa in cm/year? To calculate this divide the distance (in centimeters) between the two islands by the difference in their ages.

11. Examine the headings of the measurements that you took for the previous two questions. The headings indicate the direction the Pacific Plate is moving over the hot spot. How does the direction of motion of the Pacific Plate during the last 1.1 million years differ from direction of movement between 4.7 and 7.2 million years ago? The direction of plate movement in the last 1.1 million years_____.

    a. shows no change   b. has become more southerly   c. has become more northerly

12. Zoom out and examine the dozens of sunken volcanoes out past Nihoa, named the Emperor Seamounts. As one of these volcanic islands on the Pacific Plate moves off the hotspot it becomes inactive, or extinct, and the island begins to sink as it and the surrounding tectonic plate cool down. The speed the islands are sinking can be estimated by measuring the difference in elevation between two islands and dividing by the difference in their ages (this method assumes the islands were a similar size when they were active). Calculate how fast the Hawaiian Islands are sinking, by using the ages and elevations of Maui and Nihoa.

13. Using the speed you calculated in the previous question (and ignoring possible changes in sea level), when will the Big Island of Hawaii sink below the surface of the ocean? Divide the current maximum elevation of the Big Island by the rate you calculated in the previous question.

14. Now zoom out to ~4000 miles eye altitude and look at the chain of Hawaiian Islands again. Notice the chain continues for thousands of miles up to Aleutian Islands (between Alaska and Siberia). Examine the northernmost sunken volcano (50 49 16.99N 167 16 36.12E) in this chain. Where was that volcano located when it was still active, erupting, and above the surface of the ocean?

    a. 50 49 16.99N 167 16 36.12E   b 52 31 48.72N 166 25 43.14W

    c. 27 45 49.27N 177 10 08.75W   d. 19 28 15.23N 155 19 14.43W

15. The rock that most closely resembles the composition of continental crust based on the description in the previous section is:

    a. A   b. B   c. C   d. D

INTRODUCTORY GEOLOGY   PLATE TECTONICS

16. Based on the choice you made for question 15, what is the density of the rocks that make up continental crust? Please give your answer in grams/milliliter.

17. The rock that most closely resembles the composition of oceanic crust based on the description in the previous section is:

    a. A   b. B   c. C   d. D

18. Based on the choice you made for question 17, what is the density of the rocks that make up oceanic crust? Please give your answer in grams/milliliter.

19. Remember, because of isostasy the denser plate will be lower than the less dense plate. If oceanic and continental crust collided, based on their densities the _____ crust would sink below the _____ crust.

    a. continental; oceanic   b. oceanic; continental

20. According to the geothermal gradient, rocks buried 75 km beneath the surface would normally be at what temperature?

    At 75 km depth, rocks will be heated to about _____ degrees Celsius.

    a. 1500   b. 1250   c. 1000   d. 750

21. According to the geothermal gradient, rocks at 500 degrees Celsius will be buried how deep?

    At 500 degrees Celsius, rocks will be buried to about _____ km depth.

    a. 8   b. 12.5   c. 20   d. 27

22. What is the physical state of a dry mantle rock at point X?

    a. Completely melted   b. Starting to melt   c. Completely solid

23. What happens when the lithosphere at point X is heated to 1500 °C?

    a. No change   b. Starts to crystallize   c. Starts to melt

INTRODUCTORY GEOLOGY   PLATE TECTONICS

24. At what depth will the dry mantle rock at point X begin to melt if it is uplifted closer to Earth's surface and its temperature remains the same?

    a. 35 km   b. 25 km   c. 18 km   d. 12 km

25. What would happen to the mantle rock at point X if water is added to it?

    a. No change   b. Starts to crystallize   c. Starts to melt

26. Which of the following places represent a Wadati-Benioff zone?

    a. 10°S, 110°W   b. 0°, 0°   c. 15°S, 180°   d. 30°N, 75°E

27. The Wadati-Benioff zone is associated with which type of plate boundary?

    a. Divergent   b. Convergent (Continent-Continent)

    c. Convergent (Continent-Ocean or Ocean-Ocean)   d. Transform

28. Examine the path of the river that feeds into and flows out of Quail Lake. What direction is the North American plate moving in comparison to the Pacific Plate at this location?

    a. East   b. West

29. Given that San Francisco is located on the North American Plate and Los Angles is located on the Pacific Plate, are these two cities getting closer together or farther apart over time?

    a. Closer   b. Farther

30. Type "15 19 48.78 S 75 12 03.41 W" into the Google Earth Search bar. What type of tectonic plates are present?

    a. Ocean- Ocean   b. Ocean- Continent   c. Continent- Continent

31. What type of plate tectonic boundary is present?

    a. Transform   b. Convergent   c. Divergent

32. Type "6 21 49.68 S 29 35 37.87 E" into the Google Earth Search bar. What type of process is going on at this location?

    a. Seafloor spreading   b. Continental rifting   c. Subduction

33. What features would you expect to occur at this type of boundary?

    a. Earthquakes and a trench   b. Volcanoes and a valley

    c. Mountains and landslides   d. Earthquakes and offset rivers

34. Type "28 04 27.04N 86 55 26.84E" into the Google Earth Search bar. What type of tectonic plates are present?

    a. Ocean- Ocean   b. Ocean- Continent   c. Continent- Continent

35. What type of plate tectonic boundary is present?

    a. Transform   b. Convergent   c. Divergent

36. Type "46 55 25.66 N 152 01 25.17 E" into the Google Earth Search bar. What type of tectonic plates are present? Make sure to zoom out to get a good view of the relevant features.

    a. Ocean- Ocean   b. Ocean- Continent   c. Continent- Continent

37. What features would you expect to occur at this type of boundary?

    a. Volcanos, earthquakes and a trench   b. Volcanoes and a linear valley

    c. Mountains and landslides   d. Earthquakes and offset rivers

38. Type "43 41 07.81 N 128 16 56.29 W" into the Google Earth Search bar. What type of tectonic plates are present? Hint- make sure to re-read the section on plate boundaries before answering!

    a. Ocean- Ocean   b. Ocean- Continent   c. Continent- Continent

39. What type of plate tectonic boundary is at this exact location?

    a. Transform   b. Convergent   c. Divergent

40. This plate boundary isn't as simple as the previous examples, meaning another nearby plate boundary directly influences it. Zoom out and examine the area, what other type of boundary is nearby?

    a. Transform   b. Convergent   c. Divergent

41. Go back to the location in Google Earth that you examined for question 36 (46 55 25.66 N 152 01 25.17 E). Which of the three proposed plate tectonic mechanisms would NOT occur at this location?

    a. Slab pull b. Ridge push c. Slab suction

5 Water
Randa Harris

5.1 INTRODUCTION

Think how many times a day you take water for granted – you assume the tap will be flowing when you turn on your faucet, you expect rainfall to water your lawn, and you may count on water for your recreation. Not only is water necessary for many of life's functions, it is also a considerable geologic agent. Water can sculpt the landscape dramatically over time both by carving canyons as well as depositing thick layers of sediment. Some of these processes are slow and result in landscapes worn down over time. Others, such as floods, can be dramatic and dangerous.

What happens to water during a rainstorm? Imagine that you are outside in a parking lot with grassy areas nearby. Where does the water from the parking lot go? Much of it will run off as sheet flow and eventually join a stream. What happens to the rain in the grassy area? Much of it will infiltrate, or soak into the ground. We will deal with both surface and ground water in this lab. Both are integral parts of the water cycle, in which water gets continually recycled through the atmosphere, to the land, and back to the oceans. This cycle, powered by the sun, operates easily since water can change form from liquid to gas (or water vapor) quickly under surface conditions.

Both surface and ground water are beneficial for drinking water, industry, agriculture, recreation, and commerce. Demand for water will only increase as population increases, making it vital to protect water sources both above and below ground.

5.1.1 Learning Outcomes

After completing this chapter, you should be able to:

- Understand how streams erode, transport, and deposit sediment
- Know the different stream drainage patterns and understand what they indicate about the underlying rock
- Explain the changes that happen from the head to the mouth of a stream
- Understand the human hazards associated with floods

- Know the properties of groundwater and aquifers
- Understand the distribution of groundwater, including the water table
- Learn the main features associated with karst topography
- Understand the challenges posed by karst topography

5.1.2 Key Terms

- Aquifer
- Discharge
- Drainage Basin
- Drainage Divide
- Drainage Pattern
- Floodplain
- Karst Topography
- Natural Levee
- Permeability
- Porosity
- Recurrence Interval
- Stream Gradient

5.2 STREAMFLOW AND PARTS OF A STREAM

The running water in a stream will erode (wear away) and move material within its channel, including dissolved substances (materials taken into solution during chemical weathering). The solid sediments may range in size from tiny clay and silt particles too small for the naked eye to view up to sand and gravel sized sediments. Even boulders have been carried by large flows. The smaller particles kept in suspension by the water's flow are called suspended load. Larger particles typically travel as bed load, stumbling along the stream bed (Figure 5.1). While the dissolved, suspended, and bed loads may travel long distances (ex. from the headwaters of the Mississippi River in Minnesota to the Gulf of Mexico at New Orleans), they will eventually settle out, or deposit. These stream deposited sediments, called alluvium, can be deposited at any time, but most often occur during flood events. To more effectively transport sediment, a stream needs energy. This energy is mostly a function of the amount of water and its velocity, as more (and larger) sediment can be car-

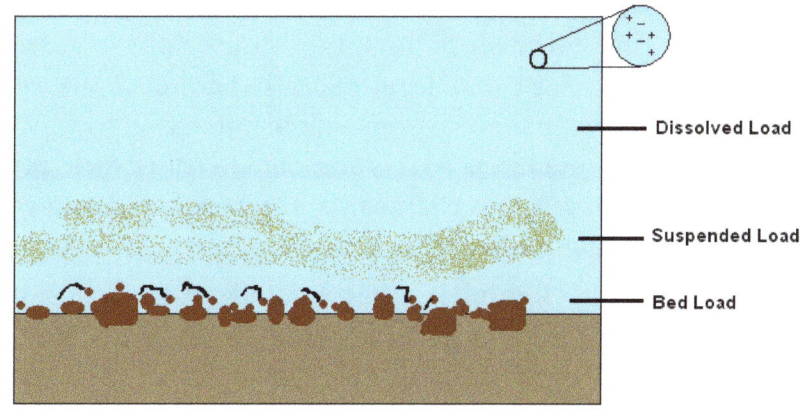

Figure 5.1 | An illustration depicting dissolved, suspended, and bed load.
Author: User "PSUEnviroDan"
Source: Wikimedia Commons
License: Public Domain

ried by a fast-moving stream. As a stream loses its energy and slows down, material will be deposited.

Under normal conditions, water will remain in a stream channel. When the amount of water in a stream exceeds it banks, the water that spills over the channel will decrease in velocity rapidly due to the greater friction on the water. As it drops velocity, it will also drop the larger sandy material it is carrying right along the channel margins, resulting in ridges of sandy alluvium called **natural levees** (Figure 5.2). As numerous flooding events occur, these ridges build up under repeated deposition. These levees are part of a larger landform known as a **floodplain**. A floodplain is the relatively flat land adjacent to the stream that is subject to flooding during times of high discharge (Figure 5.2).

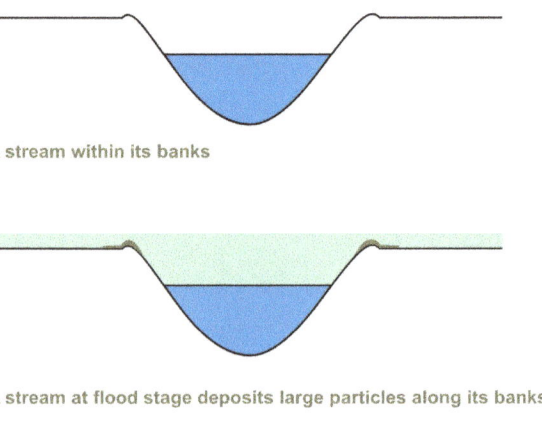

A stream within its banks

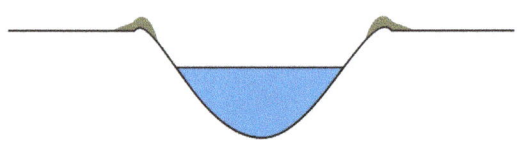

A stream at flood stage deposits large particles along its banks

After many floods, natural levees have built up along stream banks

Figure 5.2 | After many floods, natural levees have built up along stream banks.
Author: Julie Sandeen
Source: Wikimedia Commons
License: CC BY-SA 3.0

5.2.1 Stream Drainage Basins and Patterns

The **drainage basin** of a stream includes all the land that is drained by one stream, including all of its tributaries (the smaller streams that feed into the main stream). You are in a drainage basin right now. Do you know which one? You can find out on the internet. Go to the Environmental Protection Agency's webpage (epa.gov) and search for Surf Your Watershed to find out. The higher areas that separate drainage basins are called **drainage divides.** For North America, the continental divide in the Rocky Mountains separates water that drains to the west to the Pacific Ocean from water that drains to the east to the Gulf of Mexico.

As water flows over rock, it is influenced by it. Water wants to flow in the area of least resistance, so it is attracted to softer rock, rather than hard, resistant rock. This can result in characteristic patterns of drainage. Some of the more common **drainage patterns** include:

- **Dendritic** – this drainage pattern indicates uniformly resistant bedrock that often includes horizontal rocks. Since all the rock is uniform, the water is not attracted to any one area, and spreads out in a branching pattern, similar to the branches of a tree.

- **Trellis** – this drainage pattern indicates alternating resistant and non-resistant bedrock that has been deformed (folded) into parallel ridges and valleys. The water is attracted to the softer rock, and appears much like a rose climbing on a trellis in a garden.

- **Radial** – this drainage pattern forms as streams flow away from a central high point, such as a volcano, resembling the spokes in a wheel.
- **Rectangular** – this drainage pattern forms in areas in which rock has been fractured or faulted which created weakened rock. Streams are then attracted to the less resistant rock and create a network of channels that make right-angle bends as they intersect these breaking points. This pattern will often look like rectangles or squares.
- **Deranged** – this drainage pattern does not follow the rules. It consists of a random pattern of stream channels characterized by irregularity. It indicates that the drainage developed recently and has not had time to form one of the other drainage patterns yet.

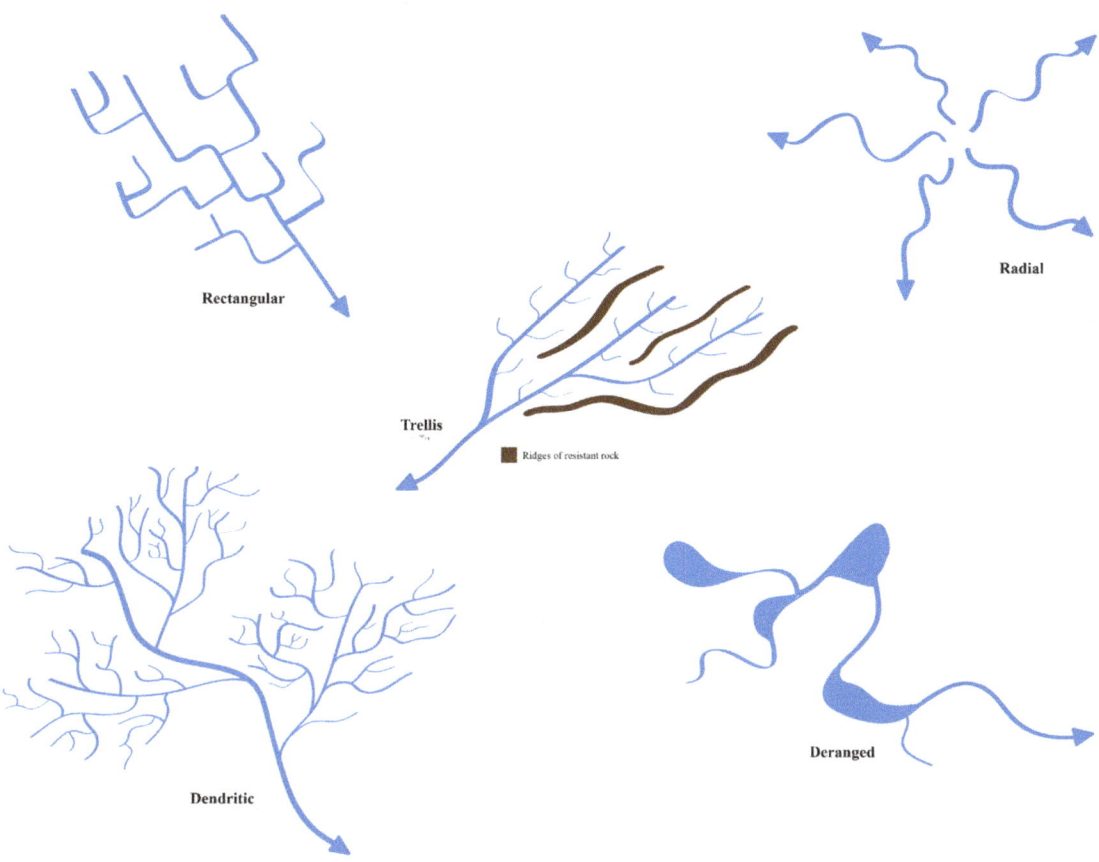

Figure 5.3 | Drainage patterns.
Author: Corey Parson
Source: Original Work
License: CC BY-SA 3.0

5.3 STREAM GRADIENT AND THE CYCLE OF STREAM EROSION

Stream gradient refers to the slope of the stream's channel, or rise over run. It is the vertical drop of the stream over a horizontal distance. You have dealt with gradient before in Topographic Maps. It can be calculated using the following equation:

$$\text{Gradient} = (\text{change in elevation}) / \text{distance}$$

Let's calculate the gradient from A to B in Figure 5.4 below. The elevation of the stream at A is 980', and the elevation of the stream at B is 920'. Use the scale bar to calculate the distance from A to B. Gradient = (980' – 920') / 2 miles, or 30 feet/mile.

Stream gradients tend to be higher in a stream's headwaters (where it originates), and lower at their mouth, where they discharge into another body of water (such as the ocean). **Discharge** measures stream flow at a given time and location, and specifically is a measure of the volume of water passing a particular point in a given period of time. It is found by multiplying the area (width multiplied by depth) of the stream channel by the velocity of the water, and is often in units of cubic feet (or meters) per second. Discharge increases downstream in most rivers, as tributaries join the main channel and add water.

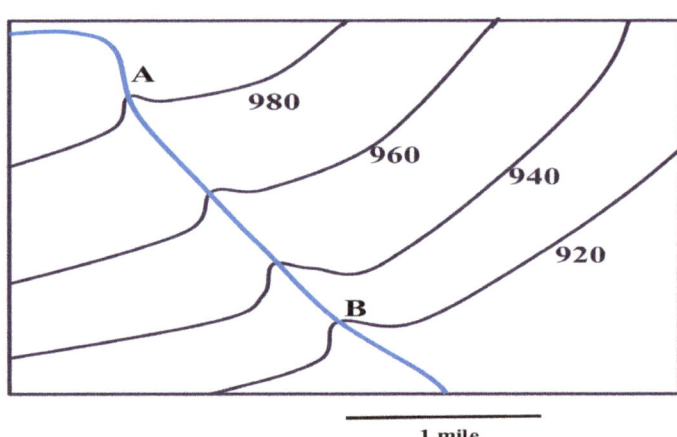

Figure 5.4 | Gradient calculation.
Author: Randa Harris
Source: Original Work
License: CC BY-SA 3.0

Sediment load (the amount of sediment carried by the stream) also changes from headwaters to mouth. At the headwaters, tributaries quickly carry their load downstream, combining with loads from other tributaries. The main river then eventually deposits that sediment load when it reaches base level. Sometimes in this process of carrying material downstream, the sediment load is large enough that the water is not capable of supporting it, so deposition occurs. If a stream becomes overloaded with sediment, braided streams may develop, with a network of intersecting channels that resembles braided hair. Sand and gravel bars are typical in braided streams, which are common in arid and semiarid regions with high erosion rates. Less commonly seen are straight streams, in which channels remain nearly straight, naturally due to a linear zone of weakness in the underlying rock. Straight channels can also be man-made, in an effort at flood control.

Streams may also be meandering, with broadly looping meanders that resemble "S"-shaped curves. The fastest water traveling in a meandering stream travels from outside bend to outside bend. This greater velocity and turbulence lead to more erosion on the outside bend, forming a featured called a cut bank. Erosion on this bank is offset by deposition on the opposite bank of the stream, where slower moving water allows sediment to settle out. These deposits are called point bars.

As meanders become more complicated, or sinuous, they may cut off a meander, discarding the meander to become a crescent-shaped oxbow lake. Check out Figure 5.6 to see the formation of an oxbow lake.

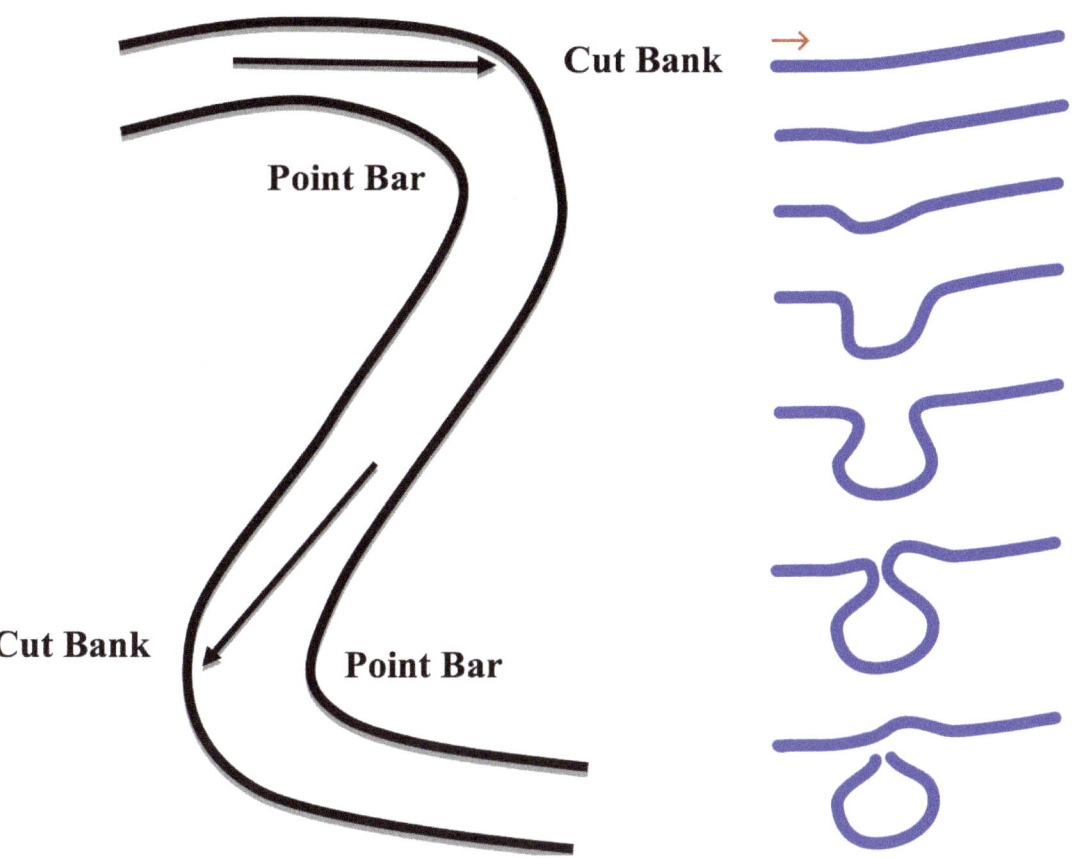

Figure 5.5 | (left) Parts of a meandering stream. The S-curves are meanders. The arrows within the stream depict where the fastest water flows. That water erodes the outside bank, creating a steep bank called the cut bank. The slowest water flows on the inside of the meander, slow enough to deposit sediment and create the point bar.
Author: Randa Harris
Source: Original Work
License: CC BY-SA 3.0

Figure 5.6 | (right) Formation of an oxbow lake. A meander begins to form and is cut off, forming the oxbow.
Author: User "Maksim"
Source: Wikimedia Commons
License: CC BY-SA 3.0

Even though streams are not living, they do go through characteristic changes over time as they change the landscape. The ultimate goal of a stream is to reach base level (the low elevation at which the stream can no longer erode its channel–

often a lake or other stream; ultimate base level is the ocean). While trying to reach this goal, the stream will experience the cycle of stream erosion, which consists of these stages:

- **Youthful (early) stage** – these streams are downcutting their channels (vertically eroding); literally they are picking up sediment from the bottom of their channels in an effort to decrease their elevation. The land surface will be above sea level, and these streams form deep V-shaped channels.
- **Mature (middle) stage** – these streams experience both vertical (downcutting) and lateral (meandering) erosion. The land surface is sloped, and streams begin to form floodplains (the flat land around streams that are subject to flooding).
- **Old age (late) stage** – these streams focus on lateral erosion and have very complicated meanders and oxbow lakes. The land surface is near base level.

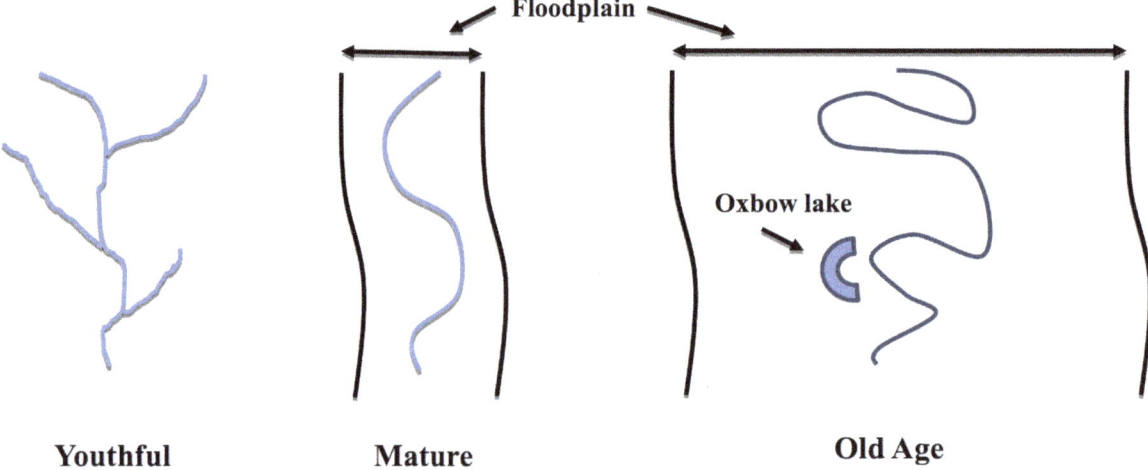

Figure 5.7 | Streams displaying the youthful, mature, and old age stages within the cycle of stream erosion. Note that the youthful stream does not have a floodplain.
Author: Randa Harris
Source: Original Work
License: CC BY-SA 3.0

An interruption may occur in this cycle. If a stream suddenly begins to downcut again, if sea level dropped (so base level dropped) or if the area around it was uplifted (think building mountains), then the stream would become rejuvenated. If the rejuvenated stream was in the old age stage, it will begin to form a deep V-shaped channel within that complicated meandering pattern that it has. This creates a neat geologic feature called an entrenched meander (Figure 5.8).

Figure 5.8 | Entrenched meanders along the San Juan River, Goosenecks State Park, Utah.
Author: User "Finetooth"
Source: Wikimedia Commons
License: CC BY-SA 3.0

5.4 LAB EXERCISE
Part A – Drainage Patterns

Map 5.1, located on the next page, is a portion of a topographic map from Royal Gorge, Colorado, courtesy of the USGS. Study the map and answer the questions below.

1. What type of stream drainage pattern is present on this map? This may be easier to determine by examining the tributaries to the main stream.

   a. dendritic   b. trellis   c. radial   d. rectangular

2. Based on the drainage pattern type, does the bedrock underlying this area consist of rocks uniformly resistant to erosion or rocks alternating between resistant and non-resistant layers?

   a. uniformly resistant bedrock   b. alternately resistant and non-resistant bedrock

3. Are the rocks likely tilted and folded or horizontal?

   a. tilted and folded   b. horizontal

4. Are streams in this area downcutting or laterally eroding?

   a. downcutting   b. laterally eroding

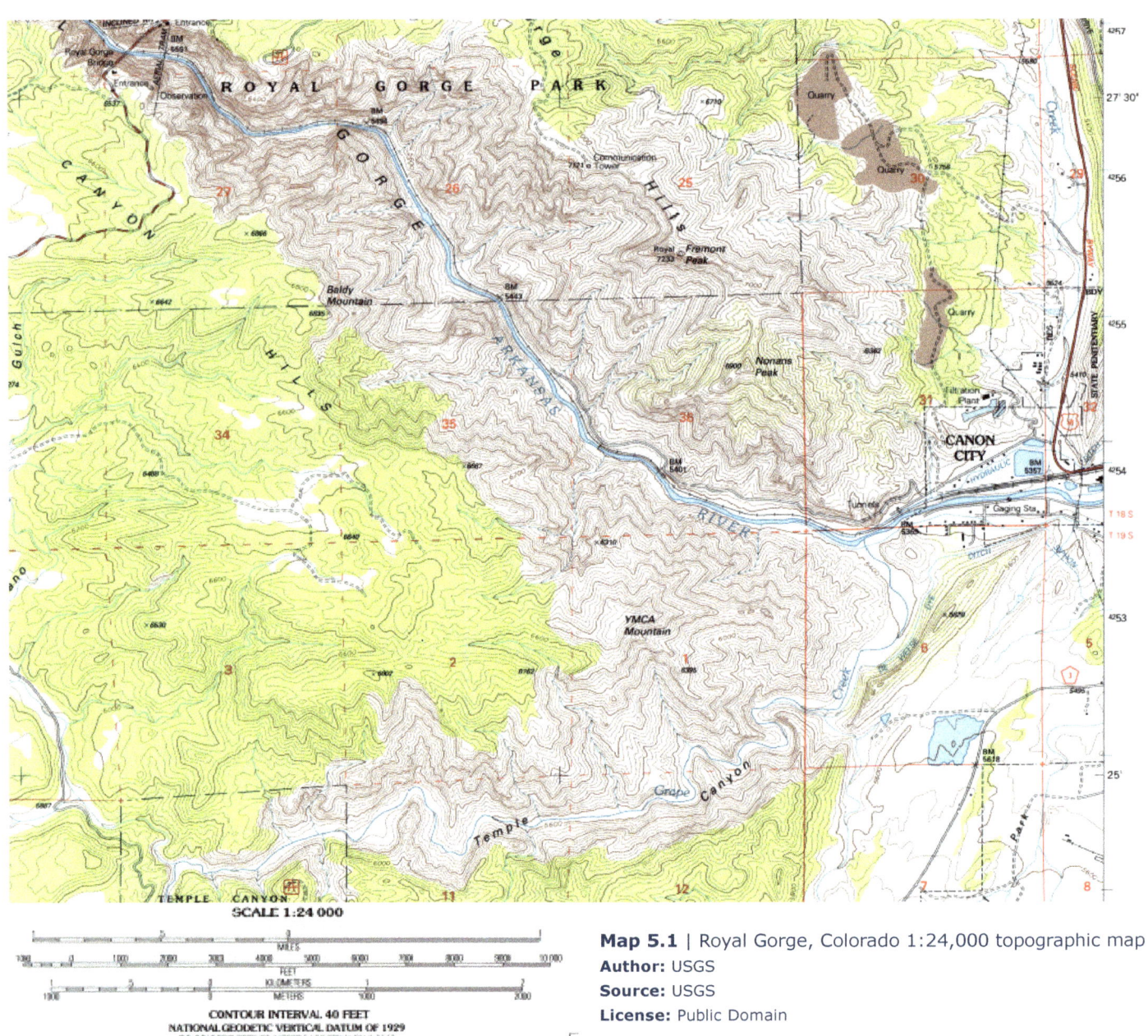

Map 5.1 | Royal Gorge, Colorado 1:24,000 topographic map
Author: USGS
Source: USGS
License: Public Domain

5. In which stage of the cycle of stream erosion is this area?

   a. old age   b. mature   c. youthful

Map 5.2 is the southeast portion of the map above, magnified to show Grape Creek. Use this map to answer question 6.

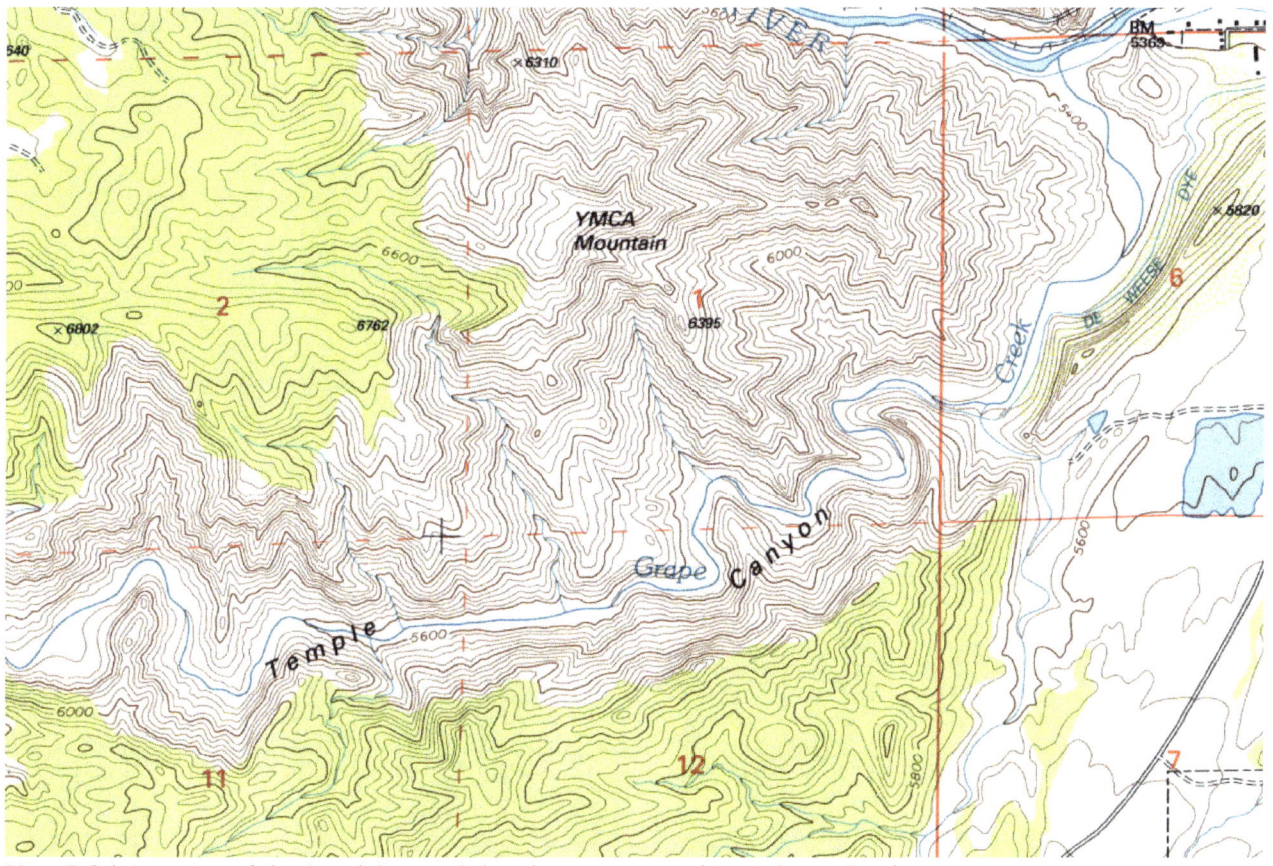

Map 5.2 | A portion of the Royal Gorge, Colorado map, zoomed in at Grape Creek
Author: USGS
Source: USGS
License: Public Domain

6. Calculate the stream gradient of Grape Creek in ft/mile. Use the index contour just above the "m" in Temple Canyon for your initial spot. Measure the gradient to the index contour past the word Creek (the last contour before it reaches Arkansas River). The distance between these areas is ~1.6 miles (measured along the curving distance of the non-magnified stream). What is the gradient?

   a. 15'/mile   b. 100'/mile   c. 125'/mile   d. 200'/mile

Use the Omaha North, Nebraska-Iowa Map 5.3 on the next page to answer questions 7-11.

7. Observe the stream on the Omaha N, Nebraska-Iowa quadrangle. In which stage of the cycle of stream erosion is this area?

   a. old age   b. mature   c. youthful

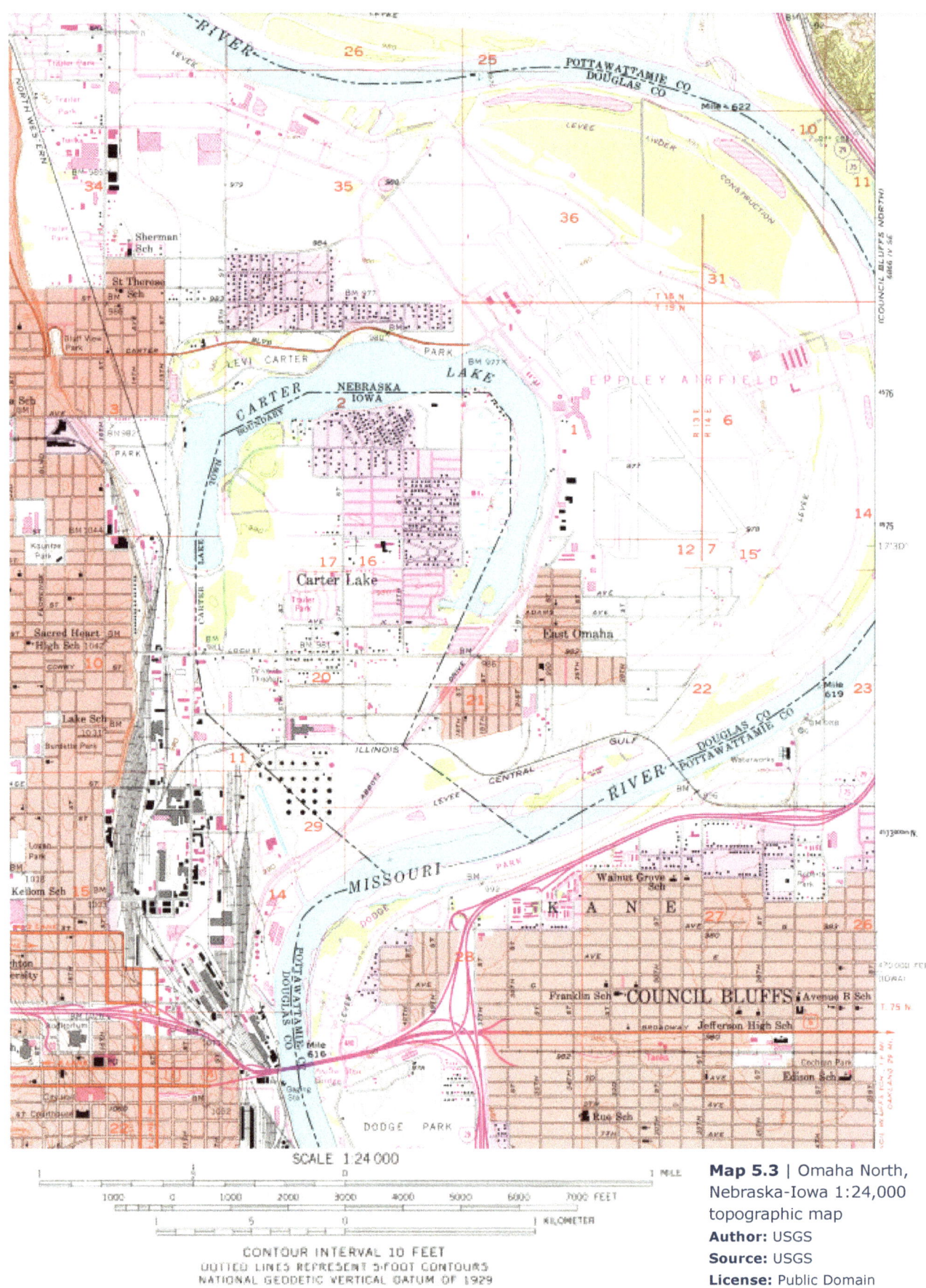

Map 5.3 | Omaha North, Nebraska-Iowa 1:24,000 topographic map
Author: USGS
Source: USGS
License: Public Domain

8. Compare the contour intervals from the Royal Gorge, Colorado map (map 5.1) to the Omaha N, Nebraska-Iowa map. Would you expect the gradient of the Missouri River in Nebraska to be greater or less than the gradient that you calculated for the Grape River in Colorado?

   a. less than the Grape River gradient   b. greater than the Grape River gradient

9. Locate the state boundary between Nebraska and Iowa along the Missouri River. Why does the boundary depart from the river channel?

   a. when the boundary was created, Iowans wanted the Carter Lake area in their state

   b. the boundary follows the course of the river at the time that it was drawn; the river has since moved

   c. none of the above

10. What is the term for the geologic feature called Carter Lake?

    a. entrenched meander   b. oxbow lake

    c. cutbank   d. point bar

11. Was Carter Lake cut off before or after the state boundary between Nebraska and Iowa was drawn?

    a. before   b. after

5.5 FLOODING

Flooding is a common and serious problem in our nation's waterways. Flood stage is reached when the water level in a stream overflows its banks. Floodplains are popular sites for development, with nice water views, but are best left for playgrounds, golf courses, and the like. Have you ever heard someone say that a flood was a 1 in 100 year flood? Does that mean that a flood of similar magnitude will occur every 100 years? No, it only means that, on average, we can expect a flood of this size or greater to occur within a 100 year peri-

Figure 5.9 | USGS officials monitoring the flooding of Sweetwater Creek over I-20 in September, 2009.
Author: USGS
Source: USGS
License: Public Domain

od. One cannot predict that it will occur in a particular year, only that each year has a 1 in 100 chance of having a flood of that magnitude. It also does not mean that only 1 flood of that size can occur within 100 years.

In order to better understand stream behavior, the U.S. Geological Survey has installed thousands of stream gauges throughout the country, locations with a permanent water level indicator and recorder. Data from these stations can be used to make flood frequency curves, which are useful in making flood control decisions. In western Georgia, a dramatic flood event occurred in September, 2009 that resulted in 11 fatalities, over $200 million in property damage, and closed Interstate 20 for a day. Rain fell from September 16-22, with a particularly intense period on September 20th. Use information below from a stream gauge located on Sweetwater Creek near Austell, Georgia, to create a flood frequency graph.

5.6 LAB EXERCISE

Part B – Recurrence Intervals

Data from the chart below was collected at the USGS site, and includes the 20 largest discharge events for Sweetwater Creek at station 02337000 from January 1, 2008 – May 1, 2015, excluding the dramatic 2009 flood (we will learn more about it later). In order to create a flood frequency graph, first the recurrence interval must be calculated (one is calculated below for an example). A **recurrence interval** refers to the average time period within which a given flood event will be equaled or exceeded once. To calculate it, first determine the rank of the flood, with a 1 going to the highest discharge event and a 20 going to the lowest discharge event. Calculate the recurrence interval using the following equation:

$$RI = (n+1) / m$$

where RI = Recurrence Interval (yrs)
n = number of years of record (in this case, 8)
m = rank of flood

Peak Discharge Date	Discharge (cfs – cubic ft/sec)	Rank	Recurrence Interval
8/27/2008	5,140		
3/2/2009	2,360		
10/13/2009	3,290		
11/12/2009	6,120	1	9
12/3/2009	2,860		
12/10/2009	2,170		
12/19/2009	3,830		

Peak Discharge Date	Discharge (cfs – cubic ft/sec)	Rank	Recurrence Interval
12/26/2009	2,650		
1/25/2010	2,500		
2/6/2010	3,680		
3/12/2010	3,600		
3/10/2011	2,350		
4/17/2011	3,100		
2/24/2013	2,060		
2/27/2013	2,190		
5/6/2013	3,610		
12/23/2013	3,790		
4/8/2014	4,170		
1/5/2015	3,970		
4/20/2015	2,940		

Now that you have completed the chart, plot the discharge against the recurrence interval on the following graph. Please note that the x-axis (for recurrence interval) is in logarithmic scale and you may need to estimate where the data points fall. A logarithmic scale is *non-linear*, based on orders of magnitude. Each mark on the x-axis is the previous mark multiplied by a value. Use a straight edge to draw a best fit line (a straight line along the graph that shows the general direction that the group of points seem to be heading – it doesn't have to hit every point on the graph) through the graph when you are done, and then answer the following questions. Make sure your best fit line continues to the end of the graph.

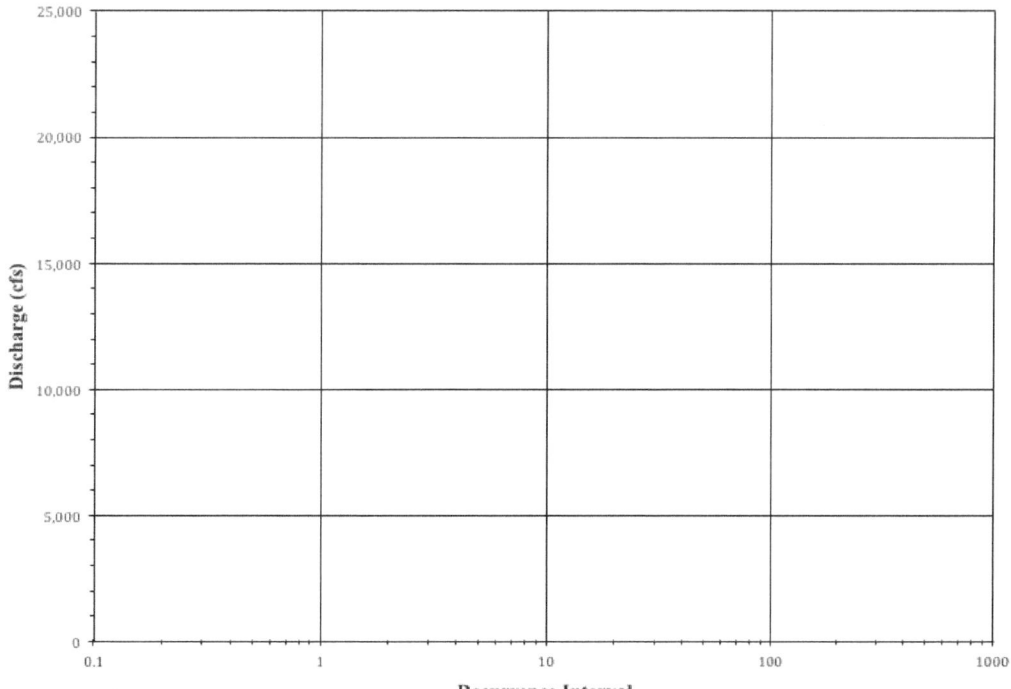

12. On which date did a flood event have a recurrence interval of 0.5?

    a. 2/27/2013   b. 10/13/2009   c. 3/10/2011   d. 4/20/2015

13. Of the following dated flood events, which one would you expect to happen more often?

    a. 8/27/2008   b. 2/24/2013   c. 2/6/2010   d. 12/23/2013

14. Observe your best fit line. What *approximate* discharge would be associated with a 50 year recurrence interval?

    a. 2,000 cfs   b. 4,750 cfs   c. 8,500 cfs   d. 14,000 cfs

15. Flood stage, or bankfull stage, on Sweetwater Creek occurs at a discharge of ~4,500 cfs. According to your best fit line, what is the recurrence interval of such a discharge?

    a. 0.5 years   b. 3 years   c. 25 years   d. 50 years

16. During the flood event of 9/23/2009, the discharge measured at this gaging station was 21,200 cfs. Note where this would plot on your graph. Would the recurrence interval for this flood plot at:

    a. 100 years   b. 300 years   c. 700 years   d. longer than 1,000 years

17. Is it possible that a flood with a similar discharge to that of the event from 9/23/2009 could happen again in the next 20 years?

    a. Yes   b. No

5.7 GROUNDWATER

It is best not to envision groundwater as underground lakes and streams (which only occasionally exist in caves), instead think of groundwater slowly seeping from one miniscule pore in the rock to another. Have you ever been to the beach and dug a hole, only to have it fill with water from the base? If so, you had reached the water table, the boundary between the unsaturated and saturated zones. Rocks and soil just beneath the land's surface are part of the unsaturated zone, and pore spaces in them are filled with air. Once the water table is reached, then rocks and soil pore spaces are filled with water, in the saturated zone.

The water table is said to mimic topography, in that it generally lies near the surface of the ground (often tens of feet below the surface, though this can vary greatly with location). The water table rises with hills and sinks with valleys, often discharging into streams. The water table receives additional inputs as rainfall infiltrates into the ground, called recharge. Its position is dynamic – during droughts the water table will lower and during wet times, it rises.

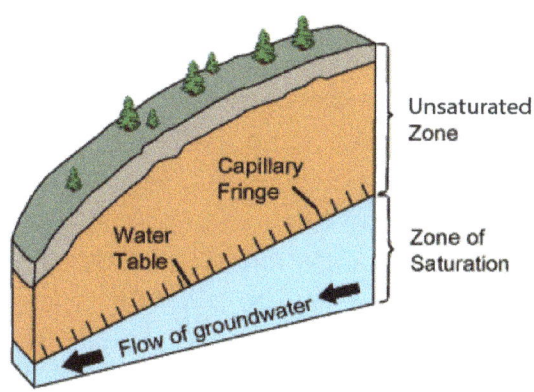

Figure 5.10 | The water table is the boundary between the unsaturated zone and saturated zone.
Author: USGS
Source: Wikimedia Commons
License: Public Domain

Two important properties of groundwater that influence its availability and movement are porosity and permeability. **Porosity** refers to the open or void space within the rock. It is expressed as a percentage of the volume of open space compared to the total rock volume. Porosity will vary with rock type. Many rocks with tight interlocking crystals (such as igneous and metamorphic rocks) will have low porosity since they lack open space. Sedimentary rocks composed of well sorted sediment tend to have high porosity because of the abundant spaces between the grains that make them up. To imagine this, envision a room filled from floor to ceiling with basketballs (similar to a rock composed completely of sand grains). Now add water to the room. The room will be able to hold a good deal of water, since the basketballs don't pack tightly due to their shape. That would be an example of high porosity.

Permeability refers to the ability of a geologic material to transport fluids. It depends upon the porosity within the rock, but also on the size of the open space and how interconnected those open spaces are. Even though a material is porous,

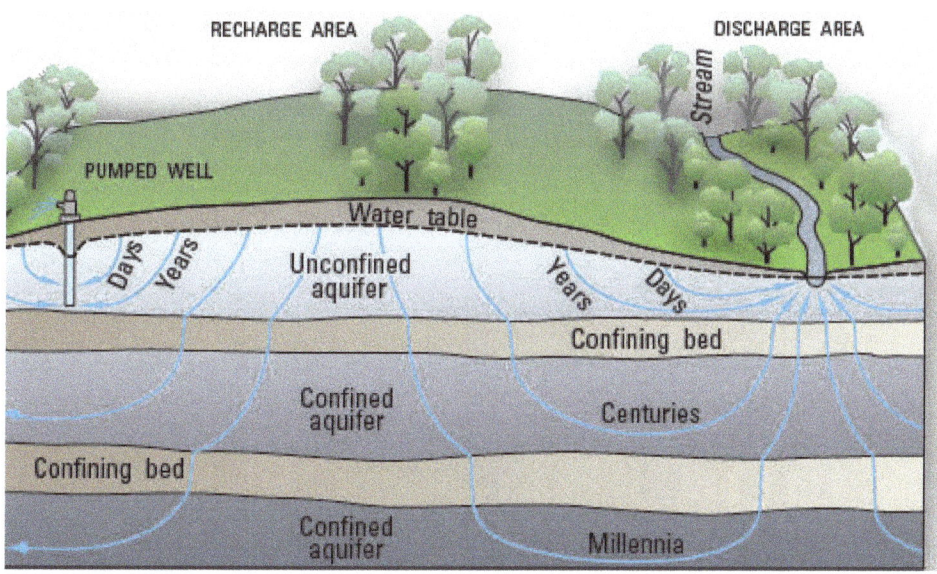

Figure 5.11 | Groundwater flow in both confined and unconfined aquifers.
Author: USGS
Source: USGS
License: Public Domain

if the open spaces aren't connected, water won't flow through it. Rocks that are permeable make good **aquifers**, geologic units that are able to yield significant water. Sedimentary rocks such as sandstone and limestone are good aquifers. Rocks that are impermeable make confining layers and prevent the flow of water. Examples of confining layers would be sedimentary rocks like shale (made from tiny clay and silt grains) or un-fractured igneous or metamorphic rock. In an unconfined aquifer, the top of the aquifer is the water table.

Groundwater generally flows from areas of higher elevation to lower elevation in the shallow subsurface. Note the flow paths in Figure 5.11. Approximately 20% of the water used in the United States is groundwater, and this water has the potential to become contaminated, mostly from sewage, landfills, industry, and agriculture. The movement of groundwater helps spread the pollutants, making containment a challenge.

5.8 LAB EXERCISE
Part C – Groundwater Flow

Many gas stations use underground storage tanks (UST) to store fuel below the ground (you have likely seen a tanker truck at a gas station filling up the UST). These UST's could leak, and gasoline could possibly reach the water table. In the diagram below, a business using a well has detected gasoline in their groundwater. To detect the source of the potential leak, contour the water table's surface and determine its flow path. There are several gas stations in the diagram, and each has the potential to have the leaking UST. Seven monitoring wells are installed in the area, and you have data about the water table elevation within each well. Using that data (in elevation above sea level), contour this map as you would any other. Add the water table elevations to the map, and using pencil, contour the groundwater elevations using a contour interval of 2 feet. Noting that the gas should flow with the groundwater, determine the direction of groundwater flow and note the most likely gas station to be the source of the gasoline leak (in a real world scenario, once the likely culprit was determined, more monitoring wells would be installed and they would be tested for gasoline residue).

Table 5.1

Monitoring Well	Water table elevation (feet)
1	794'
2	790'
3	788'
4	786'
5	786'
6	783'
7	780'

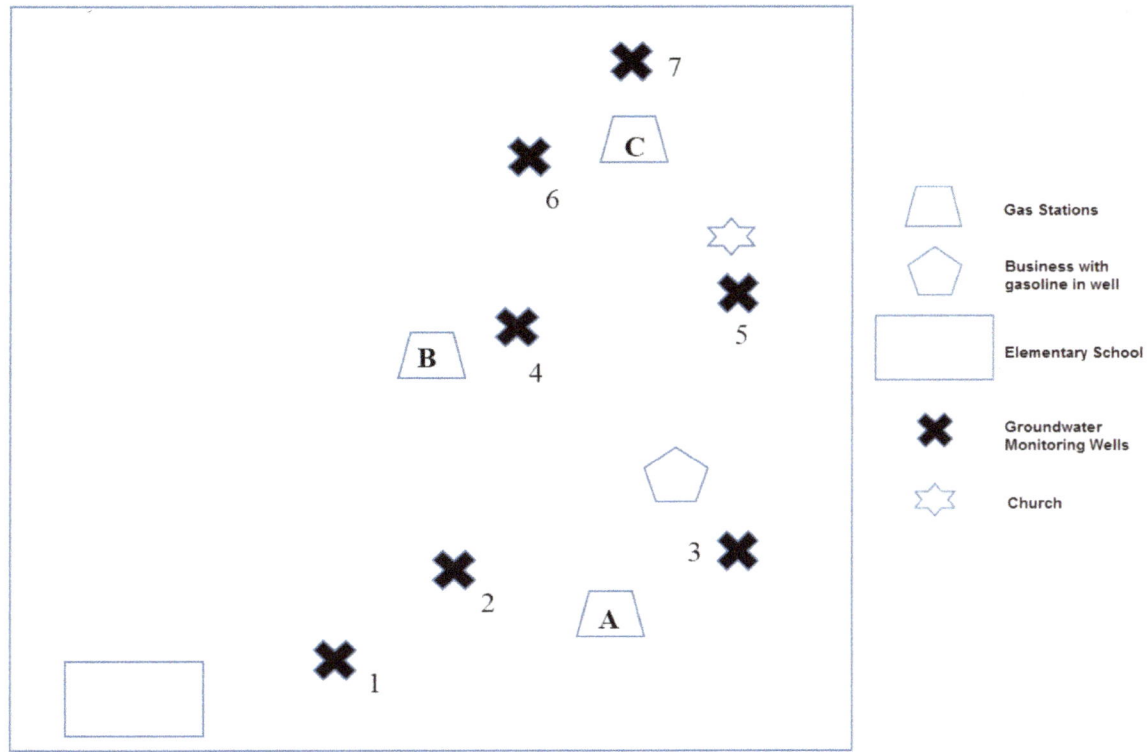

Figure 5.12 | Groundwater contamination exercise.
Author: Randa Harris
Source: Original Work
License: CC BY-SA 3.0

18. Which gas station is the most likely source of the gasoline leak?

    a. Station A   b. Station B   c. Station C

19. Is the school likely to be at risk of contamination from this same leak?

    a. Yes   b. No

20. Is the church likely to be at risk of contamination from this same leak?

    a. Yes   b. No

5.9 KARST TOPOGRAPHY

The sedimentary rock limestone is composed of the mineral calcite, which is water soluble, meaning it will dissolve in water that is weakly acidic. In humid areas where limestone is found, water dissolves away the rock, forming large cavities and depressions which vary in size and shape. As more dissolution occurs, the caves become unstable and collapse, creating sinkholes. These broad, crater-like depressions are typical of **karst topography**, named after the Karst region in

Slovenia. Karst topography is characterized by sinkholes, sink lakes (sinkholes filled with water), caves, and disappearing streams (surface streams that disappear into a sinkhole). Living in karst topography poses its challenges, and approximately one fourth of Americans in the lower 48 states live in these regions. Sinkholes can appear rather rapidly and cause great damage to any structures above them.

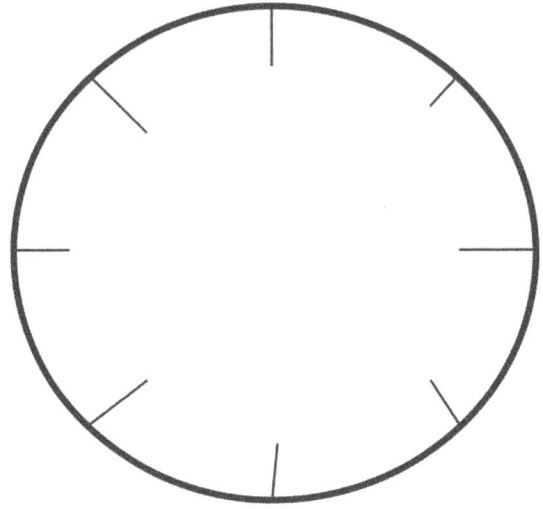

Figure 5.13 | Appearance of a sinkhole on a topographic map.
Author: Randa Harris
Source: Original Work
License: CC BY-SA 3.0

5.10 LAB EXERCISE

Part D – Karst topography

Use the Mammoth Cave, Kentucky map (Map 5.4 located at the end of this chapter) to answer the following questions. Though it is old (1922), the general geology and landforms in this area have not changed.

21. Locate Little Sinking Creek in the southern portion of the map, north of Hwy. 68 and south of the Edmonson County Line. In which direction does it flow?

    a. south   b. north   c. southeast   d. northwest

22. Follow the creek along its path. Where does it wind up?

    a. Along Hwy. 68   b. It disappears underground

23. Find Sloans Crossing. It is south of Mammoth Cave. What is the benchmark elevation at Sloans Crossing?

    a. 600'   b. 630'   c. 800'   d. 834'

INTRODUCTORY GEOLOGY   WATER

24. Now look farther south of Sloans Crossing at Hwy. 31W. Look closely at the topography south of highway, and it changes abruptly. What feature(s) can you observe south of Hwy. 31?

    a. sinkholes   b. disappearing streams

    c. generally lower land surface elevations   d. all of the above

25. Keeping the abrupt topography change in mind, which of the following is true?

    a. In the northern portion of the map, the area is underlain by limestone

    b. In the southern portion of the map, the area is underlain by limestone

26. Locate the Louisville and Nashville Railroad line just south of Hwy. 31. Would this be an easy location to maintain a railroad?

    a. Yes   b. No

Part E – Google Earth

The exercises that follow use Google Earth. For each question (or set of questions) paste the location that is given into the "Search" box. Examine each location at multiple eye altitudes and differing amounts of tilt. For any measurements that use the ruler tool, this can be accessed by clicking on the ruler icon above the image.

Search for 7 26 16.49S 75 00 00.06W and zoom out to an eye altitude of 15 miles.

27. How would one describe this river?

    a. Straight   b. Meandering   c. Low sinuosity   d. Braided

28. In this stream, erosion is occurring on the _____ because _____, while deposition is occurring on the _____ because _____.

    a. point bars; the fastest velocity water flows to this point; cut banks; the slowest velocity water flows to this point

    b. point bars; the slowest velocity water flows to this point; cut banks; the fastest velocity water flows to this point

    c. cut banks; the fastest velocity water flows to this point; point bars; the slowest velocity water flows to this point

    d. cut banks; the slowest velocity water flows to this point; point bars; the fastest velocity water flows to this point

INTRODUCTORY GEOLOGY   WATER

Search for 63 55 55.23N 17 01 07.14W and zoom out to an eye altitude of 10,000 feet.

29. How would one describe this river?

　　a. Straight　　b. Meandering　　c. Low sinuosity　　d. Braided

30. What factors control the course of this river?

　　a. Steep gradient and high discharge

　　b. Low gradient and low discharge

　　c. Low gradient and abundant sediment supply

　　d. Steep gradient and low sediment supply

Search for 38 01 12.18N 121 43 20.02W and zoom out to an eye altitude of 30,000 feet.

31. The river in this area has a rather particular pattern, what geologic process caused this?

　　a. a meander eroded through its bank and created an oxbow lake

　　b. the river is in a karst terrain and disappeared into the ground

　　c. the river is following patterns, likely faults, in the underlying bedrock

　　d. during a flood the river breached the natural levee flowing into the floodplain

32. Zoom out and examine the surrounding area, what geological hazards are likely in the area?

　　a. sinkholes

　　b. flooding of urban areas

　　c. erosion and subsidence

　　d. none of the above

Search for 41 24 30.77N 122 11 46.23W and zoom to an eye altitude 50,000 feet.

33. What type of drainage pattern is present in this area?

　　a. Trellis　　b. Dendritic　　c. Rectangular　　d. Radial　　e. Deranged

34. What does this type of drainage pattern indicate about the area?

    a. rocks in the area are homogeneous and/or flat lying

    b. rocks in the area are alternating resistant and non-resistant, forming parallel ridges and valleys

    c. stream channels radiate outward like wheel spokes from a high point

    d. stream channels flow randomly with no relation to underlying rocks or structure

Search for 36 45 41.23N 85 18 34.22W and zoom out to an eye altitude of 25,000 ft. Although there is no stream present today, the stream channels can be clearly seen.

35. What type of drainage pattern was present in this area?

    a. Trellis   b. Dendritic   c. Rectangular   d. Radial   e. Deranged

36. What does this type of drainage pattern indicate about the area?

    a. rocks in the area are homogeneous and/or flat lying

    b. rocks in the area are alternating resistant and non-resistant, forming parallel ridges and valleys

    c. stream channels radiate outward like wheel spokes from a high point

    d. stream channels flow randomly with no relation to underlying rocks or structure

37. In what direction was the main river flowing?

    a. west   b. east   c. north   d. south

Search for 28 38 01.92N 81 22 44.78W and zoom out to an eye altitude of 13,000 ft.

38. How were these lakes formed?

    a. they were man-made – all are dammed

    b. they are formed by large rivers in the area

    c. as sinkholes, as underlying soluble rock was dissolved and areas collapsed

    d. they are impact structures that filled with water

39. What type of bedrock is present in this area?

    a. limestone   b. sandstone   c. gneiss   d. granite   e. chert

Now let's travel to Mars! Mars has shown evidence of water, and NASA and others have been studying Mars intensely in recent years. Let's look at a few features on Mars. In the location toolbar across the top of Google Earth, locate the image that looks like the planet Saturn. Click the button to bring up several location options and select Mars. In the Layers box on the left, notice Global Maps. You may have to click on it to expand it. For today, we will use the highest quality images found in the Visible Images layer. The other layers are interesting, but won't be used today.

In the Search tab, type in Noctis Labyrinthus and zoom to ~300 miles.

40. These features you are seeing are linear valleys. Assume that water flowed through these valleys at some time. What type of drainage pattern would this area represent?

    a. Trellis   b. Dendritic   c. Rectangular   d. Radial   e. Deranged

41. Think about the drainage pattern you selected in the previous answer – what does this tell you about the underlying rocks?

    a. the rocks are probably fractured

    b. the rocks are uniformly resistant

    c. the rocks are part of a topographic high, like a mountain

    d. the rocks are alternately resistant and non-resistant

In the Search tab, type in Warrego Valles and zoom to ~120 miles.

42. Notice the general shape of this feature. It is thought to have formed by the runoff of either precipitation or groundwater. What type of drainage does this appear to be?

    a. Trellis   b. Dendritic   c. Rectangular   d. Radial   e. Deranged

5.11 STUDENT RESPONSES

1. What type of stream drainage pattern is present on this map? This may be easier to determine by examining the tributaries to the main stream.

    a. dendritic   b. trellis   c. radial   d. rectangular

2. Based on the drainage pattern type, does the bedrock underlying this area consist of rocks uniformly resistant to erosion or rocks alternating between resistant and non-resistant layers?

    a. uniformly resistant bedrock   b. alternately resistant and non-resistant bedrock

3. Are the rocks likely tilted and folded or horizontal?

    a. tilted and folded   b. horizontal

4. Are streams in this area downcutting or laterally eroding?

    a. downcutting   b. laterally eroding

5. In which stage of the cycle of stream erosion is this area?

    a. old age   b. mature   c. youthful

6. Calculate the stream gradient of Grape Creek in ft/mile. Use the index contour just above the "m" in Temple Canyon for your initial spot. Measure the gradient to the index contour past the word Creek (the last contour before it reaches Arkansas River). The distance between these areas is ~1.6 miles (measured along the curving distance of the non-magnified stream). What is the gradient?

    a. 15'/mile   b. 100'/mile   c. 125'/mile   d. 200'/mile

7. Observe the stream on the Omaha N, Nebraska-Iowa quadrangle. In which stage of the cycle of stream erosion is this area?

    a. old age   b. mature   c. youthful

8. Compare the contour intervals from the Royal Gorge, Colorado map (map 5.1) to the Omaha N, Nebraska-Iowa map. Would you expect the gradient of the Missouri River in Nebraska to be greater or less than the gradient that you calculated for the Grape River in Colorado?

    a. less than the Grape River gradient   b. greater than the Grape River gradient

9. Locate the state boundary between Nebraska and Iowa along the Missouri River. Why does the boundary depart from the river channel?

   a. when the boundary was created, Iowans wanted the Carter Lake area in their state

   b. the boundary follows the course of the river at the time that it was drawn; the river has since moved

   c. none of the above

10. What is the term for the geologic feature called Carter Lake?

    a. entrenched meander   b. oxbow lake

    c. cutbank   d. point bar

11. Was Carter Lake cut off before or after the state boundary between Nebraska and Iowa was drawn?

    a. before   b. after

12. On which date did a flood event have a recurrence interval of 0.5?

    a. 2/27/2013   b. 10/13/2009   c. 3/10/2011   d. 4/20/2015

13. Of the following dated flood events, which one would you expect to happen more often?

    a. 8/27/2008   b. 2/24/2013   c. 2/6/2010   d. 12/23/2013

14. Observe your best fit line. What *approximate* discharge would be associated with a 50 year recurrence interval?

    a. 2,000 cfs   b. 4,750 cfs   c. 8,500 cfs   d. 14,000 cfs

15. Flood stage, or bankfull stage, on Sweetwater Creek occurs at a discharge of ~4,500 cfs. According to your best fit line, what is the recurrence interval of such a discharge?

    a. 0.5 years   b. 3 years   c. 25 years   d. 50 years

INTRODUCTORY GEOLOGY   WATER

16. During the flood event of 9/23/2009, the discharge measured at this gaging station was 21,200 cfs. Note where this would plot on your graph. Would the recurrence interval for this flood plot at:

    a. 100 years   b. 300 years   c. 700 years   d. longer than 1,000 years

17. Is it possible that a flood with a similar discharge to that of the event from 9/23/2009 could happen again in the next 20 years?

    a. Yes   b. No

18. Which gas station is the most likely source of the gasoline leak?

    a. Station A   b. Station B   c. Station C

19. Is the school likely to be at risk of contamination from this same leak?

    a. Yes   b. No

20. Is the church likely to be at risk of contamination from this same leak?

    a. Yes   b. No

21. Locate Little Sinking Creek in the southern portion of the map, north of Hwy. 68 and south of the Edmonson County Line. In which direction does it flow?

    a. south   b. north   c. southeast   d. northwest

22. Follow the creek along its path. Where does it wind up?

    a. Along Hwy. 68   b. It disappears underground

23. Find Sloans Crossing. It is south of Mammoth Cave. What is the benchmark elevation at Sloans Crossing?

    a. 600'   b. 630'   c. 800'   d. 834'

24. Now look farther south of Sloans Crossing at Hwy. 31W. Look closely at the topography south of highway, and it changes abruptly. What feature(s) can you observe south of Hwy. 31?

    a. sinkholes   b. disappearing streams

    c. generally lower land surface elevations   d. all of the above

25. Keeping the abrupt topography change in mind, which of the following is true?

    a. In the northern portion of the map, the area is underlain by limestone

    b. In the southern portion of the map, the area is underlain by limestone

26. Locate the Louisville and Nashville Railroad line just south of Hwy. 31. Would this be an easy location to maintain a railroad?

    a. Yes   b. No

27. How would one describe this river?

    a. Straight   b. Meandering   c. Low sinuosity   d. Braided

28. In this stream, erosion is occurring on the _____ because _____, while deposition is occurring on the _____ because _____.

    a. point bars; the fastest velocity water flows to this point; cut banks; the slowest velocity water flows to this point

    b. point bars; the slowest velocity water flows to this point; cut banks; the fastest velocity water flows to this point

    c. cut banks; the fastest velocity water flows to this point; point bars; the slowest velocity water flows to this point

    d. cut banks; the slowest velocity water flows to this point; point bars; the fastest velocity water flows to this point

29. How would one describe this river?

    a. Straight   b. Meandering   c. Low sinuosity   d. Braided

30. What factors control the course of this river?

    a. Steep gradient and high discharge

    b. Low gradient and low discharge

    c. Low gradient and abundant sediment supply

    d. Steep gradient and low sediment supply

INTRODUCTORY GEOLOGY   WATER

31. The river in this area has a rather particular pattern, what geologic process caused this?

   a. a meander eroded through its bank and created an oxbow lake

   b. the river is in a karst terrain and disappeared into the ground

   c. the river is following patterns, likely faults, in the underlying bedrock

   d. during a flood the river breached the natural levee flowing into the floodplain

32. Zoom out and examine the surrounding area, what geological hazards are likely in the area?

   a. sinkholes

   b. flooding of urban areas

   c. erosion and subsidence

   d. none of the above

33. What type of drainage pattern is present in this area?

   a. Trellis   b. Dendritic   c. Rectangular   d. Radial   e. Deranged

34. What does this type of drainage pattern indicate about the area?

   a. rocks in the area are homogeneous and/or flat lying

   b. rocks in the area are alternating resistant and non-resistant, forming parallel ridges and valleys

   c. stream channels radiate outward like wheel spokes from a high point

   d. stream channels flow randomly with no relation to underlying rocks or structure

35. What type of drainage pattern was present in this area?

   a. Trellis   b. Dendritic   c. Rectangular   d. Radial   e. Deranged

36. What does this type of drainage pattern indicate about the area?

   a. rocks in the area are homogeneous and/or flat lying

   b. rocks in the area are alternating resistant and non-resistant, forming parallel ridges and valleys

   c. stream channels radiate outward like wheel spokes from a high point

   d. stream channels flow randomly with no relation to underlying rocks or structure

37. In what direction was the main river flowing?

   a. west   b. east   c. north   d. south

38. How were these lakes formed?

   a. they were man-made – all are dammed

   b. they are formed by large rivers in the area

   c. as sinkholes, as underlying soluble rock was dissolved and areas collapsed

   d. they are impact structures that filled with water

39. What type of bedrock is present in this area?

   a. limestone   b. sandstone   c. gneiss   d. granite   e. chert

40. These features you are seeing are linear valleys. Assume that water flowed through these valleys at some time. What type of drainage pattern would this area represent?

   a. Trellis   b. Dendritic   c. Rectangular   d. Radial   e. Deranged

41. Think about the drainage pattern you selected in the previous answer – what does this tell you about the underlying rocks?

   a. the rocks are probably fractured

   b. the rocks are uniformly resistant

   c. the rocks are part of a topographic high, like a mountain

   d. the rocks are alternately resistant and non-resistant

42. Notice the general shape of this feature. It is thought to have formed by the runoff of either precipitation or groundwater. What type of drainage does this appear to be?

   a. Trellis   b. Dendritic   c. Rectangular   d. Radial   e. Deranged

Map 5.4 | (appearing on following page) Mammoth Cave, Kentucky 1:24,000 topographic map
Author: USGS
Source: USGS
License: Public Domain

Climate Change
Bradley Deline

6.1 INTRODUCTION

Climate is an average of the long-term weather patterns across a geographic area, which is a complicated metric controlled by factors within the lithosphere, atmosphere, cryosphere, hydrosphere, biosphere, and anthrosphere as well as factors beyond our own planet. It is helpful to separate out humans from other life (anthrosphere verses biosphere) for several reasons, primarily because many of our activities are unique amongst life (industrialization) and it is helpful in understanding our role in climate change. Therefore, the science examining past, current, and future climate is extremely complex and interdisciplinary. You may not think of climate as a geological field of study, but the history of climate is recorded within rocks, the current climate is altered by geologic events, and future climate will be influenced by our use of geological resources such as fossil fuels. In addition to the complex nature of this subject, it is also one, if not the most, important scientific fields of study both in terms of understanding the dynamics and implications of future climate change as well as attempting to combat or mitigate the potential effects.

Though the basic science behind climate and climate change has been well studied to a point of near consensus within the scientific community, there is still significant debate amongst the broader population. This is likely related to many factors beyond science including economics, politics, the portrayal of the science by the media, and the overall public's scientific literacy. Gaining a better understanding of this issue is difficult given the enormous wealth of information and disparity in scientific literacy. This lab will explore this issue by examining climate data as well as how we, as scientists or scientific minded citizens, make interpretations and conclusions regarding data, how it is presented, and how it relates to our understanding of the world around us.

6.1.1 Learning Outcomes

After completing this chapter, you should be able to:
- Describe the climate system and how different variables are related

- Discuss how ancient climate patterns are reconstructed
- Plot, interpret, and explain the patterns in climate proxy data focusing on the sea ice extent in the North and South Poles
- Describe how heat is transported across the earth and how this can relate to local climate
- Describe the information needed to make conclusions regarding scientific patterns and how climate models should be constructed

6.1.2 Key Terms

- Albedo
- Climate Proxies
- Climate System
- Greenhouse Gases
- Ice Extent
- Negative Feedback
- Ocean Gyres
- Positive Feedback

6.2 THE CLIMATE SYSTEM

As was previously mentioned, climate is the long-term weather pattern across a region. It is important to emphasize the long-term portion of the definition to establish that climate is different from weather. Weather is the local and short-term patterns in temperature, humidity, precipitation, atmospheric pressure, wind, and other meteorological variables. As you well know, weather fluctuates throughout the day, week, month, and year such that it is difficult to see any trends beyond the random noise in the system. If you take a long-term view of weather we can begin to see patterns across time and geography that help to better understand and identify the factors that influence the climate system. **The climate system** is the interconnected network of variables that influence the earth's climate, which includes components from the lithosphere, atmosphere, hydrosphere, cryosphere, biosphere, anthrosphere, and solar system.

The heat that feeds this system comes from two primary sources. First, there is heat radiating from the Earth itself, which is coming from the decay of radioactive material and residual heat from the formation of the earth. This heat is not distributed equally, with more heat escaping in areas where the crust is thinner, such as divergent boundaries. More significantly, the earth receives heat from solar radiation. Again, this heat is not distributed equally across the earth's surface and the amount of energy received is related to the angle at which

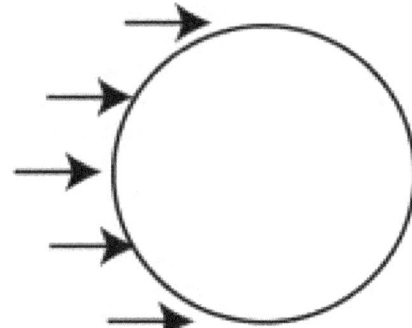

Figure 6.1 | The Earth's shape influences the angle at which the sun's rays hit the surface from perpendicular at the equator to parallel at the poles. This creates large climate differences across the Earth.
Author: Bradley Deline
Source: Original Work
License: CC BY-SA 3.0

the solar radiation hits the surface of the earth (Figure 6.1). If the solar radiation hits perpendicular to the surface more heat is absorbed than if it hits at an oblique angle, which is why the tropics are warmer than the poles.

The material on the Earth's surface is also important in that materials react differently to solar radiation. Some materials, normally dark in color, absorb and reradiate heat, most of which is retained at the surface of the planet. You are likely familiar with this if you have ever walked barefoot on dark concrete or asphalt in the summer. Other materials, that are shiny or light in color, reflect the solar radiation off the Earth's surface. Materials such as snow or ice are particularly effective at reflecting solar radiation. This is the reason Arctic explorers must use eye protection to avoid snow blindness. The proportion of solar radiation that is reflected off the Earth's surface is called **albedo**, which can vary depending on the type of ground cover. For instance the Earth's albedo is higher when it is covered with large expanses of glacial ice and thus the amount of sunlight absorbed and the temperature measured are lower.

Once heat is reradiated off of the Earth's surface it travels up into the atmosphere. Certain gases in Earth's atmosphere, called **greenhouse gases**, allow sunlight to pass but absorb terrestrial energy and radiate it in all directions including back to the surface of the Earth. These gases, such as water vapor, carbon dioxide, and methane represent a tiny, though important fraction of the material in the atmosphere. Different greenhouse gases vary both in how effectively they absorb and reradiate energy and their relative proportions in the atmosphere, such that a higher concentration of potent greenhouse gases can retain more thermal energy within the atmosphere. The rest of the reflected and radiated energy escapes from the atmosphere and dissipates into space.

Several factors can influence the simplified version of the climate system described above. The amount of radiation produced from the sun varies over time. In addition, the shape of our orbit around the sun varies over time from more circular to more elliptical because of the gravitational influences from other planets in our solar system. The angle at which solar radiation hits the planets' surface is influenced by the tilt and wobble of the Earth's axis. The distribution of water, ice, snow, vegetation, and other materials on the Earth's surface control the Earth's albedo and can change over time. The proportion of greenhouse gases can change dramatically depending on the rate of plate motion as well as the amount of volcanism, photosynthesis, weathering of rocks, burning of fossil fuels, and many other factors (to examine trends on climate and carbon dioxide levels see Figure 6.5 later in the lab).

The efficiency of the transportation of heat across the surface of the Earth also influences climate. Heat is transferred across the surface of the planet by wind, ocean currents, and storms. Therefore, the position of the continents as well as air and ocean currents affect climate and can change over time. The components in our atmosphere are also important, including water vapor and aerosols (dust). Water vapor in the atmosphere is a greenhouse gas, can reflect incoming solar radiation, and is the source of precipitation. Aerosols can come from the Earth's surface, ash from volcanoes, and the burning of fossil fuels and can alter climate by reflecting incoming solar radiation before it reaches the Earth's surface. There are numerous additional factors that also have some level of effect on the climate system.

At this point you should be able to recognize the complexity of the climate system based on the number of variables and how those variables can change over time. It is also important to recognize that all of these variables are connected. For instance, if a volcano erupts it adds some thermal energy to the climate system; it produces aerosols that block solar radiation from hitting the Earth's surface, and produces greenhouse gases that retain heat. Notice that these factors do not all influence climate in a consistent way. The change of one variable as the result of another is called feedback, which can be either positive or negative (Figure 6.2).

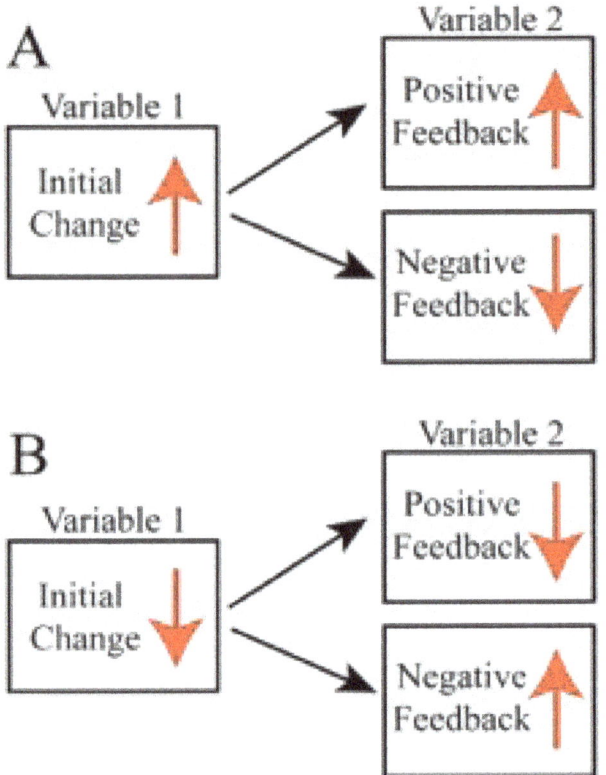

Figure 6.2 | A simple diagram showing the relationship between two variables with either positive or negative feedback following a positive (A) or negative (B) initial change.
Author: Bradley Deline
Source: Original Work
License: CC BY-SA 3.0

Positive feedback reinforces the initial change no matter the direction of that change. For instance, if the Earth warms, ice melts and reduces the albedo, which causes even more warming. This can also occur in the opposite direction, if the Earth cools, ice forms and increase the albedo, which causes more cooling. **Negative feedback** counteracts the initial change no matter the direction of the initial change. For instance, if the Earth warms, more area becomes arid resulting in an increase in the amount of dust in the atmosphere, which reflects solar radiation causing cooling. Again the opposite works, if the Earth cools, less area is arid resulting in a decrease in the amount of dust in the atmosphere, which causes warming. Therefore, an understanding of the climate system requires an identification of all of the important climate variables, how they are related to each other, the speed at which they can change, and the magnitude and direction of change of each feedback loop. The ideal way to gain a better understanding of the climate system is to study it through geologic history.

6.3 CLIMATE PROXIES AND THE CLIMATE RECORD

The first method most students think about when we talk about recording climate is using a thermometer to directly measure temperature. There are actually a few problems reconstructing climate patterns this way, including that a thermometer gives a very local signal and more importantly, thermometers are a relatively recent invention. Given that direct observations do not give us the long-term trends needed to establish climate change or patterns, we must look at a natural recorder of climate called a **climate proxy**. As climate changes it affects the deposition of sedimentary rock, rock chemistry, and fossil organisms that scientists can detect in order to reconstruct ancient climate patterns, in a field called paleoclimatology. An individual climate proxy may not give a clear signal of global climate for a couple of reasons. First, proxies show a history of the area in which they were formed, not of an entire region. Second, an individual proxy, which may have a long or a short record, can record the short-term variability of weather events. And third, most climate proxies are influenced by multiple factors. For instance, the thickness of tree rings (dendrochronology) is a wonderful proxy for temperature. Trees grow more in warmer years (producing thicker rings) and less in colder years (producing thinner rings). However, a tree could also grow slowly because of a drought or because of an infestation of pests even if it was a warm year.

If all of the individual proxies show local patterns, with some degree of weather related noise, and possibly influenced by other factors, how do we then reconstruct long-term global temperature records? The answer lies in increasing the size of the dataset. If temperature is the most important variable influencing the proxies, and we combine hundreds to thousands of individual proxy records, an overarching pattern emerges from the noise. Again, an individual proxy record may be contrary to the overall trend, but that is expected since a local region can have a cold winter in the midst of an overall hot year for the planet. To illustrate this consider the following: say we want to reconstruct overall economic patterns over the past few hundred years in the United States of America. We could examine lots of proxies for economic growth, such as employment, the stock market, individual wealth, or rates of home ownership to name a few. If we only looked at one of these proxies we likely would not get a clear picture of change. Also, if we only looked at Macon, Georgia, for example, we would be unlikely to see a trend that mimics the entire country. For instance if a new factory opened outside of Macon, GA that would be a huge economic benefit for the city, but not for the country overall. Again, the more data we have, whether it is for economics, or climate, or any other complex system, the clearer the signal becomes over the local and random noise.

Figure 6.3 | Short segment of an ice core that records ancient climate patterns.
Author: Ludovic Brucker, NASA
Source: Wikimedia Commons
License: Public Domain

One of the most commonly used climate proxy is the measurement of oxygen isotopes. As you may remember from chapter one, isotopes are atoms of the same element that differ in their weights because of differences in the number of neutrons in the nucleus of the atom. Multiple isotopes of oxygen are stable, meaning they do not radioactively decay over time. Oxygen has two stable isotopes that occur in a constant ratio on Earth. However, certain minerals (like calcite or ice) prefer one of the isotopes over the other within their crystal structure (a slightly larger or smaller atom fits better). This preference results in a ratio of oxygen isotopes that is different from the ratio found in other materials; this difference is called fractionation. The amount of fractionation in oxygen isotopes is temperature dependent, such that the mineral calcite has a different ratio of oxygen isotopes if it was formed in near freezing versus warm water temperatures. Using oxygen isotopes we can get climate records from many different sources, including coral, clams and other mollusks, the skeletons of single-celled organisms, and ice cores to name a few. Ice cores (as shown in Figure 6.3) can contain a wealth of climate data in addition to temperature data from oxygen isotopes, such as air bubbles that record the levels of greenhouse gases, concentrations of windblown aerosols, and ash from volcanic eruptions.

Other proxies include the extent of glacial sediment, sea level curves, pollen (palynology), and fossils. For instance, climatologists have used several features within fossil plants to reconstruct climate, largely because these organisms are sensitive to climate. These proxies include the thickness of tree rings, the shape of the leaves (toothier leaves are more common in colder climates), and the density of pores on leaf surfaces (more pores are needed with lower concentrations of carbon dioxide, which is necessary for photosynthesis).

As was mentioned before, by combining hundreds to thousands of individual climate records we can start to gain insight into overall climate trends. For instance, the Intergovernmental Panel on Climate Change (IPCC) and the National Oceanic and Atmospheric Administration (NOAA) regularly compile

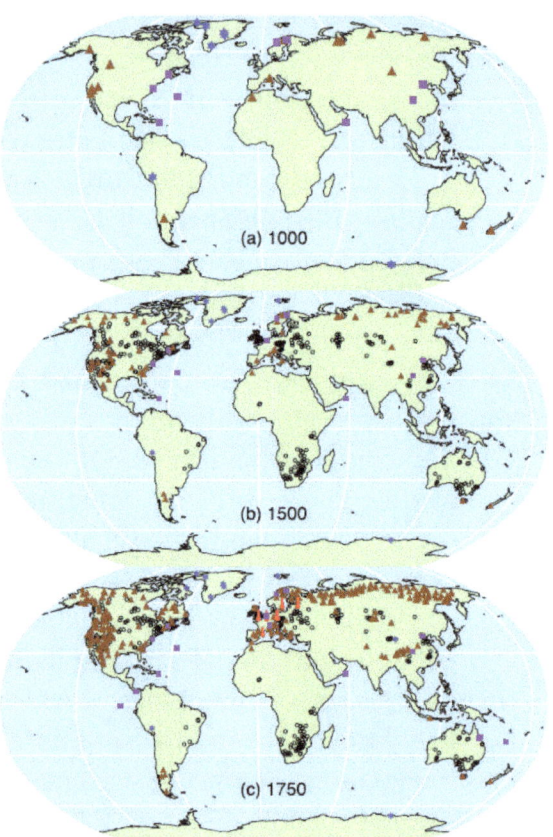

Figure 6.4 | Distribution of individual climate proxies used in the construction of Figure 5. Image from the National Ocean and Atmospheric Administration.
Author: NOAA
Source: NOAA
License: Public Domain

multiple types of proxy records from across the world (Figure 6.4) to reconstruct climate patterns (Figure 6.5). The accuracy of the climate records very much depends on the time frame being considered, with more certainty in the patterns of the recent past (Cenozoic) and less the further back in geologic time we are examining.

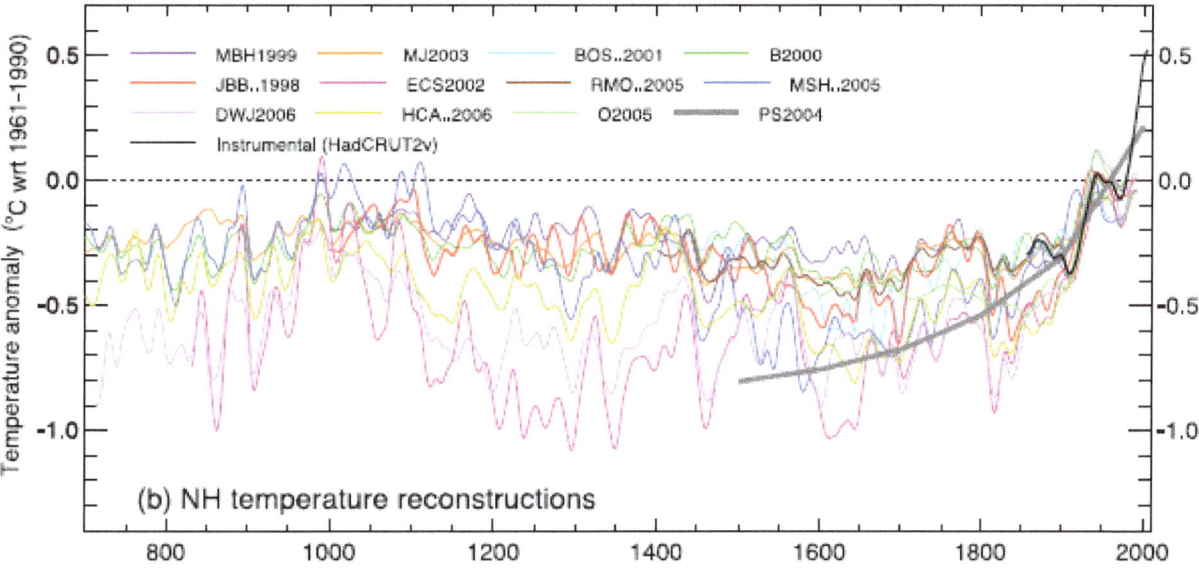

Figure 6.5 | Climate reconstruction over the last 1300 years using multiple climate proxies (different colored lines) from "Climate change 2007: the physical science basis"; Contribution of Working Group I to the Fourth Assessment Report of the Intergovernmental Panel on Climate Change Jansen, E. J. Overpeck, K. R. Briffa, J. C. Duplessy, F. Joss, V. Masson-Delmotte, D. Olago, B. Otto-Bliesner, W. R. Peltier, S. Rahmstorf, R. Ramesh, D. Raynaud, D. Rind, O. Solomina, R. Villalba, and D. Zhang.
Author: NOAA
Source: Wikimedia Commons
License: Public Domain

The climate proxy we will focus on for this lab is the extent of sea ice coverage on the North and South Polar ice sheets. This is an easy proxy to assess from satellite images and is measured as the size of the ice sheet in million square kilometers. This proxy isn't a perfect indicator of global climate change, but it is easy to understand that a warming of the Earth is likely to cause a decrease in the amount of ice at the poles and thus a decrease in the **ice extent**, while a cooling event will cause an increase in ice production. Ice extent is simply the amount of geographic area covered by a glacier as measured from satellites.

There is debate surrounding the interpretation of individual proxies and the resulting climate records, which largely stems from the economic and political aspects of climate change. This current lab was constructed following a discussion with a student regarding the information presented in several climate articles. The discussion focused on how scientific data is presented to the public and how we should make conclusions based on presented data. In considering information that we are presented with it is important to consider 1) the source of the data, 2) how the data was collected, 3) how the data is presented, and, most importantly 4) what are the reasonable conclusions you should make from the data independent of the opinions expressed alongside of the data.

6.4 LAB EXERCISE

As with Chapter 4, you will be expected to input your answers to this lab in several ways. There will be a couple of multiple-choice questions, but for the majority of the lab you will write your answers in the provided text box. This allows you to show your work in the questions requiring calculations as well as allowing you to answer open-ended questions thoroughly with multiple sentences. You will be expected to use correct grammar and complete sentences in your answers.

Materials

All of the data provided in the lab comes from the National Snow and Ice Data Center (NSIDC), housed at the University of Colorado. The data presented in this lab can be freely downloaded from them at nsidc.org. The original discussion was focused on a widely circulated though unattributed article entitled "Antarctic Sea Ice for March 2010 Significantly Greater Than 1980" published in April 2010 (climatechangehoax.com). For the dataset presented in this article as well as the following datasets, do the following before answering each set of questions: graph the data, connect the data points to better see the pattern through time, estimate the best fit line, and calculate the slope of the best fit line. Make sure to think about the data before you estimate the line of best fit so that your line falls within the data and is consistent with the trend in the data. An example of this process is shown in Figure 6.6.

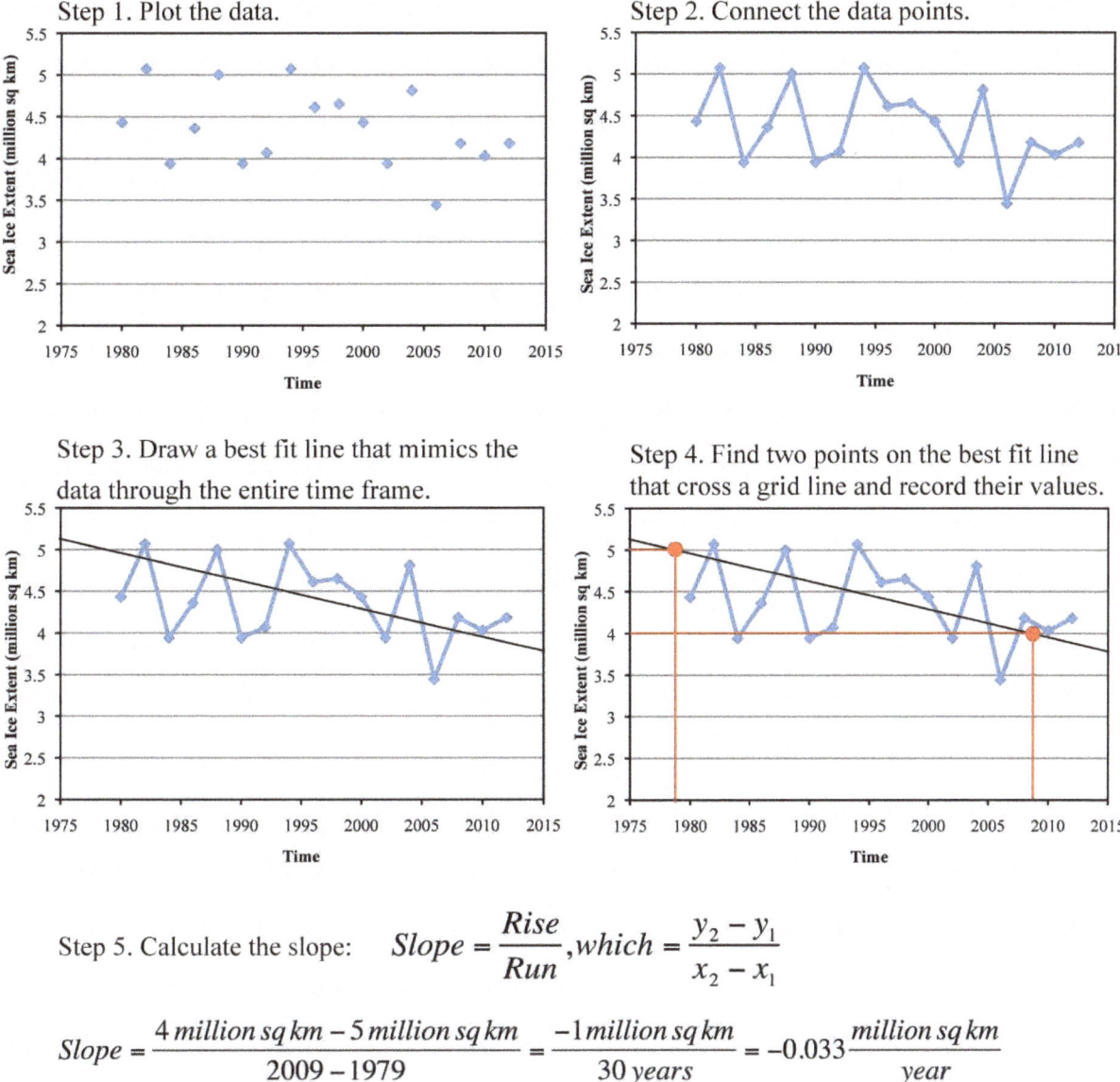

Figure 6.6 | Steps needed to analyze the ice extent data to visualize patterns: place a line of best fit, and calculate the slope of that line.
Author: Bradley Deline
Source: Original Work
License: CC BY-SA 3.0

Part A – Original Data

First, we want to take a closer look at the data presented in the original article ("Antarctic Sea Ice for March 2010 Significantly Greater Than 1980") and interpret the data. Feel free to read the article, but it isn't needed to complete the assignment or understand the patterns it is presenting. The data, which is included to the left of the graph below, is the extent of Antarctic sea ice in millions of square kilometers as measured in March of 1980 and 2010. This data is accurate and is consistent with data that can be downloaded from NSIDC. To do this, follow the instructions in Figure 6.6. In this case, steps two and three are the same and you can easily calculate slope using the original data.

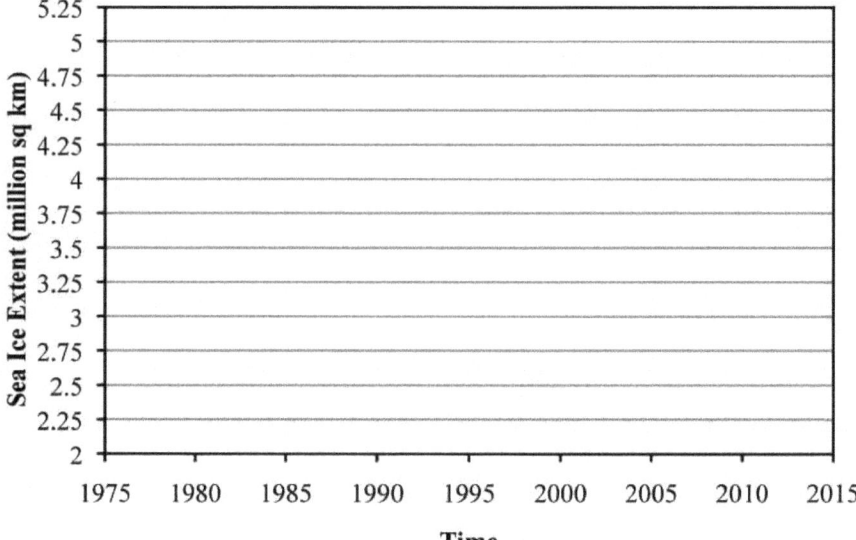

Sea Ice Extent

2010   4.0 million square km
1980   3.5 million square km.

1. Based exclusively on the data provided and your graph, what conclusion would you make regarding climate change?

    a. Sea ice is expanding, which indicates an increase in temperature

    b. Sea ice is expanding, which indicates a decrease in temperature

    c. Sea ice is contracting, which indicates an increase in temperature

    d. Sea ice is contracting, which indicates a decrease in temperature

2. What is slope of the line of best fit for this data?

    a. 0.008 million square kilometers per year

    b. 0.05 million square kilometers per year

    c. -0.05 million square kilometers per year

    d. 0.017 million square kilometers per year

    e. -0.17 million square kilometer per year

    f. 0.033 million square kilometer per year

3. Even though the above data is accurate, give and explain two reasons why this dataset might lead you to an incorrect conclusion regarding global climate change.

Part B – South Pole Sea Ice Extent

Below is an expanded dataset showing Antarctic sea ice (Figure 6.7) extent measured during March 1980 through 2012 again downloaded from NSIDC. Only the even numbered years are presented, but the addition of odd years does not alter the trend in the data. Following the instructions in Figure 6.6, graph the data, draw a line of best fit, and calculate the slope of the line.

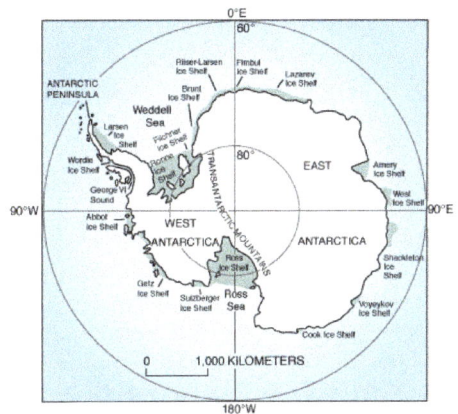

Figure 6.7 | Map of Antarctica showing the extent of the polar ice cap and the extent of the floating ice shelves.
Author: USGS
Source: Wikimedia Commons
License: Public Domain

year	Ice extent In million square km
1980	3.54
1982	4.8
1984	3.93
1986	4.08
1988	4.35
1990	4.36
1992	4.07
1994	5.07
1996	4.61
1998	4.65
2000	4.43
2002	3.94
2004	4.81
2006	3.44
2008	5.74
2010	4.03
2012	5.00

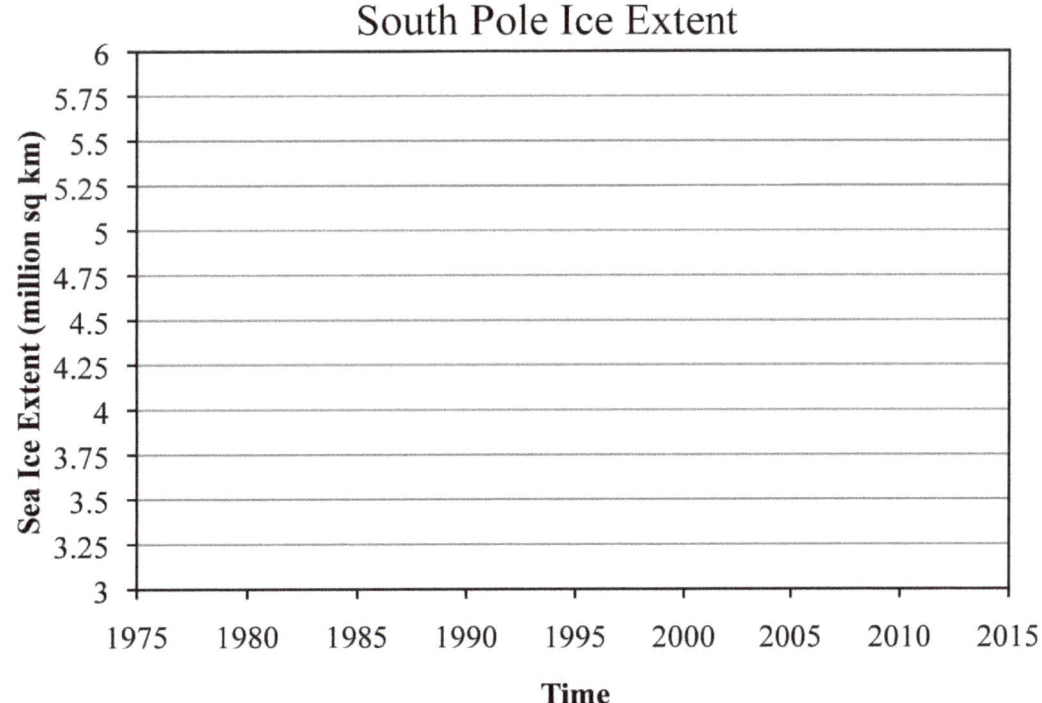

4. What is the slope of the line of best fit you estimated for this data set? Make sure to show your work.

5. What conclusion about climate change could you make from this dataset? How does your result for the extended dataset compare to the results from the data presented in the article (Part A)?

Part C – North Pole Sea Ice Extent

Next, we will examine the ice extent patterns of the northern Arctic polar ice sheet that is located around Greenland (Figure 6.8). The ice extent data is from March 1980 through 2012, for even numbered years, again downloaded from NSIDC. Following the instructions in Figure 6.6, graph the data, draw a line of best fit, and calculate the slope of the line.

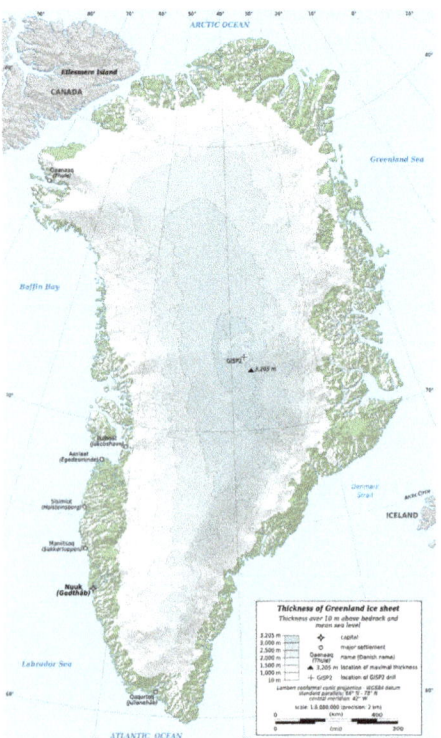

Figure 6.8 | Map of Greenland showing the extent of the polar ice cap.
Author: Eric Gaba
Source: Wikimedia Commons
License: CC BY-SA 3.0

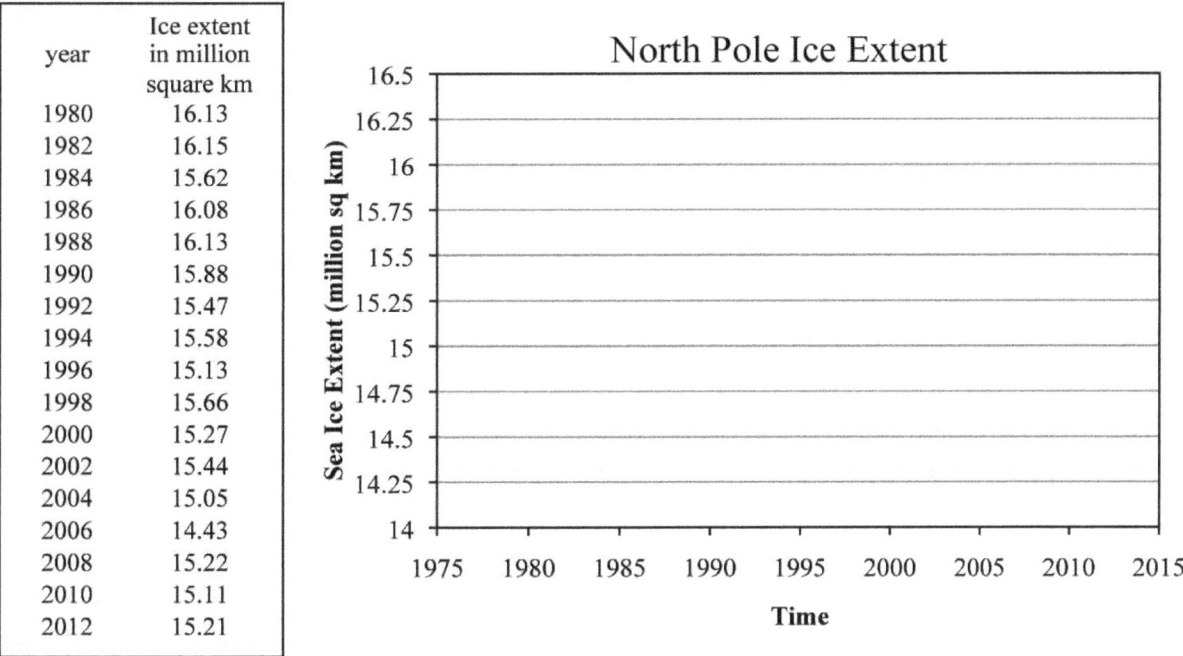

year	Ice extent in million square km
1980	16.13
1982	16.15
1984	15.62
1986	16.08
1988	16.13
1990	15.88
1992	15.47
1994	15.58
1996	15.13
1998	15.66
2000	15.27
2002	15.44
2004	15.05
2006	14.43
2008	15.22
2010	15.11
2012	15.21

6. What is the slope of the line of best fit you estimated for this data set? Make sure to show your work.

7. What conclusion on climate change could you make from this dataset? How does your result for the North Pole compare to that of the South Pole (Part B)?

6.5 HEAT TRANSPORT AND OCEAN CURRENTS

As was mentioned earlier in the lab, the tropics are warmer than the poles because of differences in the angle at which solar radiation impacts the Earth (Figure 6.1). Very little solar radiation reaches higher latitude areas because the solar radiation comes in almost parallel to the Earth's surface. Therefore, most of the thermal energy at higher latitudes comes from the movement of heat from the tropics. Heat is transported across the Earth's surface through wind currents, storms, and ocean currents. In particular, large circular ocean currents, called gyres, appear to have a significant impact on the geographic distribution of heat on Earth and large-scale climate change in Earth's history. These currents are particularly effective in melting polar ice in that they melt the sea ice from below. In addition, an examination of ocean current patterns will assist in explaining the patterns of sea ice extent you graphed earlier as the earth is warming (Part B and C).

6.6 LAB EXERCISE

Materials

A visualization of the ocean currents can be seen by downloading the file "ocean_currents.kml" either from your course's website or directly from the Science on a Sphere page from NOAA (sos.noaa.gov/kml/). Once you download the file you can open it within Google Earth. Once the file loads (which may take a few minutes), click on the wrench in the upper left corner of the screen to the left of the NOAA logo (Figure 6.9). Check the "loop animation" box and slide the cursor for animation speed to an intermediate position, then click OK. Then click the toggle animation icon (Figure 6.9) to start the animation. Watch the movement of the currents as they flow, making sure to examine the flow in different parts of the world and zoom out to get a broad prospective of the flow of the ocean currents around the world.

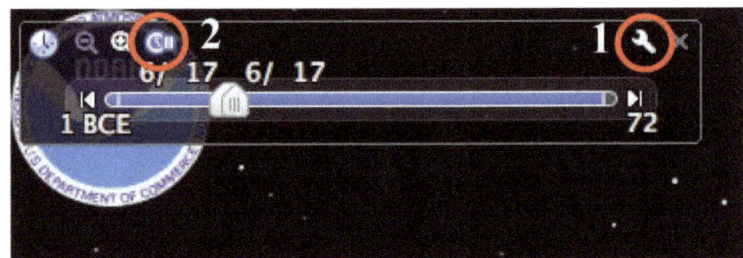

Figure 6.9 | Directions to visualize ocean currents in Google Earth. 1: Wrench icon to apply the correct settings. 2: Toggle animation icon to start the animation.
Author: Google
Source: Google Earth
License: Fair Use

If you have difficulty loading the file you can also access the visualization at: http://sos.noaa.gov/Datasets/dataset.php?id=130#. To access the visualization of the ocean currents click on "Interactive Sphere" and hit play, make sure to spin the globe to see the currents in the different oceans.

Part D – Ocean Currents and Heat Transport

8. Examine and describe the ocean currents flowing in the Atlantic Ocean from the equator to the North Pole starting from Brazil (6 08 54.55S 35 58 24.09W). Do these currents reach the northern polar ice sheet? If the tropics become warmer, how would this affect the northern ice sheet?

9. Examine and describe the ocean currents flowing in the Pacific Ocean from the equator to the South Pole starting from Somalia (0 32 32.23N 44 09 15.47E). Do these currents reach the southern polar ice sheet? If the tropics become warmer, how would this affect the southern ice sheet?

10. Based on your answer to questions eight and nine, explain why we see different trends in the sea ice extent in the south (Part B) and north poles (Part C).

11. How might the changes you saw in the previous exercises relate to global albedo, sea level, ocean salinity, and temperature?

Part E – Conclusions

Making conclusions, let alone policy decisions, regarding any complex system such as climate and how it is changing is difficult. You can see how completely accurate data can be misrepresented as well as how accurate data out of context (ice extent without understanding ocean currents) may lead you to an incorrect conclusion. It is important to base any conclusion on rational, accurate, complete, and in context data, rather than data that has been poorly collected or misrepresented. Most importantly, it is important to come to your own conclusions regarding data rather than being swayed by the opinion of the author presenting the data, which is true both in science as well as many other aspects of your life.

12. Do you think any conclusions (if any) regarding climate change should be made based on the data presented in this assignment?

13. What data do you think are needed to make a conclusion regarding climate change?

6.7 ADDITIONAL RESOURCES

To learn more about the science behind climate change and the scientific communities position on anthropogenic climate change see the following:

National Aeronautics and Space Administration (NASA) climate Site:
http://climate.nasa.gov/

National Oceanic and Atmospheric Administration Climate Site:
http://www.noaa.gov/climate.html

Geological Society of America's Position Statement on Climate Change:
https://www.geosociety.org/gsa/positions/position10.aspx

American Metrological Society's Position Statement on Climate Change:
https://www.ametsoc.org/ams/index.cfm/about-ams/ams-statements/statements-of-the-ams-in-force/climate-change/

American Chemical Society's Position Statement on Climate Change:
https://www.acs.org/content/acs/en/policy/publicpolicies/sustainability/globalclimatechange.html?_ga=2.227911517.1368062825.1497897808-1921393895.1497558864

American Physical Society's Position Statement on Climate Change:
http://www.aps.org/policy/statements/climate/

American Association for the Advancement of Science's Position Statement on Climate Change:
http://www.aaas.org/sites/default/files/migrate/uploads/aaas_climate_statement.pdf

6.8 STUDENT RESPONSES

1. Based exclusively on the data provided and your graph, what conclusion would you make regarding climate change?

    a. Sea ice is expanding, which indicates an increase in temperature

    b. Sea ice is expanding, which indicates a decrease in temperature

    c. Sea ice is contracting, which indicates an increase in temperature

    d. Sea ice is contracting, which indicates a decrease in temperature

2. What is slope of the line of best fit for this data?

    a. 0.008 million square kilometers per year

    b. 0.05 million square kilometers per year

    c. -0.05 million square kilometers per year

    d. 0.017 million square kilometers per year

    e. -0.17 million square kilometer per year

    f. 0.033 million square kilometer per year

3. Even though the above data is accurate, give and explain two reasons why this dataset might lead you to an incorrect conclusion regarding global climate change.

4. What is the slope of the line of best fit you estimated for this data set? Make sure to show your work.

5. What conclusion about climate change could you make from this dataset? How does your result for the extended dataset compare to the results from the data presented in the article (Part A)?

6. What is the slope of the line of best fit you estimated for this data set? Make sure to show your work.

7. What conclusion on climate change could you make from this dataset? How does your result for the North Pole compare to that of the South Pole (Part B)?

8. Examine and describe the ocean currents flowing in the Atlantic Ocean from the equator to the North Pole starting from Brazil (6 08 54.55S 35 58 24.09W). Do these currents reach the northern polar ice sheet? If the tropics become warmer, how would this affect the northern ice sheet?

9. Examine and describe the ocean currents flowing in the Pacific Ocean from the equator to the South Pole starting from Somalia (0 32 32.23N 44 09 15.47E). Do these currents reach the southern polar ice sheet? If the tropics become warmer, how would this affect the southern ice sheet?

10. Based on your answer to questions eight and nine, explain why we see different trends in the sea ice extent in the south (Part B) and north poles (Part C).

11. How might the changes you saw in the previous exercises relate to global albedo, sea level, ocean salinity, and temperature?

12. Do you think any conclusions (if any) regarding climate change should be made based on the data presented in this assignment?

13. What data do you think are needed to make a conclusion regarding climate change?

7 Matter and Minerals
Randa Harris

7.1 INTRODUCTION

Have you used a mineral yet today? While many people may initially say no, answer these questions: Have you brushed your teeth? Have you eaten anything that might contain salt? Did you put on make-up this morning, or do you have painted fingernails or toenails? Have you used a cellphone? What about a car, bike, or public transportation? If you have done any of those things, you have used at least one mineral, and in many cases you have used a great number of minerals. Minerals are very useful and common in everyday products, but most people do not even realize it.

A **mineral** is defined as a naturally occurring, inorganic solid with a definite chemical composition and a characteristic crystalline structure. Let's break that definition down. By naturally occurring, it means that anything man has created, like the beautiful synthetic bismuth in Figure 7.1, does not count as a mineral. To be an inorganic solid, the mineral must not be composed of the complex carbon molecules that are characteristic of life and must be in the solid state, rather than vapor or liquid. This means that water, a liquid, is not a mineral, while ice, a solid, would be (as long as it is not man-made). A definite chemical composition refers to the chemical formula of a mineral. For most minerals, this does not vary (ex. halite is NaCl), though some minerals have a range of compositions, since one element can substitute for another of similar size and charge (ex. olivine is $(Mg,Fe)_2SiO_4$, and its magnesium and iron content can vary). The atoms within minerals are lined up in an orderly fashion, so that the characteristic crystalline structure is just an outward manifestation of the internal atomic arrangement.

Figure 7.1 | Synthetic bismuth
Author: Philippe Giabbanelli
Source: Wikimedia Commons
License: CC BY-SA 3.0

Minerals are not only important for their many uses, but also as the building blocks of rocks. In this lab, you will lay the foundation for all the future rock labs in the course. Correct mineral identification is critical in geology, so work through this lab carefully. There are several thousand minerals, but we will focus on only eighteen of the most common ones.

7.1.1 Learning Outcomes

After completing this chapter, you should be able to:
- Know the definition of a mineral
- Understand the many different physical properties of minerals, and how to apply them to mineral identification
- Be able to distinguish mineral cleavage from mineral fracture
- Identify 18 minerals

7.1.2 Key Terms

- Cleavage
- Crystal Form
- Fracture
- Hardness
- Luster
- Mineral
- Specific Gravity
- Streak
- Tenacity

7.2 PHYSICAL PROPERTIES

Identifying a mineral is a little like playing detective. Minerals are identified by their physical properties. For example, look at Figure 7.2. How would you describe it? You may say that it is shiny, gold, and has a particular shape. Each of these descriptions is actually a physical property (shiny=luster, gold=color, shape=crystal form). Physical properties can vary within the same minerals, so caution should be applied. For example, color is a property that is not a very realistic diagnostic tool in many cases. Quartz is a mineral that comes in a variety of colors, as evidenced by Figure 7.3. Occasionally color can be helpful, as in the case of the mineral olivine. Olivine is said to be "olive green" (a light to dark green)

Figure 7.2 | Describe this mineral.
Author: Randa Harris
Source: Original Work
License: CC BY-SA 3.0

Figure 7.3 | Examples of the different varieties of quartz (jasper, rose quartz, smoky quartz, agate, amethyst, citrine, and petrified wood), demonstrating the difficulty of identifying this mineral.
Author: Randa Harris
Source: Original Work
License: CC BY-SA 3.0

Figure 7.4 | The mineral olivine is "olive green."
Author: Randa Harris
Source: Original Work
License: CC BY-SA 3.0

as seen in Figure 7.4. Make sure you use caution when using color to help identify minerals. We will cover each of the physical properties in detail to help you identify the minerals.

7.2.1 Hardness

Hardness refers to the resistance of a mineral to being scratched by a different mineral or other material and is a product of the strength of the bonds between the atoms of a mineral. Whatever substance does the scratching is harder; the item scratched is softer. Hardness is based off a scale of 1 to 10 created by a mineralogist named Friedrich Mohs (Figure 7.5). Mohs' scale lists ten minerals in order of relative hardness. Each mineral on the scale can scratch a mineral of lower number. Your mineral kit comes with several items of a known hardness. The glass plate has a hardness of 5.5, the iron nail has a hardness of 4, the copper wire has a hardness of 3, and your fingernail has a hardness of 2.5. If you can scratch a mineral, then it would be softer than your fingernail, so therefore its hardness would be <2.5. When trying to scratch a surface, use force, but be cautious with the glass plate. **ALWAYS lay the glass plate on a flat surface rather than holding it in your hand in case it breaks.** Do not confuse mineral powder with a scratch – use your finger to feel for a groove created by a scratch (mineral powder is left behind when a soft mineral scratches a harder surface). Materials of similar hardness have difficulty scratching each other, so that, for example, your fingernail may not be able to always scratch biotite mica, which has a hardness of 2.5.

Number	Mineral	Hardness of Test Kit Items
1	Talc	(softest mineral)
2	Gypsum	2.5 – Fingernail
3	Calcite	3 – Copper Wire
4	Fluorite	4 – Nail
5	Apatite	5.5 – Glass Plate
6	Orthoclase Feldspar	
7	Quartz	
8	Topaz	
9	Corundum	
10	Diamond	(hardest mineral)

Figure 7.5 | Mohs' Scale of Hardness

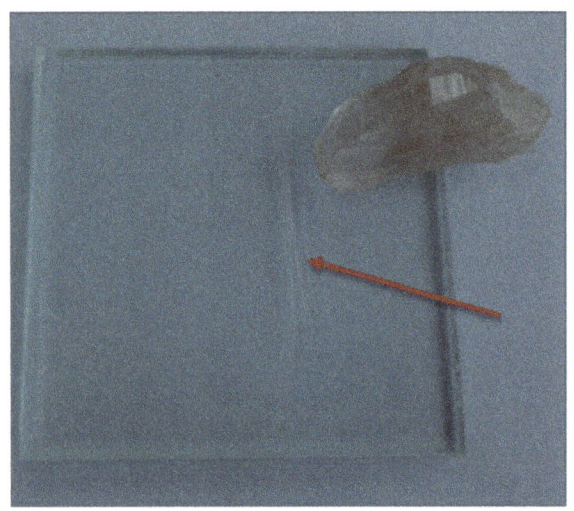

Figure 7.6 | An example of a scratch made by the mineral quartz on a streak plate. The red arrow is pointing to the scratch. Quartz, therefore, is harder than glass.
Author: Randa Harris
Source: Original Work
License: CC BY-SA 3.0

Figure 7.7 | An example of a scratch made by a fingernail on the mineral gypsum. The red arrow is pointing to the scratch. Gypsum, therefore, is softer than a fingernail.
Author: Randa Harris
Source: Original Work
License: CC BY-SA 3.0

7.3 LAB EXERCISE

Materials

Your HOL Lab Kit contains 18 numbered mineral samples, separated into 3 bags (labeled as Mineral Bag 1, 2, or 3). Use these instructions to test and identify them. You will test for different properties after learning about them, then work on identification at the end of the lab. The HOL kit has been specifically tailored to this class – make sure that you are using the kit required by this class, as other rock and mineral kits will not work. Images will be provided of the correct kit; make sure that you closely compare your kit to the images so that you are working with the correct samples. Empty the contents of the testing kit. It will contain:

a. A 3" copper wire

b. Glass plate (wrapped in paper) – this will be used in testing hardness

c. Zinc coated nail

d. Unglazed porcelain plate (wrapped in paper) – this will be used as a streak plate

e. Hydrochloric acid

f. Magnifying glass (10x). To use this, hold it very close to your eye and bring the sample near the glass until it is in focus (approximately one inch from your eye).

g. Gloves and protective goggles (for use with the acid)

Take out Minerals Bag 1 and lay the six mineral samples out on a white sheet of paper. It should appear like Figure 7.8. We will first examine hardness from these six samples, and will answer more questions about them later in the lab. Look closely at each of the minerals, using the hand lens to observe them. In this bag, you have the following minerals (not listed in order): Microcline (also called Potassium Feldspar), Fluorite, Quartz, Olivine, Talc, and Selenite (also called Gypsum). They are numbered 1-6.

You need to experiment with each sample to test for its hardness and use Figure 7.5 for reference. Remember that hardness is determined by scratching the mineral (or using the mineral to scratch something else). First, decide which minerals have a hardness greater than 5.5 (the hardness of glass). Lay the glass on a flat surface (not in your hand), then try to scratch it with each mineral. Bare down hard with the mineral, much like trying to leave a scratch on a car with a key. Table 7.1 is given at the end of this lab for you to make notations about each mineral. Note that you do not have to fill in every physical property for every mineral (that would be very time-consuming with 18 samples). Just fill in the properties you are asked about as you work. Note now on the table which minerals have a hardness greater than 5.5. You may also test samples by using materials to scratch them. The copper wire has a hardness of 3. Any mineral that it can scratch will have a hardness less than 3. You can further refine this by using your fingernail (only natural fingernails work for this). Your fingernail has a hardness of 2.5, so if the copper wire scratches a mineral and your fingernail also scratches it, you know its hardness must be <2.5. The zinc coated nail has a hardness of 4. Also use it to scratch the minerals. Minerals may also be used to scratch each other. For example, if you have two minerals that have a hardness of <2.5, you can see if one will scratch the other. Then you know it is harder, since it did the scratching.

Figure 7.8 | The six minerals (#1-6) in Minerals Bag 1 in the HOL kit.
Author: Randa Harris
Source: Original Work
License: CC BY-SA 3.0

1. Sample 1: What is this sample's hardness?

    a. harder than glass

    b. softer than glass but harder than nail

    c. softer than nail but harder than copper

    d. softer than copper but harder than a fingernail

    e. softer than a fingernail

2. Sample 2: What is this sample's hardness?

    a. harder than glass

    b. softer than glass but harder than nail

    c. softer than nail but harder than copper

    d. softer than copper but harder than a fingernail

    e. softer than a fingernail

3. Sample 3: What is this sample's hardness?

    a. harder than glass

    b. softer than glass but harder than nail

    c. softer than nail but harder than copper

    d. softer than copper but harder than a fingernail

    e. softer than a fingernail

4. Sample 4: What is this sample's hardness?

    a. harder than glass

    b. softer than glass but harder than nail

    c. softer than nail but harder than copper

    d. softer than copper but harder than a fingernail

    e. softer than a fingernail

5. Sample 5: What is this sample's hardness?

   a. harder than glass

   b. softer than glass but harder than nail

   c. softer than nail but harder than copper

   d. softer than copper but harder than a fingernail

   e. softer than a fingernail

6. Sample 6. What is this sample's hardness?

   a. harder than glass

   b. softer than glass but harder than nail

   c. softer than nail but harder than copper

   d. softer than copper but harder than a fingernail

   e. softer than a fingernail

7.4 CRYSTAL FORM

This property refers to the geometric shape that a crystal naturally grows into, and is a reflection of the orderly internal arrangement of atoms within the mineral. If minerals have space to grow when they are developing, they will display their **crystal form**. These ideal growth conditions do not always occur, however, so many minerals do not display their ideal crystal form due to crowded conditions during growth. Examples of crystal form are shown in Figure 7.9.

Cube

Hexagonal Prism

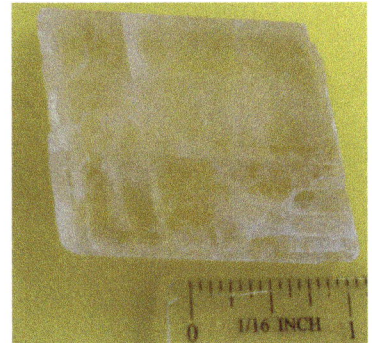
Rhombohedron

Octahedron (8 faces)

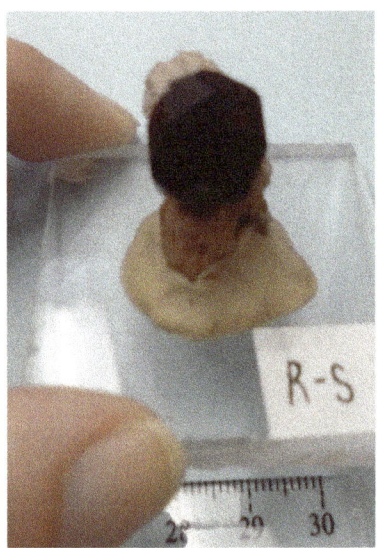
Dodecahedron (12 faces)

Figure 7.9 | Examples of crystal form
Author: Randa Harris
Source: Original Work
License: CC BY-SA 3.0

7.5 CLEAVAGE

As minerals are broken (such as with a rock hammer, for example), some may cleave, or break, along smooth flat planes known as **cleavage**. These flat surfaces are parallel to directions of weakness within the crystal. All the bonds among the atoms within a mineral may not be of the same strength, so that when a mineral is broken, it breaks along these zones of weakness. This results in flat cleavage planes. Minerals with perfect cleavage break along a smooth, flat plane, while those with poor cleavage break in a more irregular fashion. Some minerals do not contain zones of weakness either because all of the bonds are the same strength or the weaker bonds are not aligned within a plane. If this is the case it will not have cleavage, but rather breaks in a random and irregular fashion. Make sure to distinguish cleavage from crystal form. Crystal form occurs as a mineral *grows*, while cleavage only forms as a mineral *breaks*. See Figure 7.10 for the main types of cleavage and an example of each.

# of Cleavages & Direction	Cleavage Name	Example
0 (none) – mineral fractures	No cleavage planes	
1	Basal cleavage – flat sheets	
2 – cleavages at or near 90°	Prismatic cleavage – rectangular cross-sections	
2 – cleavages not at 90°	Prismatic cleavage – parallelogram cross-sections	

3 – cleavages at 90°	Cubic cleavage – cubes	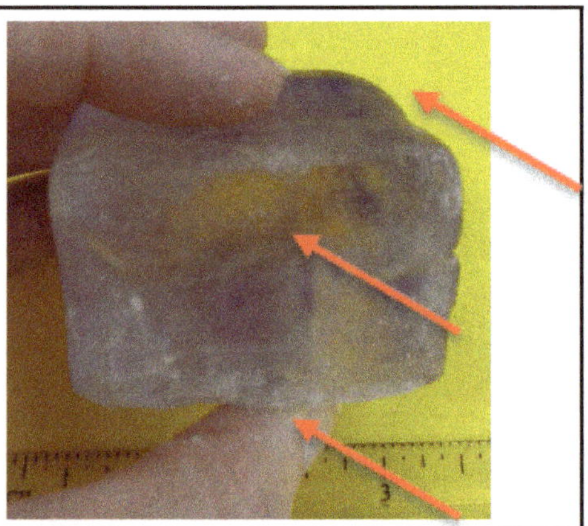
3 – cleavages not at 90°	Rhombohedral cleavage – rhombs	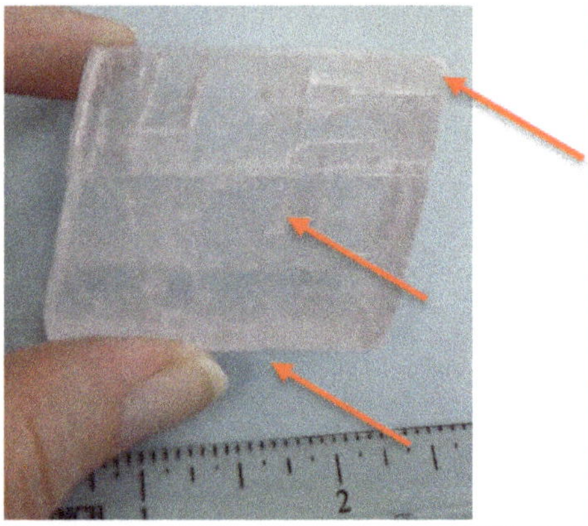
4	Octahedral cleavage	

Figure 7.10 | Chart with the main types of cleavage, along with picture examples. Red arrows are pointing to the different cleavage planes in each picture.
Author: Randa Harris
Source: Original Work
License: CC BY-SA 3.0

A mineral may have one or more cleavage planes. Planes that are parallel are considered the same direction of cleavage and should only count as one. One direction of cleavage is termed basal cleavage. Minerals that display this cleavage will break off in flat sheets. Two directions of cleavage is termed prismatic, while three directions of cleavage at 90° is referred to as cubic. A mineral with four directions of cleavage is termed octahedral. With 2 or more cleavage planes present, it is important to pay attention to the angle of the cleavage planes. To determine the angle of cleavage, look at the intersection of cleavage planes. Commonly, cleavage planes will intersect at 60°, 90° (right angles), or 120°. Be cautious when you see a flat surface on a mineral – not every flat surface is a cleavage plane. Crystal faces can be flat, but remember they form as a mineral grows, while cleavage forms as a mineral breaks. The crystal form of quartz is a hexagonal prism, with nice flat sides. But when quartz is hit with a rock hammer, it breaks in an irregular fashion and does not exhibit cleavage. Also use caution when trying to distinguish the minerals pyroxene and amphibole. Both minerals are black or greenish-black, with similar hardness, making them difficult to tell apart. You must observe the cleavage angles to tell them apart. Cleavage angles in pyroxene are near 90°, so expect it to look boxy and form right angles. Cleavage angles in amphibole are 60° and 120°, so expect a more bladed or pyramid like appearance (Figure 7.11).

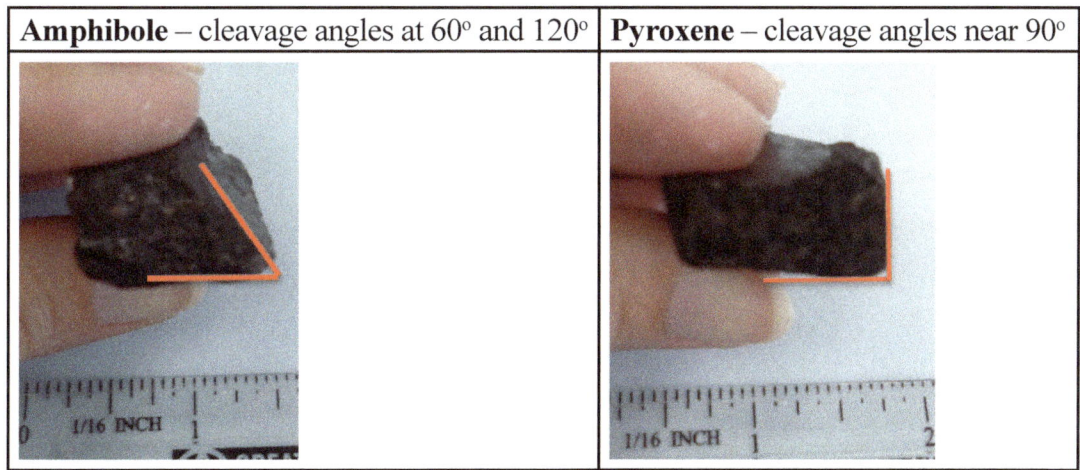

Figure 7.11 | Comparison of cleavage angles between amphibole and pyroxene.
Author: Randa Harris
Source: Original Work
License: CC BY-SA 3.0

7.6 FRACTURE

When minerals do not break along cleavage planes, but rather break irregularly, they are said to **fracture**. Commonly fracture surfaces are either uneven or conchoidal, a ribbed, smoothly curved surface similar to broken glass (Figure 7.12).

Figure 7.12 | This piece of igneous rock called obsidian has been hit with a hammer and is displaying conchoidal fracture.
Author: Randa Harris
Source: Original Work
License: CC BY-SA 3.0

7.7 LAB EXERCISE

Take out Minerals Bag 2 and lay the six mineral samples out on a white sheet of paper. It should appear like Figure 7.13. We will first examine cleavage and fracture, along with hardness, from these six samples, and will answer more questions about them later in the lab. Look closely at each of the minerals, using the hand lens to observe them. In this bag, you have the following minerals (not listed in order): Pyroxene, Muscovite Mica, Halite, Amphibole, Calcite, and Biotite Mica. They are numbered 7-12.

Figure 7.13 | The six minerals (#7-12) in Minerals Bag 2 in the HOL kit.
Author: Randa Harris
Source: Original Work
License: CC BY-SA 3.0

7. Sample 7: This sample has:

    a. no cleavage (it fractures)   b. 1 cleavage plane

    c. 2 cleavage planes at 90°   d. 3 cleavage planes at 90°

    e. 4 cleavage planes

8. Sample 8: This sample has:

    a. no cleavage (it fractures)   b. 1 cleavage plane

    c. 2 cleavage planes not at 90° d. 3 cleavage planes not at 90°

    e. 4 cleavage planes

9. Sample 9: This sample has:

    a. no cleavage (it fractures)   b. 1 cleavage plane

    c. 2 cleavage planes not at 90°   d. 3 cleavage planes at 90°

    e. 4 cleavage planes

10. Sample 10: This sample has:

    a. no cleavage (it fractures)   b. 1 cleavage plane

    c. 2 cleavage planes not at 90°   d. 2 cleavage planes at 90°

    e. 4 cleavage planes

11. Sample 11: This sample has:

    a. no cleavage (it fractures)   b. 1 cleavage plane

    c. 2 cleavage planes not at 90°   d. 2 cleavage planes at 90°

    e. 3 cleavage planes at 90°

12. Sample 11: What is this sample's hardness?

    a. harder than glass

    b. softer than glass but harder than nail

    c. softer than nail but harder than copper

    d. softer than copper but harder than a fingernail

    e. softer than a fingernail

13. Sample 12: This sample has:

    a. no cleavage (it fractures)   b. 1 cleavage plane

    c. 2 cleavage planes not at 90°   d. 2 cleavage planes at 90°

    e. 3 cleavage planes not at 90°

14. Sample 12: What is this sample's hardness?

    a. harder than glass

    b. softer than glass but harder than nail

    c. softer than nail but harder than copper

    d. softer than copper but harder than a fingernail

    e. softer than a fingernail

7.8 LUSTER

Luster refers to the appearance of the reflection of light from a mineral's surface. It is generally broken into two main types: metallic and non-metallic. Minerals with a metallic luster have the color of a metal, like silver, gold, copper, or brass (Figure 7.14). While minerals with a metallic luster are often shiny, not all shiny minerals are metallic. Make sure you look for the color of a metal, rather than for just a shine. Minerals with non-metallic luster do not appear like metals. They may be vitreous (glassy), earthy (dull), waxy (similar to a candle's luster), greasy (oily), or other types (Figure 7.15).

Figure 7.14 | Examples of the metallic luster of pyrite, also known as "fool's gold."
Author: Randa Harris
Source: Original Work
License: CC BY-SA 3.0

Figure 7.15 | Examples of different types of non-metallic lusters.
Author: Randa Harris
Source: Original Work
License: CC BY-SA 3.0

7.9 STREAK

Streak is an easily detectable physical property. It refers to the color left behind on an unglazed piece of porcelain when a mineral is rubbed along its surface. A streak plate is included in your rock and mineral kit to test this property. Often a mineral will have a streak of a different color than the color of the mineral (for example, pyrite has a dark gray streak, Figure 7.16). Some minerals will have a white streak, which is difficult to see along the white streak plate. If you rub a mineral along the streak plate and do not see an obvious streak, wipe your finger along the streak plate. A mineral with a white streak will leave a white powder behind that will rub on your finger (Figure 7.17).

Figure 7.16 | An example of the dark gray streak left behind when pyrite is rubbed along a streak plate.
Author: Randa Harris
Source: Original Work
License: CC BY-SA 3.0

Figure 7.17 | An example of the white streak (on finger) left behind when fluorite is rubbed along a streak plate.
Author: Randa Harris
Source: Original Work
License: CC BY-SA 3.0

7.10 SPECIAL PHYSICAL PROPERTIES

Several minerals have unique properties that aid in their identification. **Tenacity** refers to the way a mineral resists breakage. If a mineral shatters like glass, it is said to be brittle (like quartz), while minerals that can be hammered are malleable (like copper, Figure 7.18). Minerals may be elastic, in which they are flexible and bend like a plastic comb, but return to their original shape (like mica, Figure 7.19). Sectile minerals are soft like wax, and can be separated with a knife (like gypsum).

Some minerals react when dilute hydrochloric acid is placed on them. Carbonate

Figure 7.18 | Copper, which can be hammered into thin sheets, is malleable.
Author: Randa Harris
Source: Original Work
License: CC BY-SA 3.0

minerals (minerals that include CO_3 in their chemical formula) will effervesce or fizz when acid is applied to them. When you test a mineral with acid, be cautious and use just a drop of the acid. Use your magnifying glass to look closely for bubbles (Figure 7.20). The acid is very dilute and will not burn your skin or clothing, but wash your hands after use (gloves and goggles are provided). Also make sure that you rinse with water and wipe off the acid from the minerals that you test.

Minerals may be magnetic, and this property is simply tested by seeing if your nail is attracted to a mineral. Magnetite is an example of a magnetic mineral. The mineral halite is simply table salt, so it will taste salty. Graphite is used in pencils, and makes a nice smudge when rubbed along paper. Talc will feel soapy when touched.

Specific gravity is the ratio of a mineral's weight to the weight of an equal volume of water. A mineral with a specific gravity of 2 would weigh twice as much as water. Most minerals are heavier than water, and the average specific gravity for all minerals is approximately 2.7. Some minerals are quite heavy, such as pyrite with a specific gravity of 4.9-5.2, native copper, with a specific gravity of 8.8-9.0, and native gold at 19.3, which makes panning useful for gold, as the heavy mineral stays behind as you wash material out of the pan.

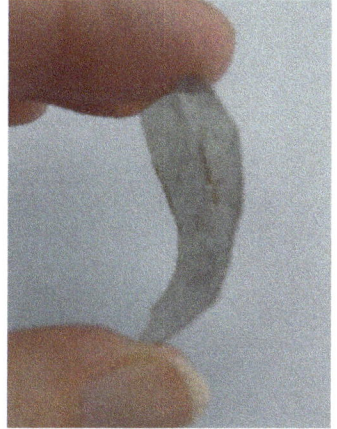

Figure 7.19 | Muscovite mica, which bends but returns to its original shape, is elastic.
Author: Randa Harris
Source: Original Work
License: CC BY-SA 3.0

Figure 7.20 | Note the effervescing acid bubbles at the red arrow on this piece of calcite.
Author: Randa Harris
Source: Original Work
License: CC BY-SA 3.0

7.11 LAB EXERCISE

Take out Minerals Bag 3 and lay the six mineral samples out on a white sheet of paper. It should appear like Figure 7.21. We will first examine several properties, including streak, from these six samples, and will answer more questions about them later in the lab. Look closely at each of the minerals, using the hand lens to observe them. In this bag, you have the following minerals (not listed in order): Magnetite, Graphite, Copper, Sulfur, Hematite, and Pyrite. They are numbered 13-18.

Figure 7.21 | The six minerals (#13-18) in Minerals Bag 3 in the HOL kit.
Author: Randa Harris
Source: Original Work
License: CC BY-SA 3.0

15. Sample 13: What is the streak of this sample?

    a. dark gray streak
    b. white streak
    c. reddish brown streak
    d. pale yellow streak

16. Sample 13: What is the luster of this sample?

    a. non-metallic, vitreous
    b. non-metallic, earthy
    c. non-metallic, greasy
    d. non-metallic, waxy
    e. metallic

17. Sample 14: What is the streak of this sample?

    a. dark gray to black streak
    b. white streak
    c. reddish brown streak
    d. pale yellow streak

18. Sample 14: Which other item(s) is/are characteristic(s) of this sample?

    a. stains the fingers

    b. harder than glass

    c. greasy feel

    d. both a and b

    e. both a and c

19. Sample 15: What is the streak of this sample?

    a. dark gray to black streak

    b. white streak

    c. reddish brown streak

    d. pale yellow streak

20. Sample 16: What is the luster of this sample?

    a. non-metallic, vitreous

    b. non-metallic, earthy

    c. non-metallic, greasy

    d. non-metallic, waxy

    e. metallic

21. Sample 16: Due to its appearance, this sample has often been confused with native gold, a mineral with a hardness of 2.5-3. How does its hardness compare with that of gold?

    a. Sample 16 is harder than gold.

    b. Sample 16 is softer than gold.

22. Sample 17: What is the streak of this sample?

    a. dark gray to black streak

    b. white streak

    c. reddish brown streak

    d. pale yellow streak

23. Sample 17: What another unique property does this sample have?

    a. effervescence in acid

    b. it is magnetic

    c. it tastes salty

    d. it feels soapy

    e. it writes on paper

24. Sample 18. Examine this entire sample closely with a hand lens. What is the luster of this sample?

   a. non-metallic, vitreous   b. non-metallic, earthy

   c. non-metallic, greasy   d. non-metallic, waxy

   e. metallic

Now that you have had practice at detecting the properties of your 18 mineral samples, take the next step of identifying each sample and answering the questions below. Use the Mineral Identification Chart (Figure 7.22) to help you.

25. Sample 1: What is this sample?

   a. Microcline   b. Fluorite   c. Quartz

   d. Olivine   e. Talc   f. Selenite (Gypsum)

26. Sample 2: What is this sample?

   a. Microcline   b. Fluorite   c. Quartz

   d. Olivine   e. Talc   f. Selenite (Gypsum)

Luster	Hardness	Cleavage	Other Properties	Mineral Name
Non-Metallic	> Glass	Poor Cleavage	Red-brown, black, silver in color. H=6. St=red-brown	Hematite
			Olive-green in color. H=6. St=white. Commonly granular	Olivine
			Variety of colors. H=7. Conchoidal fracture. Vitreous luster.	Quartz
		Clearly Shows Cleavage	Black to greenish black in color. H=6. C=2 planes at ~60° and 120°. Elongated crystals.	Amphibole
			Tan-pink, white, green in color. H=6. C=2 planes at 90°.	Microcline
			Black to greenish black in color. H=6. C=2 planes at ~90°. Short stubby crystals.	Pyroxene
	< Glass	Poor Cleavage	Dark gray to black in color. H=1. Greasy feel - will smudge fingers.	Graphite
			Yellow in color. H=1.5-2.5. St=white to yellow.	Sulfur
			White to green in color. H=1. Soapy feel.	Talc
		Clearly Shows Cleavage	Brown to black in color. H=2.5. C=1 perfect. Breaks into thin sheets that are elastic.	Biotite Mica
			White to transparent in color. H=3. C=3 rhombohedral. Strong effervescence in acid.	Calcite
			Transparent, yellow, purple, green in color. H=4. C=4 - octahedral.	Fluorite
			Transparent to white in color. H=2. C=3, though 2 directions may be difficult to see.	Gypsum
			White to transparent in color. H=2.5. C=3, cubic. Tastes salty.	Halite
			Transparent, light brown, to yellow in color. H=2.5. C=1 perfect. Breaks into thin sheets that are elastic.	Muscovite Mica
Metallic	> Glass	Poor Cleavage	Black in color. H=6. St=black. Strongly magnetic.	Magnetite
			Brass-yellow in color. H=6.5. St=dark gray.	Pyrite
	< Glass	Poor Cleavage	Copper-red in color. Tarnishes to green or black in air. H=2.5-3. St=copper-red.	Copper

Figure 7.22 | Mineral Identification Chart
Author: Randa Harris
Source: Original Work
License: CC BY-SA 3.0

27. Sample 3: What is this sample?

    a. Microcline   b. Fluorite   c. Quartz

    d. Olivine   e. Talc   f. Selenite (Gypsum)

28. Sample 4: What is this sample?

    a. Microcline   b. Fluorite   c. Quartz

    d. Olivine   e. Talc   f. Selenite (Gypsum)

29. Sample 4: What other unique property does this sample have?

    a. effervescence in acid   b. it is magnetic

    c. it tastes salty   d. it feels soapy

    e. it writes on paper

30. Sample 5: What is this sample?

    a. Microcline   b. Fluorite   c. Quartz

    d. Olivine   e. Talc   f. Selenite (Gypsum)

31. Sample 6: What is this sample?

    a. Microcline   b. Fluorite   c. Quartz

    d. Olivine   e. Talc   f. Selenite (Gypsum)

32. Sample 7: What is this sample?

    a. Pyroxene   b. Muscovite Mica   c. Halite

    d. Amphibole   e. Calcite   f. Biotite Mica

33. Sample 7: What other unique property does this sample have?

    a. effervescence in acid   b. it is magnetic   c. it tastes salty

    d. it feels soapy   e. it writes on paper

INTRODUCTORY GEOLOGY MATTER AND MINERALS

34. Sample 8: What is this sample?

   a. Pyroxene   b. Muscovite Mica   c. Halite

   d. Amphibole   e. Calcite   f. Biotite Mica

35. Sample 8: What other unique property does this sample have?

   a. effervescence in acid   b. it is magnetic

   c. it tastes salty   d. it feels soapy

   e. it writes on paper

36. Sample 9: What is this sample?

   a. Pyroxene   b. Muscovite Mica   c. Halite

   d. Amphibole   e. Calcite   f. Biotite Mica

37. Sample 10: What is this sample?

   a. Pyroxene   b. Muscovite Mica   c. Halite

   d. Amphibole   e. Calcite   f. Biotite Mica

38. Sample 10: Test this tenacity of this mineral by trying to bend it. Which way does it behave?

   a. sectile   b. malleable   c. elastic   d. brittle

39. Sample 11: What is this sample?

   a. Pyroxene   b. Muscovite Mica   c. Halite

   d. Amphibole   e. Calcite   f. Biotite Mica

40. Sample 12: What is this sample?

   a. Pyroxene   b. Muscovite Mica   c. Halite

   d. Amphibole   e. Calcite   f. Biotite Mica

41. Sample 13: What is this sample?

   a. Magnetite   b. Graphite   c. Copper

   d. Sulfur   e. Hematite   f. Pyrite

42. Sample 14: What is this sample?

   a. Magnetite   b. Graphite   c. Copper

   d. Sulfur   e. Hematite   f. Pyrite

43. Sample 15: What is this sample?

   a. Magnetite   b. Graphite   c. Copper

   d. Sulfur   e. Hematite   f. Pyrite

44. Sample 16: What is this sample?

   a. Magnetite   b. Graphite   c. Copper

   d. Sulfur   e. Hematite   f. Pyrite

45. Sample 17: What is this sample?

   a. Magnetite   b. Graphite   c. Copper

   d. Sulfur   e. Hematite   f. Pyrite

46. Sample 18: What is this sample?

   a. Magnetite   b. Graphite   c. Copper

   d. Sulfur   e. Hematite   f. Pyrite

INTRODUCTORY GEOLOGY — MATTER AND MINERALS

Table 7.1 | Mineral Notation Chart – Fill in this chart as you work through the lab. An example of a mineral you do not have in your kit (#0) is included. You do not have to fill out every column for every mineral – just follow along in the lab and determine the properties you are asked about.
Author: Randa Harris
Source: Original Work
License: CC BY-SA 3.0

Mineral #	Luster	Hardness	Cleavage/Fracture	Streak	Other Notable Properties (include color when diagnostic)	Name
0	Metallic	~2.5 – may scratch fingernail	3 - cubic	gray	High specific gravity because it is heavy	Galena

Mineral #	Luster	Hardness	Cleavage/Fracture	Streak	Other Notable Properties (include color when diagnostic)	Name

Mineral #	Luster	Hardness	Cleavage/Fracture	Streak	Other Notable Properties (include color when diagnostic)	Name

Mineral #	Luster	Hardness	Cleavage/Fracture	Streak	Other Notable Properties (include color when diagnostic)	Name

7.12 STUDENT RESPONSES

The following is a summary of the questions in this lab for ease in submitting answers online.

1. Sample 1: What is this sample's hardness?

    a. harder than glass

    b. softer than glass but harder than nail

    c. softer than nail but harder than copper

    d. softer than copper but harder than a fingernail

    e. softer than a fingernail

2. Sample 2: What is this sample's hardness?

    a. harder than glass

    b. softer than glass but harder than nail

    c. softer than nail but harder than copper

    d. softer than copper but harder than a fingernail

    e. softer than a fingernail

3. Sample 3: What is this sample's hardness?

    a. harder than glass

    b. softer than glass but harder than nail

    c. softer than nail but harder than copper

    d. softer than copper but harder than a fingernail

    e. softer than a fingernail

4. Sample 4: What is this sample's hardness?

    a. harder than glass

    b. softer than glass but harder than nail

    c. softer than nail but harder than copper

    d. softer than copper but harder than a fingernail

    e. softer than a fingernail

5. Sample 5: What is this sample's hardness?

    a. harder than glass

    b. softer than glass but harder than nail

    c. softer than nail but harder than copper

    d. softer than copper but harder than a fingernail

    e. softer than a fingernail

6. Sample 6. What is this sample's hardness?

    a. harder than glass

    b. softer than glass but harder than nail

    c. softer than nail but harder than copper

    d. softer than copper but harder than a fingernail

    e. softer than a fingernail

7. Sample 7: This sample has:

    a. no cleavage (it fractures)   b. 1 cleavage plane

    c. 2 cleavage planes at 90°   d. 3 cleavage planes at 90°

    e. 4 cleavage planes

8. Sample 8: This sample has:

    a. no cleavage (it fractures)   b. 1 cleavage plane

    c. 2 cleavage planes not at 90°   d. 3 cleavage planes not at 90°

    e. 4 cleavage planes

9. Sample 9: This sample has:

    a. no cleavage (it fractures)   b. 1 cleavage plane

    c. 2 cleavage planes not at 90°   d. 3 cleavage planes at 90°

    e. 4 cleavage planes

10. Sample 10: This sample has:

    a. no cleavage (it fractures)   b. 1 cleavage plane

    c. 2 cleavage planes not at 90°   d. 2 cleavage planes at 90°

    e. 4 cleavage planes

11. Sample 11: This sample has:

    a. no cleavage (it fractures)   b. 1 cleavage plane

    c. 2 cleavage planes not at 90°   d. 2 cleavage planes at 90°

    e. 3 cleavage planes at 90°

12. Sample 11: What is this sample's hardness?

    a. harder than glass

    b. softer than glass but harder than nail

    c. softer than nail but harder than copper

    d. softer than copper but harder than a fingernail

    e. softer than a fingernail

13. Sample 12: This sample has:

    a. no cleavage (it fractures) b. 1 cleavage plane

    c. 2 cleavage planes not at 90° d. 2 cleavage planes at 90°

    e. 3 cleavage planes not at 90°

14. Sample 12: What is this sample's hardness?

    a. harder than glass

    b. softer than glass but harder than nail

    c. softer than nail but harder than copper

    d. softer than copper but harder than a fingernail

    e. softer than a fingernail

15. Sample 13: What is the streak of this sample?

    a. dark gray streak b. white streak

    c. reddish brown streak d. pale yellow streak

16. Sample 13: What is the luster of this sample?

    a. non-metallic, vitreous b. non-metallic, earthy

    c. non-metallic, greasy d. non-metallic, waxy

    e. metallic

17. Sample 14: What is the streak of this sample?

    a. dark gray to black streak b. white streak

    c. reddish brown streak d. pale yellow streak

18. Sample 14: Which other item(s) is/are characteristic(s) of this sample?

    a. stains the fingers b. harder than glass

    c. greasy feel d. both a and b

    e. both a and c

INTRODUCTORY GEOLOGY   MATTER AND MINERALS

19. Sample 15: What is the streak of this sample?

   a. dark gray to black streak   b. white streak

   c. reddish brown streak   d. pale yellow streak

20. Sample 16: What is the luster of this sample?

   a. non-metallic, vitreous   b. non-metallic, earthy

   c. non-metallic, greasy   d. non-metallic, waxy

   e. metallic

21. Sample 16: Due to its appearance, this sample has often been confused with native gold, a mineral with a hardness of 2.5-3. How does its hardness compare with that of gold?

   a. Sample 16 is harder than gold.   b. Sample 16 is softer than gold.

22. Sample 17: What is the streak of this sample?

   a. dark gray to black streak   b. white streak

   c. reddish brown streak   d. pale yellow streak

23. Sample 17: What another unique property does this sample have?

   a. effervescence in acid   b. it is magnetic

   c. it tastes salty   d. it feels soapy

   e. it writes on paper

24. Sample 18. Examine this entire sample closely with a hand lens. What is the luster of this sample?

   a. non-metallic, vitreous   b. non-metallic, earthy

   c. non-metallic, greasy   d. non-metallic, waxy

   e. metallic

25. Sample 1: What is this sample?

    a. Microcline   b. Fluorite   c. Quartz

    d. Olivine   e. Talc   f. Selenite (Gypsum)

26. Sample 2: What is this sample?

    a. Microcline   b. Fluorite   c. Quartz

    d. Olivine   e. Talc   f. Selenite (Gypsum)

27. Sample 3: What is this sample?

    a. Microcline   b. Fluorite   c. Quartz

    d. Olivine   e. Talc   f. Selenite (Gypsum)

28. Sample 4: What is this sample?

    a. Microcline   b. Fluorite   c. Quartz

    d. Olivine   e. Talc   f. Selenite (Gypsum)

29. Sample 4: What other unique property does this sample have?

    a. effervescence in acid   b. it is magnetic

    c. it tastes salty   d. it feels soapy

    e. it writes on paper

30. Sample 5: What is this sample?

    a. Microcline   b. Fluorite   c. Quartz

    d. Olivine   e. Talc   f. Selenite (Gypsum)

31. Sample 6: What is this sample?

    a. Microcline   b. Fluorite   c. Quartz

    d. Olivine   e. Talc   f. Selenite (Gypsum)

32. Sample 7: What is this sample?

    a. Pyroxene   b. Muscovite Mica   c. Halite

    d. Amphibole   e. Calcite   f. Biotite Mica

33. Sample 7: What other unique property does this sample have?

    a. effervescence in acid   b. it is magnetic   c. it tastes salty

    d. it feels soapy   e. it writes on paper

34. Sample 8: What is this sample?

    a. Pyroxene   b. Muscovite Mica   c. Halite

    d. Amphibole   e. Calcite   f. Biotite Mica

35. Sample 8: What other unique property does this sample have?

    a. effervescence in acid   b. it is magnetic   c. it tastes salty

    d. it feels soapy   e. it writes on paper

36. Sample 9: What is this sample?

    a. Pyroxene   b. Muscovite Mica   c. Halite

    d. Amphibole   e. Calcite   f. Biotite Mica

37. Sample 10: What is this sample?

    a. Pyroxene   b. Muscovite Mica   c. Halite

    d. Amphibole   e. Calcite   f. Biotite Mica

38. Sample 10: Test this tenacity of this mineral by trying to bend it. Which way does it behave?

    a. sectile   b. malleable   c. elastic   d. brittle

39. Sample 11: What is this sample?

    a. Pyroxene   b. Muscovite Mica   c. Halite

    d. Amphibole   e. Calcite   f. Biotite Mica

40. Sample 12: What is this sample?

    a. Pyroxene   b. Muscovite Mica   c. Halite

    d. Amphibole   e. Calcite   f. Biotite Mica

41. Sample 13: What is this sample?

    a. Magnetite   b. Graphite   c. Copper

    d. Sulfur   e. Hematite   f. Pyrite

42. Sample 14: What is this sample?

    a. Magnetite   b. Graphite   c. Copper

    d. Sulfur   e. Hematite   f. Pyrite

43. Sample 15: What is this sample?

    a. Magnetite   b. Graphite   c. Copper

    d. Sulfur   e. Hematite   f. Pyrite

44. Sample 16: What is this sample?

    a. Magnetite   b. Graphite   c. Copper

    d. Sulfur   e. Hematite   f. Pyrite

45. Sample 17: What is this sample?

    a. Magnetite   b. Graphite   c. Copper

    d. Sulfur   e. Hematite   f. Pyrite

46. Sample 18: What is this sample?

    a. Magnetite   b. Graphite   c. Copper

    d. Sulfur   e. Hematite   f. Pyrite

8 Igneous Rocks
Karen Tefend

8.1 INTRODUCTION

All rocks found on the Earth are classified into one of three groups: igneous, sedimentary, or metamorphic. This rock classification is based on the origin of each of these rock types, or if you prefer, based on the rock-forming process that formed the rock. The focus of this chapter will be on igneous rocks, which are the only rocks that form from what was once a molten or liquid state. Therefore, based on their mode of origin, igneous rocks are defined as those rock types that form by the cooling of magma or lava. You would be right in thinking that there is more to the classification of igneous rocks than stated in the previous sentence, as there are dozens of different igneous rocks that are considered commonplace, and dozens of more types that are less common, and also quite a few igneous rock types that are quite scarce, yet each igneous rock has a name that distinguishes it from all the rest of the igneous group of rocks. So, if they all start out as molten material (magma or lava), which must harden to form a rock, then it is logical to assume that these igneous rocks differ from one another primarily due to: 1) the original composition of the molten material from which the rock is derived, and 2) the cooling process of the molten material that ended up forming the rock. These two parameters define the classification of igneous rocks, which are simplified into the two terms: composition and texture. Igneous rock composition refers to what is in the rock (the chemical composition or the minerals that are present), and the word texture refers to the features that we see in the rock such as the mineral sizes or the presence of glass, fragmented material, or vesicles (holes) in the igneous rock.

8.1.1 Learning Outcomes

After completing this chapter, you should be able to:
- Classify igneous rock types based on color, texture, and mafic color index
- Identify, when possible, the minerals present in an igneous rock
- Determine the cooling history of the igneous rock

8.1.2 Key Terms

- Aphanitic
- Extrusive
- Felsic (Silicic)
- Ferromagnesian
- Glassy
- Intermediate
- Intrusive
- Mafic
- Nonferromagnesian
- Phaneritic
- Phenocryst
- Plutonic
- Porphyritic
- Ultramafic
- Vesicular
- Volcanic

8.2 IGNEOUS ROCK ORIGIN

8.2.1 Magma Composition

It seems like a bad joke, but before any igneous rock can form, there must be molten material known as magma produced, which means that you must first have a rock to melt to make magma in order for it to cool and become an igneous rock. Which brings more questions: what rock melted to form the magma? Was there more than one rock type that melted to form that magma? Did the rocks completely melt, or did only certain minerals inside of those rocks melt (a process known as partial melting)? Once that melted material formed, what happened to it next? Did some other process occur to change the composition of that magma, before ending up as the igneous rock that we are studying? These are just a few of the questions that a person should consider when studying the origin of igneous rocks.

Most rocks (there are very few exceptions!) contain minerals that are crystalline solids composed of the chemical elements. In your chapter on minerals you learned that the most common minerals belong to a group known as the silicate minerals, so it makes sense that magmas form from the melting of rocks that most likely contain abundant silicate minerals. However all minerals (not just the silicates) have a certain set of conditions, such as temperature, at which they can melt. Since rocks contain a mixture of minerals it is easy to see how only some of the minerals in a rock may melt, and why others stay as a solid. Furthermore, the temperature conditions are important, as only minerals that can melt at "lower" temperatures (such as 600°C) may experience melting, whereas the temperature would have to increase (for example, to 1200°C) in order for other minerals to *also* melt (remember the lower temperature minerals are still melting) and thus add their chemical components to the magma that is being generated. This brings up an important point: even if the same types of rocks are melting, we can generate different magma compositions purely by melting at different temperatures!

Once magma is generated, it will eventually start to rise upward through the Earth's lithosphere, as magma is more buoyant than the source rock that generated it. This separation of the magma from the source region will result in new thermal conditions as the magma moves away from the heated portion of the lithosphere and encounters cooler rocks, which results in the magma also cooling. As with melting, minerals also have a certain set of conditions at which they form, or crystallize, from within a cooling magma body. You would be right in thinking that the sequence of mineral crystallization is the opposite sequence of crystal melting. The sequence of mineral formation from magma has been experimentally determined by Norman L. Bowen in the early 1900's, and the now famous Bowen's reaction series appears in countless textbooks and lab manuals (Figure 8.1).

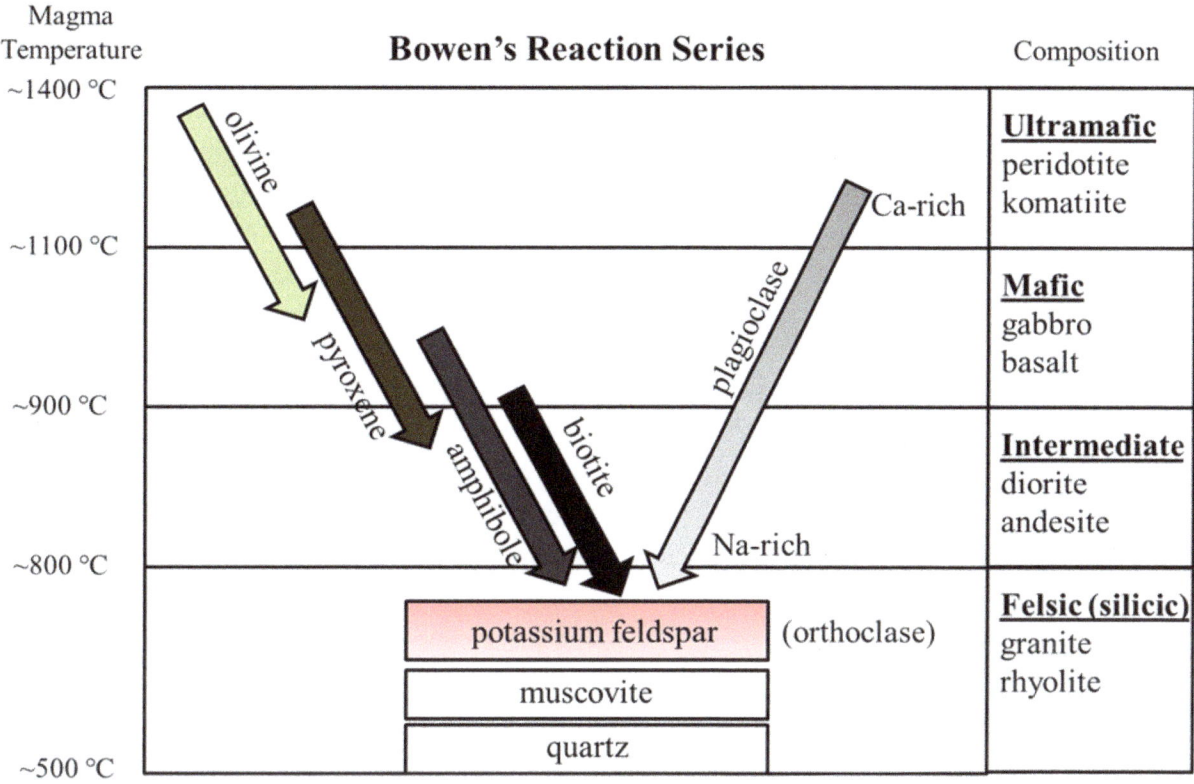

Figure 8.1 | Bowen's Reaction Series, showing the progression of mineral crystallization as magma temperatures drop from ~1400 °C to ~500 °C. Note the corresponding names for igneous rock composition (underlined and bolded) and some example rock types within each compositional group.
Author: Karen Tefend
Source: Original Work
License: CC BY-SA 3.0

This "reaction series" refers to the chemical reactions that are the formation of minerals, through chemical bonding of elements within the magma, in a sequence that is based on falling magma temperatures. Close examination of Figure 8.1 shows that the first mineral to crystallize in a cooling magma of ultramafic composition is olivine; the length of the arrow indicates the range of temperatures at which olivine can form. Once temperatures fall below this range, olivine crystals will no longer form; instead, other minerals such as pyroxene will start to crystallize (a small interval of temperatures exists where both olivine and pyroxene can crystallize). Minerals that form in cooling magma are called crystals, or **phenocrysts**; as these phenocrysts are forming, they are removing chemical elements from the magma. For example, olivine phenocrysts take magnesium (Mg) and iron (Fe) from the magma and incorporate them into their crystal structure. This behavior of mineral phenocrysts to take certain chemical elements into their structure, while excluding other elements, means that the composition of the magma must be changing as phenocrysts are forming!

There can be more than one mineral type crystallizing within the cooling magma, as the arrows in Figure 8.1 demonstrate. The minerals on the left side of Bowen's reaction series are referred to as a discontinuous series, as these minerals (olivine, pyroxene, amphibole, and biotite) all remove the iron (Fe), magnesium (Mg), and manganese (Mn) from the magma during crystallization, but do so at certain temperature ranges. These iron- and magnesium-rich minerals are referred to as **ferromagnesian** minerals (ferro = iron) and are usually green, dark gray, or black in color due to the absorption of visible light by iron and magnesium atoms. On the right side of Bowen's reaction series is a long arrow labelled plagioclase feldspar. Plagioclase crystallizes over a large temperature interval and represents a continuous series of crystallization even though its composition changes from calcium (Ca) rich to sodium (Na) rich. As the magma temperature drops and plagioclase first starts to crystallize (form), it will take in the calcium atoms into the crystal structure, but as magma temperatures continue to drop, plagioclase takes in sodium atoms preferentially. As a result, the higher temperature calcium-rich plagioclase is dark gray in color due to the high calcium content, but the lower temperature sodium-rich plagioclase is white due to the high sodium content. Finally, at the bottom of the graph in Figure 8.1, we see that three more minerals can form as temperatures continue to drop. These minerals (potassium feldspar, muscovite, and quartz) are considered to be the "low temperature minerals", as they are the last to form during cooling, and therefore first to melt as a rock is heated. The previous removal of iron and magnesium from the magma results in the formation of the latest-forming minerals that are deficient in these chemical elements; these minerals are referred to as **nonferromagnesian** minerals, which are much lighter in color. For example, the potassium-rich feldspar (also known as orthoclase) can be a pale pink or white in color. The references to mineral color are necessary, as the color of any mineral is primarily due to the chemical elements that are in the minerals, and therefore the color of an igneous rock will be dependent on the mineral content (or chemical composition) of the rock.

8.3 IGNEOUS ROCK COMPOSITION

Often added to the Bowen's reaction series diagram are the igneous rock classifications as well as example igneous rock names that are entirely dependent on the minerals that are found in them. For example, you can expect to find abundant olivine, and maybe a little pyroxene and a little Ca-rich plagioclase, in an **ultramafic** rock called peridotite or komatiite, or that pyroxene, plagioclase, and possibly some olivine or amphibole may be present in a **mafic** rock such as gabbro or basalt. You can also expect to see quartz, muscovite, potassium feldspar, and maybe a little biotite and Na-rich plagioclase in a **felsic** (or **silicic**) rock such as granite or rhyolite. Figure 8.1 demonstrates nicely that the classification of an igneous rock depends partly on the minerals that may be present in the rock, and since the minerals have certain colors due to their chemical makeup, then the rocks must have certain colors. For example, a rock composed of mostly olivine will be green in color due to olivine's green color; such a rock would be called ultramafic. A rock that has a large amount of ferromagnesian minerals in it will be a dark-colored rock because the ferromagnesian minerals (other than olivine) tend to be dark colored; an igneous rock that is dark in color is called a **mafic** rock ("ma-" comes from magnesium, and "fic" from ferric iron). An igneous rock with a large amount of nonferromagnesian minerals will be light in color, such as the silicic or felsic rocks ("fel" from feldspar, and "sic" from silica-rich quartz). So, based on color alone, we've been able to start classifying the igneous rocks.

In Figure 8.2 are examples of igneous rocks that represent the mafic and felsic rock compositions (Figures 8.2A and 8.2C, respectively), as well as an **intermediate** rock type (Figure 8.2B). Notice that the felsic rocks can have a small amount of dark-colored ferromagnesian minerals, but is predominately composed of light-colored minerals, whereas the mafic rock has a higher percentage of dark-colored ferromagnesian minerals, which results in a darker-colored rock. A rock that is considered intermediate between the mafic and felsic rocks is truly an intermediate in terms of the color and mineral composition; such a rock would have less ferromagnesian minerals than the mafic rocks, yet more ferromagnesian minerals than the felsic rocks.

Figure 8.2 | Examples of igneous rocks from the mafic (A), intermediate (B), and felsic (C) rock compositions. Photo scale on bottom is in centimeters.
Author: Karen Tefend
Source: Original Work
License: CC BY-SA 3.0

As previously mentioned, classifying rocks into one of the igneous rock compositions (ultramafic, mafic, intermediate, and felsic) depends on the minerals that each rock contains. Identification of the minerals can be difficult in rocks such as in Figure 8.2A, as the majority of minerals are dark in color and it can be difficult to distinguish each mineral. An easy method of determining the igneous rock composition is by determining the percentage of dark-colored minerals in the rock, without trying to identify the actual minerals present; this method of classification relies on a mafic color index (MCI), where the term mafic refers to any dark gray, black or green colored mineral (Figure 8.3). Igneous rocks with 0-15% dark colored minerals (or 0-15% MCI) are the felsic rocks (Figure 8.3A), igneous rocks with 46-85% MCI are the mafic rocks (Figure 8.3C), and igneous rocks with over 85% MCI are considered ultramafic (Figure 8.3D). This means that any rock with an intermediate composition or with a 16-45% MCI is an intermediate igneous rock (Figure 8.3B). Estimating the percentage of dark-colored minerals is only possible if the minerals are large enough to see; in that case a person can still recognize a mafic rock by its dark-colored appearance, and a felsic rock by its light-colored

Figure 8.3 | Examples of how igneous rocks can be classified using the Mafic Color Index (MCI), which is a visual classification based on the amount of ferromagnesian minerals in the rock: (A) The small amount of tiny black phenocrysts (biotite) gives this rock a 0-15% MCI value; (B) numerous dark phenocrysts (amphibole) gives this rock a 16-45% MCI value; (C) this rock lacks visible phenocrysts, but the black color in this rock results in a 46-85% MCI value; (D) this rock is entirely green in color due to the overwhelming amount of olivine. Any rock with this much olivine is always classified as having over 85% MCI values.
Author: Karen Tefend
Source: Original Work
License: CC BY-SA 3.0

appearance. An intermediate rock will be somewhat lighter than a mafic rock, yet darker than a felsic rock. Finally, an ultramafic rock is typically green in color, due to the large amount of green-colored olivine in the rock. Such rocks that contain minerals that are too small to see are shown in Figure 8.4; note that you can still distinguish between mafic (Figure 8.4A), intermediate (Figure 8.4B) and felsic (Figure 8.4C) by the overall color of the rock. The intermediate igneous rock in Figure 8.4B does have a few visible phenocrysts; this odd texture will be covered later in this chapter.

Figure 8.4 | Examples of igneous rocks from the mafic (A), intermediate (B), and felsic (C) rock compositions. Notice the difference in appearance between these rocks and those in Figure 8.2
Author: Karen Tefend
Source: Original Work
License: CC BY-SA 3.0

8.4 LAB EXERCISE

Part A - Igneous Rock Composition

Before attempting to answer the following questions, remove the eight rock samples from the Igneous Rocks bag in your HOL rock kit (samples I1 – I8) and place them on a clean sheet of white paper. Your samples should look identical to the samples in Figure 8.5.

Separate the 8 samples into three groups based on their color (dark, light or intermediate). As a first attempt at classifying rock compositions based on color, most students end up with three light colored rocks (felsic), two intermediate rocks, and three dark colored rocks (mafic). There actually are only two mafic, two intermediate, two felsic, and two other rocks that we have yet to discuss, but for now leave your samples as they are. Take a close look at the dark colored rocks in your mafic pile; how are they different? When you hold them closer to a light source, notice how they reflect light; one is dull looking, one has small shiny surfaces, and one is extremely shiny and smooth. The one that is very smooth and shiny is not a mafic rock, even though it is dark in color. This particular rock is an example of that rare rock type that lacks minerals (crystals); instead, it is almost entirely made of glass, and is most probably felsic in composition. Go ahead and move this rock to its own space on your piece of paper. Now take a close look at your three light-colored felsic rocks. One will have several minerals visible that you will see as different colors. The other two rocks will be fairly uniform in color; one is composed of very tiny minerals that require a microscope to view them, and the other rock seems to be fragile and light-weight. This fragile appearing rock is another example of rock that is again almost entirely glass with a very few, tiny phenocrysts; set this rock aside with your other glassy rock. Now you should have four piles, with two rocks in each pile, and you can now proceed with the following questions. You may want to refer to Figures 8.1 through 8.4 to help with identification.

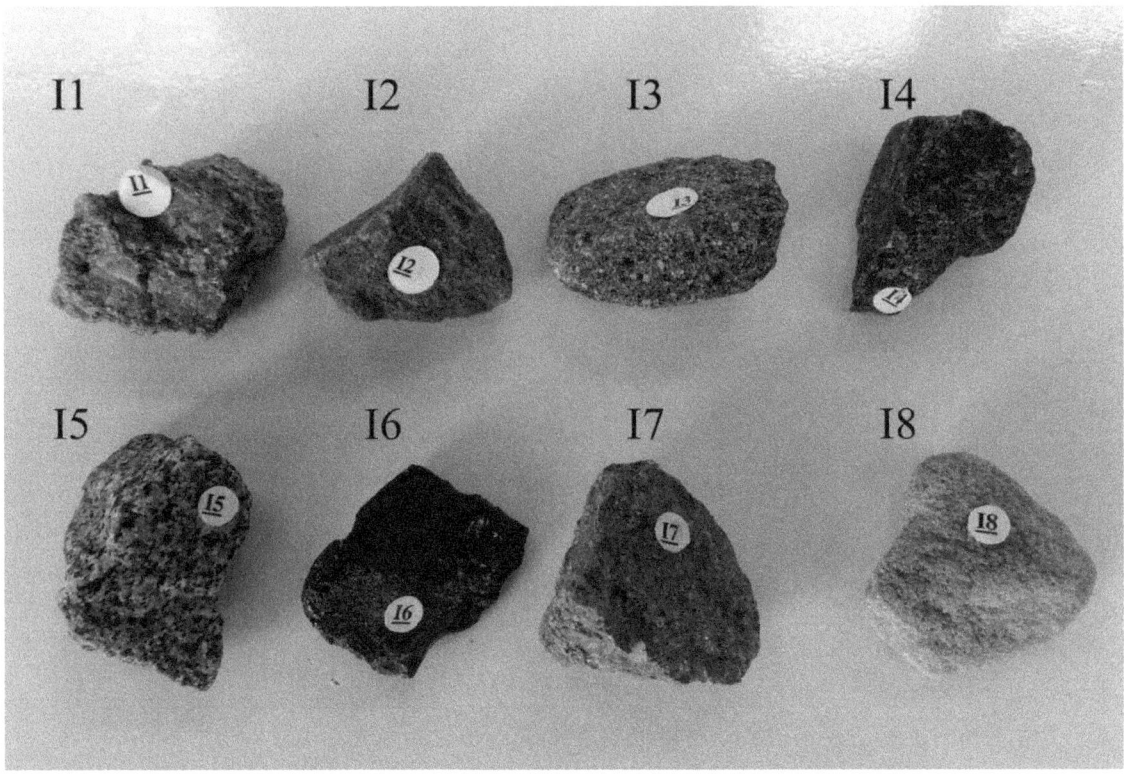

Figure 8.5 | These samples are in your rock kit, in a bag labeled "IGNEOUS ROCKS". Each sample is identified by the letter "I" for igneous, and a number 1 through 8. A hand lens is also provided in your rock kit; use the hand lens for magnified viewing.
Author: Karen Tefend
Source: Original Work
License: CC BY-SA 3.0

1. Sample I1 has what mafic color index (MCI)?

    a. 0-15% MCI   b. 16-45% MCI   c. 46-85% MCI   d. >85% MCI

2. Sample I1 has pink colored minerals present, which are most likely _____.

    a. quartz   b. biotite   c. orthoclase   d. Ca-rich plagioclase

3. The small, dark colored minerals in Sample I1 are most likely _____.

    a. quartz   b. biotite   c. orthoclase   d. Na-rich plagioclase

4. Sample I1 belongs to which igneous rock classification?

    a. ultramafic   b. mafic   c. intermediate   d. felsic

5. Sample I2 has what mafic color index (MCI)?

    a. 0-15% MCI   b. 16-45% MCI   c. 46-85% MCI   d. >85% MCI

6. Sample I2 belongs to which igneous rock classification?

    a. ultramafic   b. mafic   c. intermediate   d. felsic

7. Sample I3 has what mafic color index (MCI)?

    a. 0-15% MCI   b. 16-45% MCI   c. 46-85% MCI   d. >85% MCI

8. Sample I3 belongs to which igneous rock classification?

    a. ultramafic   b. mafic   c. intermediate   d. felsic

9. Sample I4 has what mafic color index (MCI)?

    a. 0-15% MCI   b. 16-45% MCI   c. 46-85% MCI   d. >85% MCI

10. Sample I4 belongs to which igneous rock classification?

    a. ultramafic   b. mafic   c. intermediate   d. felsic

11. Sample I5 has what mafic color index (MCI)?

    a. 0-15% MCI   b. 16-45% MCI   c. 46-85% MCI   d. >85% MCI

12. The white minerals in Sample I5 are most likely _____.

    a. quartz   b. biotite   c. orthoclase   d. plagioclase

13. Sample I7 has what mafic color index (MCI)?

    a. 0-15% MCI   b. 16-45% MCI   c. 46-85% MCI   d. >85% MCI

14. According to Figure 8.1, Sample I4 will never contain which mineral?

    a. olivine   b. pyroxene   c. amphibole   d. quartz

15. According to Figure 8.1, Sample I7 must contain which mineral?

    a. olivine   b. pyroxene   c. amphibole   d. quartz

16. According to Bowen's Reaction Series, Sample I5 (if heated) should start melting at what temperature?

    a. ~1100 °C   b. ~900 °C   c. ~800 °C   d. ~500 °C

17. According to Figure 8.1, when Sample I1 is heated, what is the first mineral to melt?

    a. olivine   b. Ca-rich plagioclase

    c. potassium feldspar   d. quartz

8.5 IGNEOUS ROCK TEXTURE

The classification of igneous rocks is based not just on composition, but also on texture. As mentioned earlier, texture refers to the features that we see in the rock such as the mineral sizes or the presence of glass, fragmented material, or vesicles (holes) in the igneous rock. We will cover mineral crystal sizes and vesicles in this section.

Since the crystals or phenocrysts form while the magma is cooling, then the size of the crystals must have something to do with the cooling process. Recall that each mineral derives its chemical composition directly from the magma, and that each mineral has a certain temperature interval during which that particular mineral can form. The chemical elements that become part of the mineral must migrate from the liquid magma to link or bond with other elements in a certain way to form the crystal structure that is unique for that mineral. What do you think will happen if the magma's temperature drops quickly, or if the magma's temperature drops slowly? Either way, the time allowed for the migration of the chemical elements to form a crystal is affected. When magma cools slowly, there is plenty of time for the migration of the needed chemical elements to form a certain mineral; that particular mineral can become quite large in size, large enough for a person to see without the aid of a microscope. As a result, this igneous rock with its visible minerals is said to have a **phaneritic** texture (phan = large). The rock samples shown in Figure 8.2 are all phaneritic rocks. Figure 8.2A is a phaneritic mafic rock called gabbro, Figures 8.2B and 8.3B are a phaneritic intermediate rock called diorite, and the rock in Figure 8.2C is a phaneritic felsic rock known as granite. If you refer back to Figure 8.1 (Bowen's reaction series) you will see that these rock names are listed on the right side of the diagram.

Magma that cools relatively quickly will have the opposite result as described above; there is less time for the migration of the chemical elements to form a mineral, and as a result the minerals will not have time to form large crystals. Therefore, many small crystals of a particular mineral will form in the magma. Igneous rocks that are composed of crystals too small to see (unless you have a microscope) are called **aphanitic** igneous rocks. Figure 8.3C, 8.4A and 8.4C are aphanitic rocks; because Figure 8.3C (and 8.4A) is dark in color, it is a mafic aphanitic rock called basalt. The felsic rock in Figure 8.4C is called rhyolite. It is important to note that basalt and gabbro are both mafic rocks and have the same composition, but one rock represents a magma that cooled fast (basalt), and the other represents a mafic magma that cooled slowly (gabbro). The same can be said for the other rock compositions: the felsic rocks rhyolite and granite have identical compositions but one cooled fast (rhyolite) and the other cooled slower (granite). The intermediate rocks diorite (Figure 8.2B) and andesite (Figure 8.4B) also represent magmas that cooled slow or a bit faster, respectively. Sometimes there are some visible crystals in an otherwise aphanitic rock, such as the andesite in Figure 8.4B. The texture of such a rock is referred to as **porphyritic**, or more accurately porphyritic-aphanitic since it is a porphyritic andesite, and all andesites are aphanitic. Two different crystal sizes within an igneous rock indicates that the cooling rate of the magma increased; while the magma was cooling slowly, larger crystals can form, but if the magma starts to cool faster, then only small crystals can form. A phaneritic rock can also be referred to as a porphyritic-phaneritic rock if the phaneritic rock contains some very large crystals (ie. the size of your thumb!) in addition to the other visible crystals. In Figure 8.6 are two porphyritic rocks: a porphyritic-aphanitic basalt, and a porphyritic-phaneritic granite.

Figure 8.6 | (A) An example of a porphyritic-aphanitic mafic rock with needle-shaped amphibole phenocrysts (arrow points to one phenocryst that is 1cm in length); No other minerals in (A) are large enough to see. (B) An example of a porphyritic-phaneritic felsic rock with large feldspars (outlined phenocryst is 3 cm length). Surrounding these large feldspars are smaller (yet still visible) dark and light colored minerals.
Author: Karen Tefend
Source: Original Work
License: CC BY-SA 3.0

Sometimes the magma cools so quickly that there isn't time to form any minerals as the chemical elements in the magma do not have time to migrate into any crystal structure. When this happens, the magma becomes a dense glass called obsidian (Figure 8.7A). By definition, glass is a chaotic arrangement of the chemical elements, and therefore not considered to be a mineral; igneous rocks composed primarily of glass are said to have a **glassy** texture. The identification of a glassy rock such as obsidian is easy once you recall the properties of glass; any thick glass pane or a glass bottle that is broken will have this smooth, curve shaped pattern on the broken edge called conchoidal fracture (this was covered in your mineral chapter). Even though obsidian is naturally occurring, and not man made, it still breaks in this conchoidal pattern. If you look closely at the obsidian in Figure 8.7A, you will see the curved (conchoidal) surfaces by noticing the shiny pattern on the rock. Obsidian appears quite dark in color regardless of its composition because it is a dense glass, and light cannot pass through this thick glass; however, if the edges of the obsidian sample are thin enough, you may be able to see through the glass.

Figure 8.7 | Igneous rocks with glassy texture: obsidian (A) and pumice (B).
Author: Karen Tefend
Source: Original Work
License: CC BY-SA 3.0

In Figure 8.7B there is another igneous rock that is also composed primarily of glass due to a very fast rate of magma cooling. This rock is called pumice, and is commonly referred to as the rock that floats on water due to its low density. The glass in this rock is stretched out into very fine fibers of glass which formed during the eruptive phase of a volcano. Because these fibers are so thin, they are easy to break (unlike the dense obsidian) and any conchoidal fractures on these fibers are too small to see without the aid of a microscope. Pumice can have any composition (felsic to mafic), but unlike obsidian the color of the pumice can be used to determine the magma composition, as felsic pumice is always light in color and mafic pumice will be dark in color. Mafic pumice with a dark grey, red or black color is also known as scoria.

8.6 IGNEOUS ROCK FORMATION—PLUTONIC VS. VOLCANIC

The different crystal sizes and presence or absence of glass in an igneous rock is primarily controlled by the rate of magma cooling. Magmas that cool below the surface of the earth tend to cool slowly, as the surrounding rock acts as an insulator, not unlike a coffee thermos. Most of us are aware that even the most expensive coffee thermos does not prevent your coffee from cooling; it just slows the rate of cooling. Magma that stays below the surface of the earth can take tens of thousands of years to completely crystallize depending on the size of the magma body. Once the magma has completely crystallized, the entire igneous body is called a pluton. Sometimes portions of these plutons are exposed at the earth's surface where direct observation of the rock is possible, and upon inspection of this "plutonic" rock, you would see that it is composed of minerals that are large enough to see without the aid of a microscope. Therefore, any igneous rock sample that is considered to have a phaneritic texture (or porphyritic-phaneritic), is also referred to as a **plutonic** rock. A plutonic rock is also called an **intrusive** rock as it is derived from magma that intruded the rock layers but never reached the earth's surface.

If magma does manage to reach the earth's surface, it is no longer insulated by the rocks around it and will therefore cool rapidly. Magma that reaches the earth's surface through a fissure or central vent (volcano) will lose some of its dissolved gas and becomes lava, and any rock that forms from lava will have an aphanitic texture due to fast cooling (or have a glassy texture due to very fast cooling). Flowing lava may continue to release gas while cooling; this is typical of mafic lava flows. If the lava hardens while these gases are bubbling out of the lava, a small hole or vesicle may form on the rock, and even though the rock is aphanitic (due to fast cooling), the name of the rock can be given the term "**vesicular**" to indicate the presence of these vesicles. For example, a basalt with vesicles is called vesicular basalt (Figure 8.8). Aphanitic rocks and rocks with glassy texture are also known as **volcanic** igneous rocks, or **extrusive** igneous rocks, as the magma was "extruded" onto the surface of the earth. Porphyritic-aphanitic rocks are also considered to be volcanic or extrusive

Figure 8.8 | An aphanitic mafic rock (basalt), with gas escape structures called vesicles. Arrow points to one vesicle that is ~1cm in diameter. This is an example of another texture type, called vesicular texture, and the name of this rock is a vesicular basalt.
Author: Karen Tefend
Source: Original Work
License: CC BY-SA 3.0

rocks; these rocks may have begun crystallizing under the earth's surface as a slowly cooling magma (as evidence by the small amount of visible phenocrysts), but this magma was extruded onto the surface as lava before finally crystallizing completely on the surface of the earth to form an extrusive igneous rock with a porphyritic texture, or more accurately, a porphyritic-aphanitic texture.

A summary of the terms used to classify the igneous rocks are provided in Figure 8.9 in order to help with the identification of the igneous rock samples provided in your rock kit. Refer to the preceding figures for further help.

Composition

Texture		Felsic	Intermediate	Mafic
		0-15% MCI	16-45% MCI	46-85% MCI
	Phaneritic visible minerals, slow cooling (plutonic)	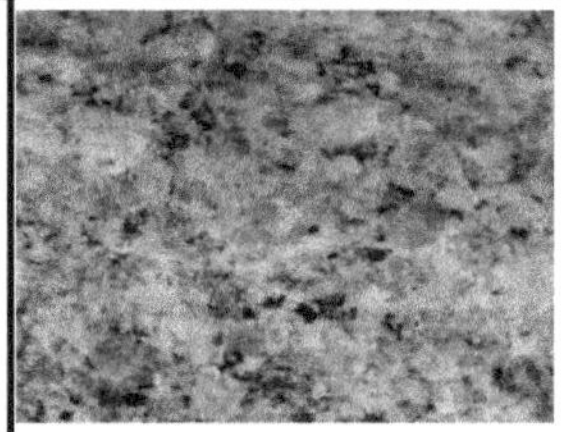 granite	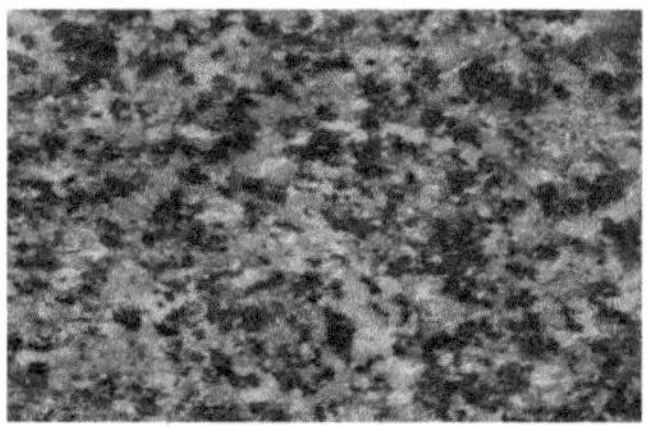 diorite	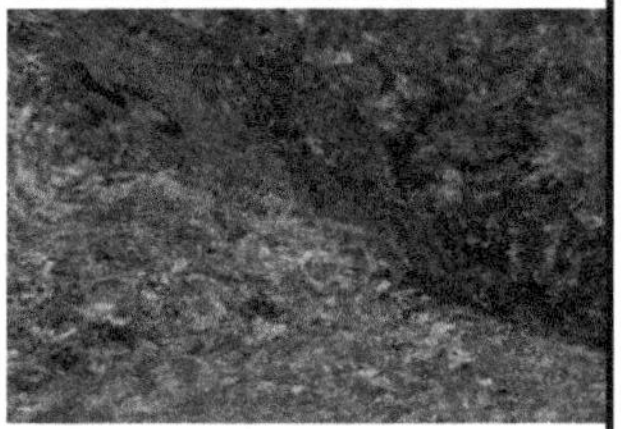 gabbro
	Aphanitic No visible crystals, fast cooling (volcanic)	rhyolite	 andesite*	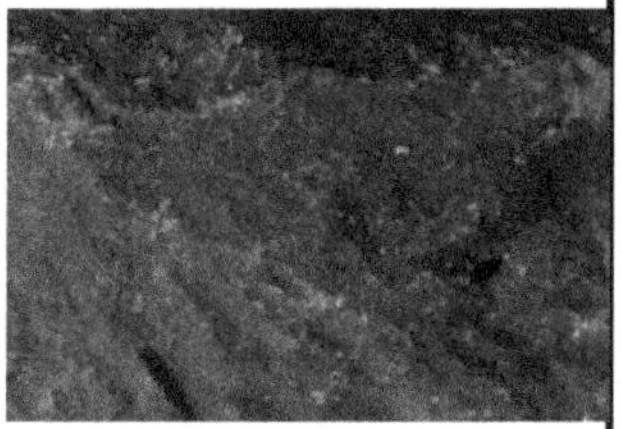 basalt
	Porphyritic	Use this term if there are visible crystals in basalt or andesite, or if there are very large crystals in granite		
	Glassy	Dense glass that is dark in color (obsidian), or Glass froth that is light or intermediate in color (pumice)		

*this andesite is porphyritic-aphanitic.

Figure 8.9 | Chart showing some common igneous rock textures and compositions. MCI = mafic color index, or the percentage of dark colored ferromagnesian minerals present. Recall that any composition can be phaneritic, aphanitic, porphyritic or glassy. Vesicular texture is not as common and is only seen in some aphanitic rocks.
Author: Karen Tefend
Source: Original Work
License: CC BY-SA 3.0

8.7 LAB EXERCISE

Part B - Igneous Rock Texture

For this exercise, you will determine the texture and rate of cooling for the igneous rock samples in your HOL kit. Refer to Figures 8.6 through 8.9 for guidance.

18. Sample I1 has what texture?

    a. phaneritic   b. aphanitic   c. porphyritic-aphanitic   d. glassy

19. How did Sample I1 form?

    a. by magma that cooled slowly

    b. by fast cooling lava

    c. by a very fast cooling of a lava

    d. by a very fast cooling magma during a volcanic eruption

    e. by magma that was cooling slowly, then as a lava that cooled quickly

20. Sample I1 is called_____.

    a. obsidian   b. pumice   c. granite   d. rhyolite

    e. diorite   f. andesite   g. gabbro   h. basalt

21. Sample I1 can also be called _____.

    a. volcanic   b. plutonic

22. Sample I2 has what texture?

    a. phaneritic   b. aphanitic   c. porphyritic-aphanitic   d. glassy

23. How did Sample I2 form?

    a. by magma that cooled slowly

    b. by fast cooling lava

    c. by a very fast cooling of a lava

    d. by a very fast cooling magma during a volcanic eruption

    e. by magma that was cooling slowly, then as a lava that cooled quickly

24. Sample I2 is called_____.

   a. obsidian   b. pumice   c. granite   d. rhyolite

   e. diorite   f. andesite   g. gabbro   h. basalt

25. Sample I2 can also be called _____.

   a. volcanic   b. plutonic

26. Sample I3 has what texture?

   a. phaneritic   b. aphanitic   c. porphyritic-aphanitic   d. glassy

27. How did Sample I3 form?

   a. by magma that cooled slowly

   b. by fast cooling lava

   c. by a very fast cooling of a lava

   d. by a very fast cooling magma during a volcanic eruption

   e. by magma that was cooling slowly, then as a lava that cooled quickly

28. Sample I3 is called_____.

   a. obsidian   b. pumice   c. granite   d. rhyolite

   e. diorite   f. andesite   g. gabbro   h. basalt

29. Sample I3 can also be called _____.

   a. volcanic   b. plutonic

30. Sample I4 has what texture?

   a. phaneritic   b. aphanitic   c. porphyritic-aphanitic   d. glassy

31. How did Sample I4 form?

   a. by magma that cooled slowly

   b. by fast cooling lava

   c. by a very fast cooling of a lava

   d. by a very fast cooling magma during a volcanic eruption

   e. by magma that was cooling slowly, then as a lava that cooled quickly

32. Sample I4 is called_____.

   a. obsidian   b. pumice   c. granite   d. rhyolite

   e. diorite   f. andesite   g. gabbro   h. basalt

33. Sample I4 can also be called _____.

   a. volcanic   b. plutonic

34. Sample I5 has what texture?

   a. phaneritic   b. aphanitic   c. porphyritic-aphanitic   d. glassy

35. How did Sample I5 form?

   a. by magma that cooled slowly

   b. by fast cooling lava

   c. by a very fast cooling of a lava

   d. by a very fast cooling magma during a volcanic eruption

   e. by magma that was cooling slowly, then as a lava that cooled quickly

36. Sample I5 is called_____.

   a. obsidian   b. pumice   c. granite   d. rhyolite

   e. diorite   f. andesite   g. gabbro   h. basalt

37. Sample I5 can also be called _____.

   a. volcanic   b. plutonic

38. Sample I6 has what texture?

    a. phaneritic   b. aphanitic   c. porphyritic-aphanitic   d. glassy

39. Sample I6 has what composition?

    a. felsic   b. intermediate   c. mafic   d. any of the above

40. How did Sample I6 form?

    a. by magma that cooled slowly

    b. by fast cooling lava

    c. by a very fast cooling of a lava

    d. by a very fast cooling magma during a volcanic eruption

    e. by magma that was cooling slowly, then as a lava that cooled quickly

41. Sample I6 is called_____.

    a. obsidian   b. pumice   c. granite   d. rhyolite

    e. diorite   f. andesite   g. gabbro   h. basalt

42. Sample I7 has what texture?

    a. phaneritic   b. aphanitic   c. porphyritic-aphanitic   d. glassy

43. How did Sample I7 form?

    a. by magma that cooled slowly

    b. by fast cooling lava

    c. by a very fast cooling of a lava

    d. by a very fast cooling magma during a volcanic eruption

    e. by magma that was cooling slowly, then as a lava that cooled quickly

44. Sample I7 is called_____.

    a. obsidian   b. pumice   c. granite   d. rhyolite

    e. diorite   f. andesite   g. gabbro   h. basalt

45. Sample I7 can also be called _____.

    a. volcanic   b. plutonic

46. Sample I8 has what texture?

    a. phaneritic   b. aphanitic   c. porphyritic-aphanitic   d. glassy

47. Sample I8 has what composition?

    a. felsic   b. intermediate   c. mafic   d. any of the above

48. How did Sample I8 form?

    a. by magma that cooled slowly

    b. by fast cooling lava

    c. by a very fast cooling of a lava

    d. by a very fast cooling magma during a volcanic eruption

    e. by magma that was cooling slowly, then as a lava that cooled quickly

49. Sample I8 is called_____.

    a. obsidian   b. pumice   c. granite   d. rhyolite

    e. diorite   f. andesite   g. gabbro   h. basalt

Part C - Google Earth

Igneous intrusions such as plutons are large magma bodies that have crystallized into solid rock. Regional uplift and erosion of the earth's surface can expose portions of these plutons, and large exposures that are over 100 km² are called batholiths, and a small portion (less than 100 km²) of a pluton exposed at the earth's surface is called a stock. The following exercises require that you use Google Earth.

50. Type the following coordinates into the search bar in Google Earth: 33 48 14.24 N 84 08 44.31 W. This is Stone Mountain, GA. Zoom in to an eye elevation of ~2400 ft for a closer look. Based on the color, and assuming no color change has occurred due to weathering at this location, what is the composition of rock?

    a. ultramafic   b. mafic   c. intermediate   d. felsic

51. Zoom out to an eye elevation of ~1500 ft so that you can see the entire mountain. Click on the ruler icon to open up the ruler function. Select the Line tab, and change the map length to kilometers. What is the longest length of exposed portion of Stone Mountain?

    a. 3000 km   b. 260 km   c. 6 km   d. 2.6 km

52. Based on your measurement of the longest dimension, Stone Mountain can be classified as:

    a. a batholith   b. a stock

53. Type the following coordinates into the search bar in Google Earth: 64 58 45.54 N 16 43 07.04 W. Zoom in to an eye elevation of ~12979 ft for a closer look. Based on the color, what is the composition of this rock?

    a. ultramafic   b. mafic   c. intermediate   d. felsic

54. Zoom out to an eye elevation of 41,566 ft. Based on the proximity of the volcanic crater (now containing a lake), the igneous rocks identified in question 53 are:

    a. extrusive   b. intrusive

8.8 STUDENT RESPONSES

1. Sample I1 has what mafic color index (MCI)?

    a. 0-15% MCI   b. 16-45% MCI   c. 46-85% MCI   d. >85% MCI

2. Sample I1 has pink colored minerals present, which are most likely _____.

    a. quartz   b. biotite   c. orthoclase   d. Ca-rich plagioclase

3. The small, dark colored minerals in Sample I1 are most likely _____.

    a. quartz   b. biotite   c. orthoclase   d. Na-rich plagioclase

4. Sample I1 belongs to which igneous rock classification?

    a. ultramafic   b. mafic   c. intermediate   d. felsic

5. Sample I2 has what mafic color index (MCI)?

    a. 0-15% MCI   b. 16-45% MCI   c. 46-85% MCI   d. >85% MCI

6. Sample I2 belongs to which igneous rock classification?

    a. ultramafic   b. mafic   c. intermediate   d. felsic

7. Sample I3 has what mafic color index (MCI)?

    a. 0-15% MCI   b. 16-45% MCI   c. 46-85% MCI   d. >85% MCI

8. Sample I3 belongs to which igneous rock classification?

    a. ultramafic   b. mafic   c. intermediate   d. felsic

9. Sample I4 has what mafic color index (MCI)?

    a. 0-15% MCI   b. 16-45% MCI   c. 46-85% MCI   d. >85% MCI

10. Sample I4 belongs to which igneous rock classification?

    a. ultramafic   b. mafic   c. intermediate   d. felsic

11. Sample I5 has what mafic color index (MCI)?

    a. 0-15% MCI   b. 16-45% MCI   c. 46-85% MCI   d. >85% MCI

12. The white minerals in Sample I5 are most likely _____.

    a. quartz　　　b. biotite　　　c. orthoclase　　　d. plagioclase

13. Sample I7 has what mafic color index (MCI)?

    a. 0-15% MCI　　b. 16-45% MCI　　c. 46-85% MCI　　d. >85% MCI

14. According to Figure 8.1, Sample I4 will never contain which mineral?

    a. olivine　　b. pyroxene　　c. amphibole　　d. quartz

15. According to Figure 8.1, Sample I7 must contain which mineral?

    a. olivine　　b. pyroxene　　c. amphibole　　d. quartz

16. According to Bowen's Reaction Series, Sample I5 (if heated) should start melting at what temperature?

    a. ~1100 °C　　b. ~900 °C　　c. ~800 °C　　d. ~500 °C

17. According to Figure 8.1, when Sample I1 is heated, what is the first mineral to melt?

    a. olivine　　b. Ca-rich plagioclase　　c. potassium feldspar　　d. quartz

18. Sample I1 has what texture?

    a. phaneritic　　b. aphanitic　　c. porphyritic-aphanitic　　d. glassy

19. How did Sample I1 form?

    a. by magma that cooled slowly

    b. by fast cooling lava

    c. by a very fast cooling of a lava

    d. by a very fast cooling magma during a volcanic eruption

    e. by magma that was cooling slowly, then as a lava that cooled quickly

20. Sample I1 is called _____.

    a. obsidian   b. pumice   c. granite   d. rhyolite

    e. diorite   f. andesite   g. gabbro   h. basalt

21. Sample I1 can also be called _____.

    a. volcanic   b. plutonic

22. Sample I2 has what texture?

    a. phaneritic   b. aphanitic   c. porphyritic-aphanitic   d. glassy

23. How did Sample I2 form?

    a. by magma that cooled slowly

    b. by fast cooling lava

    c. by a very fast cooling of a lava

    d. by a very fast cooling magma during a volcanic eruption

    e. by magma that was cooling slowly, then as a lava that cooled quickly

24. Sample I2 is called _____.

    a. obsidian   b. pumice   c. granite   d. rhyolite

    e. diorite   f. andesite   g. gabbro   h. basalt

25. Sample I2 can also be called _____.

    a. volcanic   b. plutonic

26. Sample I3 has what texture?

    a. phaneritic   b. aphanitic   c. porphyritic-aphanitic   d. glassy

27. How did Sample I3 form?

    a. by magma that cooled slowly

    b. by fast cooling lava

    c. by a very fast cooling of a lava

    d. by a very fast cooling magma during a volcanic eruption

    e. by magma that was cooling slowly, then as a lava that cooled quickly

28. Sample I3 is called_____.

    a. obsidian   b. pumice   c. granite   d. rhyolite

    e. diorite   f. andesite g. gabbro   h. basalt

29. Sample I3 can also be called _____.

    a. volcanic   b. plutonic

30. Sample I4 has what texture?

    a. phaneritic   b. aphanitic   c. porphyritic-aphanitic   d. glassy

31. How did Sample I4 form?

    a. by magma that cooled slowly

    b. by fast cooling lava

    c. by a very fast cooling of a lava

    d. by a very fast cooling magma during a volcanic eruption

    e. by magma that was cooling slowly, then as a lava that cooled quickly

32. Sample I4 is called_____.

    a. obsidian   b. pumice   c. granite   d. rhyolite

    e. diorite   f. andesite g. gabbro   h. basalt

33. Sample I4 can also be called _____.

   a. volcanic   b. plutonic

34. Sample I5 has what texture?

   a. phaneritic   b. aphanitic   c. porphyritic-aphanitic   d. glassy

35. How did Sample I5 form?

   a. by magma that cooled slowly

   b. by fast cooling lava

   c. by a very fast cooling of a lava

   d. by a very fast cooling magma during a volcanic eruption

   e. by magma that was cooling slowly, then as a lava that cooled quickly

36. Sample I5 is called_____.

   a. obsidian   b. pumice   c. granite   d. rhyolite

   e. diorite   f. andesite g. gabbro   h. basalt

37. Sample I5 can also be called _____.

   a. volcanic   b. plutonic

38. Sample I6 has what texture?

   a. phaneritic   b. aphanitic   c. porphyritic-aphanitic   d. glassy

39. Sample I6 has what composition?

   a. felsic   b. intermediate   c. mafic   d. any of the above

40. How did Sample I6 form?

   a. by magma that cooled slowly

   b. by fast cooling lava

   c. by a very fast cooling of a lava

   d. by a very fast cooling magma during a volcanic eruption

   e. by magma that was cooling slowly, then as a lava that cooled quickly

41. Sample I6 is called_____.

   a. obsidian   b. pumice   c. granite   d. rhyolite

   e. diorite   f. andesite   g. gabbro   h. basalt

42. Sample I7 has what texture?

   a. phaneritic   b. aphanitic   c. porphyritic-aphanitic   d. glassy

43. How did Sample I7 form?

   a. by magma that cooled slowly

   b. by fast cooling lava

   c. by a very fast cooling of a lava

   d. by a very fast cooling magma during a volcanic eruption

   e. by magma that was cooling slowly, then as a lava that cooled quickly

44. Sample I7 is called_____.

   a. obsidian   b. pumice   c. granite   d. rhyolite

   e. diorite   f. andesite   g. gabbro   h. basalt

45. Sample I7 can also be called _____.

   a. volcanic   b. plutonic

46. Sample I8 has what texture?

    a. phaneritic   b. aphanitic   c. porphyritic-aphanitic   d. glassy

47. Sample I8 has what composition?

    a. felsic   b. intermediate   c. mafic   d. any of the above

48. How did Sample I8 form?

    a. by magma that cooled slowly

    b. by fast cooling lava

    c. by a very fast cooling of a lava

    d. by a very fast cooling magma during a volcanic eruption

    e. by magma that was cooling slowly, then as a lava that cooled quickly

49. Sample I8 is called _____.

    a. obsidian   b. pumice   c. granite   d. rhyolite

    e. diorite   f. andesite   g. gabbro   h. basalt

50. Type the following coordinates into the search bar in Google Earth: 33 48 14.24 N 84 08 44.31 W. This is Stone Mountain, GA. Zoom in to an eye elevation of ~2400 ft for a closer look. Based on the color, and assuming no color change has occurred due to weathering at this location, what is the composition of rock?

    a. ultramafic   b. mafic   c. intermediate   d. felsic

51. Zoom out to an eye elevation of ~1500 ft so that you can see the entire mountain. Click on the ruler icon to open up the ruler function. Select the Line tab, and change the map length to kilometers. What is the longest length of exposed portion of Stone Mountain?

    a. 3000 km   b. 260 km   c. 6 km   d. 2.6 km

52. Based on your measurement of the longest dimension, Stone Mountain can be classified as:

    a. a batholith   b. a stock

53. Type the following coordinates into the search bar in Google Earth: 64 58 45.54 N 16 43 07.04 W. Zoom in to an eye elevation of ~12979 ft for a closer look. Based on the color, what is the composition of this rock?

   a. ultramafic   b. mafic   c. intermediate   d. felsic

54. Zoom out to an eye elevation of 41,566 ft. Based on the proximity of the volcanic crater (now containing a lake), the igneous rocks identified in question 53 are:

   a. extrusive   b. intrusive

9 Volcanoes
Karen Tefend

9.1 INTRODUCTION

How would you like to live on an active volcano? Surprisingly, a lot of people are living on or near active volcanoes, and many more live near volcanoes that are currently considered to be "dormant", exhibiting a volcanically-quiet period of time over the past 10,000 years, but with the potential to erupt in the future (Figure 9.1). Are they crazy (Figure 9.2)? Maybe some are, but not all volcanoes erupt explosively; for example, the type of volcano that forms the Hawaiian Islands is a type that erupts effusively, with lava running down the sides (flanks) of the volcano (Figure 9.3). Hawaiian citizens are familiar with this style of eruption, and they are also aware that their particular volcano will not erupt explosively. Explosive eruptions are more likely for the volcanoes that make up the Islands of Sumatra and Java (Indonesia); these volcanoes are very dangerous, and yet Indonesia is one of the most densely populated countries in the world. So, it would seem very important to know why some volcanoes are dangerous, and why others are not. Of the "dangerous" types that are currently dormant, will we have enough of a warning before they erupt? Is it only the eruption that we should worry about, or are there other hazards that we must be aware of? These questions will be addressed in this chapter, along with a few other questions that you might have, such as why volcanoes form in certain locations, especially in relation to plate tectonics, and the association of magma type with volcano type and eruptive style.

Figure 9.1 | An idyllic setting with rolling fields, crashing waves, and a beautiful snow-capped volcano is pictured.
Author: U.S. Fish and Wildlife Service
Source: Public Domain Images
License: Public Domain

Figure 9.2 | A not-so-idyllic photo of a volcano during an explosive eruption is pictured.
Author: Barry Voight
Source: Wikimedia Commons
License: Public Domain

Figure 9.3 | Nighttime photo of an erupting volcano; note that this particular volcanic eruption involves the flow of lava down the sides of the volcano, as opposed to the explosive eruption in the previous image, Figure 9.2.
Author: Tryfon Topalidis
Source: Wikimedia Commons
License: CC BY-SA 3.0

9.1.1 Learning Outcomes

After completing this chapter, you should be able to:
- Relate magma type with plate boundaries
- Understand why most magma crystallizes underground
- Associate volcano form with eruption type and magma type
- Recognize the hazards associated with volcanoes

9.1.2 Key Terms

- Composite volcano
- Dike
- Flood Basalt
- Lahar
- Lava Dome
- Pyroclastic Flow
- Shield Volcano
- Silica Tetrahedron
- Sill
- Viscosity

9.2 MAGMA GENERATION

In the previous chapter on igneous rocks you learned about the concept of partial melting, and in the chapter on plate tectonics you learned about the conditions necessary for mantle rocks to melt; we will review these concepts in this section.

Most magma is generated at the base of the earth's crust; Figure 9.4 is a pressure-temperature diagram similar to the one you saw in the plate tectonics chapter. On the left side of the solid black line (called the solidus) is a region where the temperature is too low for a rock to melt. On the right side of the solidus line is the region where rock will melt. Notice that the solidus line is not a vertical line going straight down, but is sloped at an angle less than vertical, demonstrating that with increasing pressure the temperature must also increase in order for a rock to melt. Now take a look at the conditions at the base of the crust, at point "X". This rock at "X" is not hot enough to melt; or it can be said that the rock at point X is under too much pressure to melt. To make this rock melt, either the temperature must increase (arrow "a"), or the pressure must decrease (arrow "b"), or we can have both hotter tem-

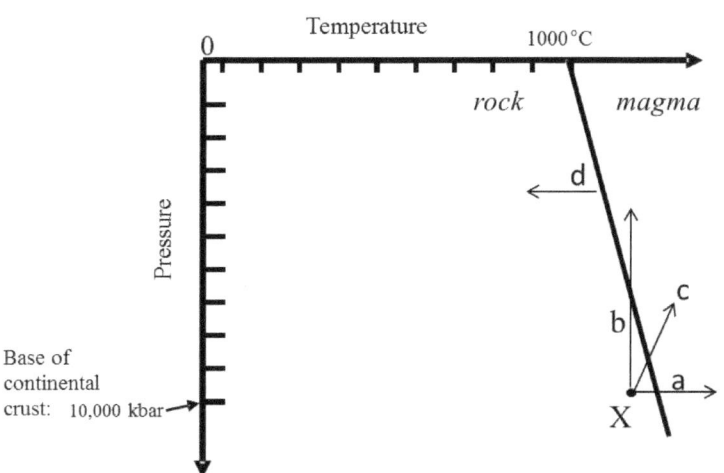

Figure 9.4 | Pressure and Temperature graph of crustal rocks. The line that separates the region of solid versus liquid rock is called the solidus. Note that a rock at point X cannot melt unless one of the following conditions occur: either a temperature increase (arrow "a"), or a pressure decrease (arrow "b"), or any combination of these two changes (arrow "c"), or by adding water to change the melting conditions which shifts the solidus line to the left (arrow "d").
Author: Karen Tefend
Source: Original Work
License: CC BY-SA 3.0

perature and lower pressure occur simultaneously (arrow "c"). Regardless of the path taken, we can make this rock X cross the solid line and become magma. The only other way we can make rock X cross the solid line and become magma is to move this line (arrow "d" on Figure 9.4); in other words, change the melting temperature of the rock. This can be done by adding water, which lowers the melting temperature of rock, and now we can make rock X melt without actually having to change the temperature and pressure conditions.

Now let us think of plate tectonics and the types of boundaries that have magma associated with them (Figure 9.5). Tectonic plates that are diverging (or pulling apart), causes the underlying region of the mantle to experience reduced pressure conditions (just like what the cheerleader at the bottom of a pyramid experiences when everyone else jumps off his back). If the mantle is already fairly warm, the decreased pressure may just be enough for magma to be produced (arrow "b" in Figure 9.5). Where tectonic plates are converging (coming together), one of the plates may subduct below the other plate; recall that subduction will only occur if the tectonic plate has an oceanic crust type. This subducting oceanic crust-topped plate will contain minerals that are hydrated (water in their crystal structure), and as the plate subducts, the hydrated minerals will become unstable and water will be released.

This water will lower the melting temperature of the mantle region directly above the subducting plate, and as a result magma is produced (arrow "d" in Figure 9.5). The last way to melt rock is to just increase the temperature of the rock; this particular melting mechanism does not have to be associated with any particular plate boundary. Instead there must be a region known as a hot spot, caused by mantle plumes (arrow "a" in Figure 9.5). Mantle plumes are thought to be generated at the core-mantle boundary, and are regions of increased temperature that can cause melting of the lithospheric region. With the lithosphere broken into several tectonic plates that have been migrating over these plumes throughout geologic time, the resulting hotspot-generated volcanoes can be found anywhere in the world.

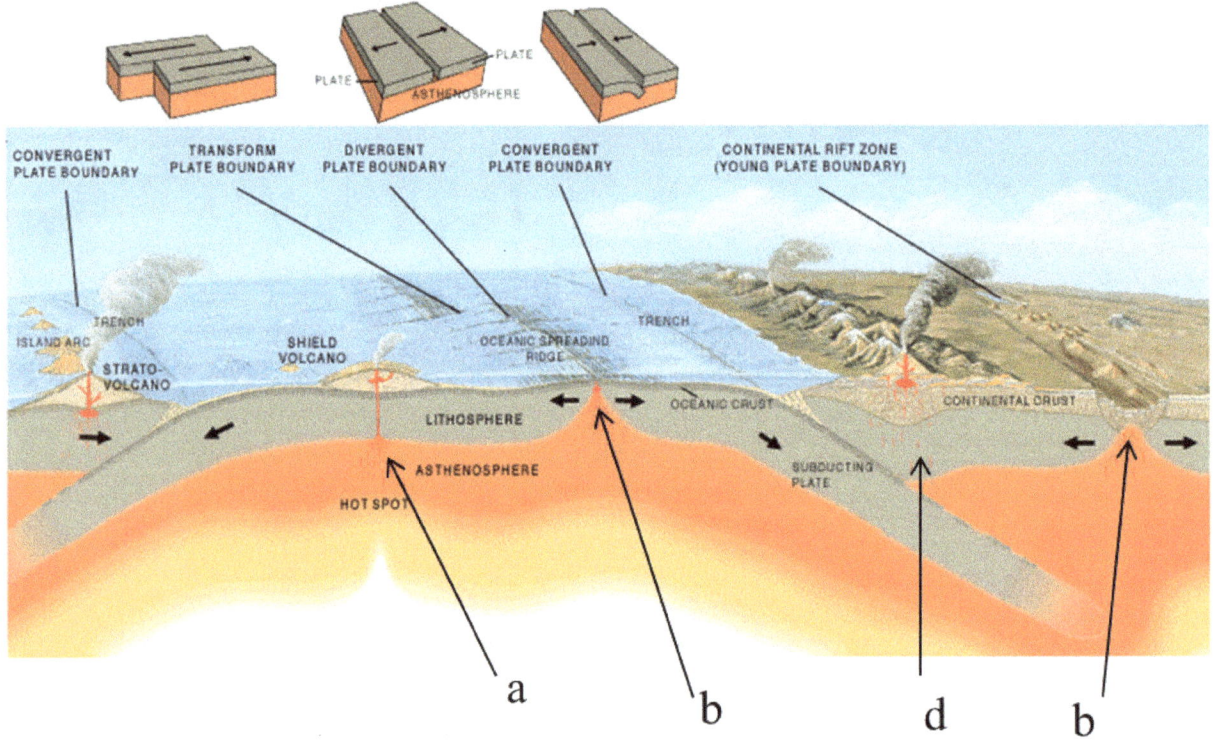

Figure 9.5 | Areas of magma generation at certain plate boundaries (b and d), or within a plate due to a hotspot (a). Arrows a, b, and d correspond to the same arrows as shown in Figure 9.1.
Author: Jose F. Vigil
Source: Wikimedia Commons
License: Public Domain

9.3 LAB EXERCISE

Part A - Magma Generation and Plate Tectonics

Refer to Figure 9.4 to help answer the questions. The exercises that follow use Google Earth. For each question (or set of questions) paste the location that is given into the "Search" box. When finding your locations in Google Earth, be sure to zoom out to higher eye elevations in order to see all of the important features of each area.

1. Type 19 53 48.36 N 155 34 58.11 W in the search bar on Google Earth, and zoom out to an eye altitude of ~615 miles. The magma that resulted in the formation of these islands was generated by what process?

    a. decreased pressure (arrow "b")

    b. the addition of water (arrow "d", which shifts the solidus to the left)

    c. increased temperature (arrow "a")

2. Type 50 16 27.25 N 29 22 05.11 W in the search bar on Google Earth, and zoom out to an eye altitude of ~2663 miles. The magma generated at this location is due to:

    a. decreased pressure (arrow "b")

    b. the addition of water (arrow "d", which shifts the solidus to the left)

    c. increased temperature (arrow "a")

3. Type 36 11 46.85 S 71 09 48.03 W in the search bar on Google Earth, and zoom out to an eye altitude of ~ 2865 miles. The magma responsible for this volcano was generated by what process?

    a. decreased pressure (arrow "b")

    b. the addition of water (arrow "d", which shifts the solidus to the left)

    c. increased temperature (arrow "a")

4. Type "San Andreas Fault" in the search bar on Google Earth; no magma is generated here, because this plate boundary is _____.

    a. a divergent plate boundary, with decompression melting occurs

    b. a convergent plate boundary, where water lowers the melting temperature of rock

    c. a transform plate boundary, where no magma is produced

9.4 MAGMATIC PROCESSES OCCURRING WITHIN THE EARTH'S CRUST

Once magma is generated by one of the mechanisms mentioned earlier (increased temperature, decreased pressure, or by adding water), the magma rises upward through the surrounding rock mainly through pre-existing fractures in the brittle lithosphere. A lot of magma stops rising upward through the continental crust because it has encountered an area in the crust that has the same density as the magma. This is a natural occurrence when matter is under the influence of gravity: denser material will sink, whereas lower density material will rise upward (think of a lava lamp containing two liquids of different densities). This magma will cool slowly underground, and crystallize completely into an igneous rock body called a pluton (Figure 9.6). A lot of these plutonic rock bodies are found along our western coast of North America and other areas where magma must move upward through thick crust. Some of this magma may reach the surface either through a fissure or through a central vent in a volcanic structure; as the majority of magma stays underground and crystallizes into plutons, the smaller amount of magma that continues upward towards the surface may eventually crystallize completely within these fissures and become a **dike** or **sill** (Figure 9.6). Dikes and sills can be found on the surface of the earth due to uplift and erosion of the surrounding rock layers; sills will be oriented parallel to the surrounding rock layers, but dikes cut across them and are therefore easy to identify on the Earth's surface.

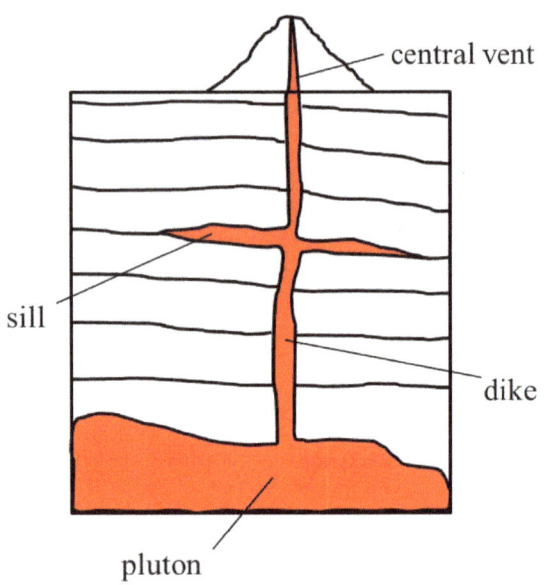

Figure 9.6 | Magma that crystallizes underground will form a pluton, an igneous rock body with a phaneritic texture. Magma can travel further upward through the crust by way of fractures through the rock layers to form a dike or magma may migrate in between rock layers and form a sill.
Author: Karen Tefend
Source: Original Work
License: CC BY-SA 3.0

Recall that magma is generated by partial melting of a source rock, and if we refer back to Bowen's Reaction Series (Igneous Rock chapter, or Figure 9.7 in this chapter), we see that minerals have different melting temperatures. If a rock undergoes partial melting, the minerals that melt are those minerals on the bottom of the Reaction Series that are rich in silica (SiO_2) but poor in iron and magnesium. This means that magma generated by partial melting will be richer in silica and contain less iron and magnesium than the source rock (the rock that partially melted). The different magma types (and the solid rock forms that they make) in Figure 9.7 are listed in order of decreasing iron and magnesium, and increasing silica, with ultramafic rocks as the most iron and magnesium rich, and silica poor, and the felsic rocks as the most deficient in iron and magnesium but containing the most silica. Therefore partial melting of a mantle rock, which has an ultramafic composition, will produce a mafic magma; likewise, partial melting of a mafic rock can produce an intermediate magma type, and partial melting of an intermediate rock can produce felsic magma.

There are other ways that felsic and intermediate magmas are produced; for example, magma that is moving upward (ascending) through the crust may melt some of those crustal rocks, which can change the magma's composition. Recall that the continental crust is the lowest density rock type of the earth, of granitic composition, and is quite thick at convergent plate boundaries (see arrow "d" in Figure 9.5); at convergent plate boundaries, magma may have to travel through this thick portion of the continental crust. At any time before the magma is completely crystallized, it can melt the surrounding crust and perhaps end up with a more silica rich composition, therefore an intermediate or even a mafic magma will become more felsic as it travels through the continental crust. Whereas most magma crystallizes below ground, a portion of it does erupt onto the surface of the earth, either as a passive lava flow or under more explosive conditions such as the May 1980 eruption of Mount St. Helens, WA. Whether the style of eruption is passive or active (explosive) depends on the magma composition.

9.5 LAB EXERCISE

Part B - Association of Magma Type with Tectonic Setting

The questions in this exercise demonstrate the control that tectonic setting has on the type of magma produced. For the Google Earth questions, copy and paste the latitude and longitude coordinates into the search bar (or just type them in).

5. A(n) _____ at the Mid-Ocean Ridge, where oceanic plates are diverging and magma is generated by partial melting of the mantle.

    a. ultramafic magma is produced   b. mafic magma is produced

    c. intermediate magma is produced   d. felsic magma is produced

6. Type 19 28 19.70 N 155 35 31.94 W in the search bar on Google Earth, and zoom out to an eye altitude of ~119 miles. This volcano is composed of:

    a. mafic rocks, because a hotspot partially melted the mantle below the oceanic crust

    b. mafic rocks, because a hotspot partially melted the oceanic crust

    c. ultramafic rocks, because a hotspot partially melted the mantle below the oceanic crust

    d. ultramafic rocks, because a hotspot partially melted the oceanic crust

7. Type 35 35 08.45 S 70 45 08.22 W in the search bar in Google Earth. This volcano formed from an intermediate magma type, because:

   a. subduction of oceanic crust beneath the continental crust occurs here

   b. continental crust is subducting, causing magma to form

   c. a hot spot is partial melting the continental crust

   d. this is the result of a divergent plate boundary

8. Type 36 40 41.62 N 108 50 17.22 W in the search bar in Google Earth. The dark colored rock that forms a straight line on the surface is most likely:

   a. a pluton of ultramafic rock

   b. a dike of ultramafic rock

   c. a sill of mafic rock

   d. a dike of mafic rock

9.6 MAGMA COMPOSITION AND VISCOSITY

In the chapter on igneous rocks, you learned that the igneous rock classification is in part based on the mineral content of the rock. For example, ultramafic rocks are igneous rocks composed primarily of olivine and a lesser amount of calcium-rich plagioclase and pyroxene, whereas quartz, muscovite and potassium feldspar are the typical minerals found in felsic rocks (Figure 9.7). We need to re-

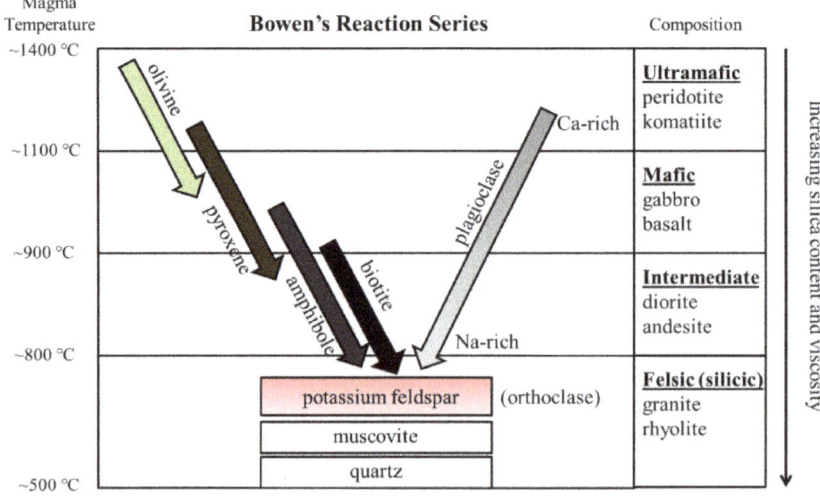

Figure 9.7 | Bowen's Reaction Series, and the magma compositions with the corresponding rock names listed on the right.
Author: Karen Tefend
Source: Original Work
License: CC BY-SA 3.0

view the mineral content of these rocks, because igneous rocks are the crystallized result of cooled magma, and the minerals that form during cooling depend on the chemical composition of the magma; for example, a mafic magma will form a mafic rock containing a large amount of the ferromagnesian minerals pyroxene and amphibole, but will not contain quartz (the mineral that is always present in felsic rocks).

Recall that most of the minerals in igneous rocks are silicate minerals, and all of the minerals shown in Bowen's Reaction Series belong to the silicate mineral group. All silicate minerals have crystal structures containing **silica tetrahedron** (a silicon atom linked to four oxygen atoms), and these silica tetrahedra can be linked in a variety of ways to form sheets, linked chains, or a 3-dimensional framework. The lower temperature minerals (quartz, muscovite and orthoclase) have more linked tetrahedra than the high temperature minerals (olivine and pyroxene). Figure 9.8A shows how the silica tetrahedra are linked to form the mineral quartz. Note that this is a 2-dimensional diagram of a 3-dimensional structure, and there are a lot more tetrahedra connected in the area above and behind the typed page. What is important about this figure is what happens to the silica tetrahedra when quartz melts; Figure 9.8B shows that even though the crystal structure is lost (the regularly repeated structure is gone), the tetrahedral links are maintained, albeit distorted. The bonds that link these tetrahedra are strong, and magma temperatures are not high enough to break these bonds. This means that mag-

○ Silicon atom

● Oxygen atom

 Tetrahedron that has apex pointed away from viewer; the 4th oxygen atom in each tetrahedra not shown.

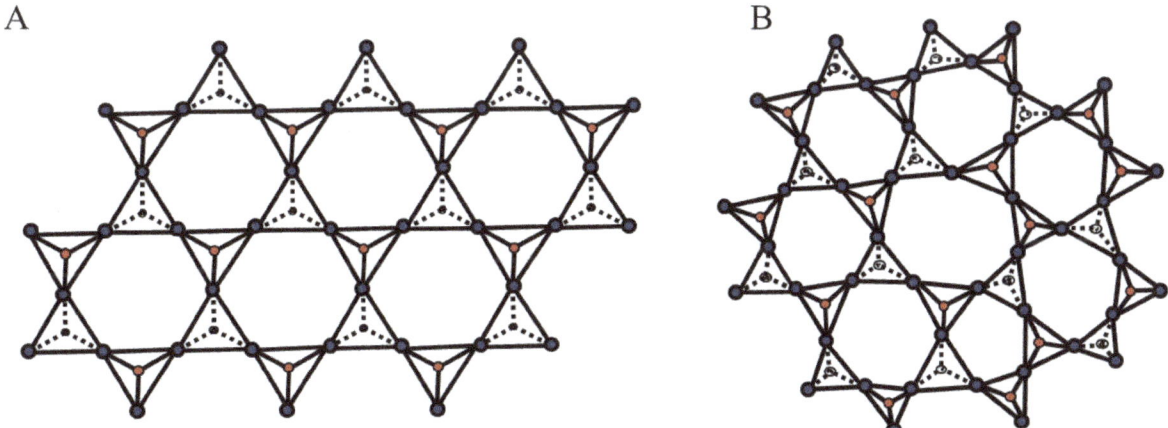

Figure 9.8 | (A) Regular repeated structure of silica tetrahedra in quartz; note that the 4th oxygen of each tetrahedra is not drawn for simpler viewing. These hexagonally arranged tetrahedral are joined to others to form a 3-dimensional framework. (B) shows how the tetrahedra stay linked when quartz melts; the arrangement of tetrahedral is irregular, but the links are maintained.
Author: Karen Tefend
Source: Original Work
License: CC BY-SA 3.0

mas that can crystallize quartz will have a lot of these tetrahedra linked in the magma, whereas mafic magmas which do not contain enough silicon to crystallize quartz, will instead crystallize minerals that have fewer linked tetrahedra.

Why does the silica content (the amount of linked tetrahedra) of magma matter so much? A large amount of linked silica tetrahedra will result in a magma or lava that is very viscous, meaning that it cannot flow easily (**viscosity** means resistance to flow). The temperature of a lava also affects the viscosity; think of how ketchup from your refrigerator flows, and how a ketchup stored in your pantry flows; of these two fluids, the colder ketchup has the higher viscosity. In the case of magmas or lavas, the hotter the lava, the easier it flows, and the less silica that is present, the lower the viscosity (see right side of the diagram in Figure 9.7). This means that mafic lavas can flow faster than intermediate or felsic lavas. The silica content of magma affects not only the shape of the volcano, but the style of eruption, or whether an eruption will be a lava that flows, or a magma that blows (up).

9.7 LAB EXERCISE

Part C - Magma Viscosity

The following questions address what factors control how fast a magma or lava can flow. The resistance to flow (viscosity) depends primarily on the magma or lava composition, and is also affected by temperature.

9. Intermediate lavas can flow _____ than mafic lavas, due to the _____ viscosity.

    a. slower, higher   b. slower, lower   c. faster, higher   d. faster, lower

10. Let's relate food items to magmas of different viscosity; if we compare how honey and water flow when poured from a container:

    a. then the honey represents felsic magma, and the water represents mafic magma

    b. then the honey represents mafic magma, and the water represents mafic magma.

11. Imagine putting the honey in the refrigerator overnight; will its viscosity be affected?

    a. yes; the viscosity will increase

    b. yes, the viscosity will decrease

    c. no. there will be no change as the composition stays the same

12. Keep the honey in mind while you answer this question: when it is first erupted, basalt lava typically erupts at 1200°C; after flowing away from the vent, the temperature falls, therefore the viscosity of the basaltic lava will:

    a. increase   b. decrease   c. stay the same

9.8 VOLCANIC LANDFORMS AND ERUPTION STYLES

The size, shape, and eruptive style of any volcano ultimately depend on the magma composition. We will focus mainly on mafic and felsic magmas as intermediate magmas have properties that are intermediate between these two types, and ignore the ultramafic magma as this type no longer forms (due to a cooler earth's interior). As mentioned earlier, mafic magmas are lower in silica and are therefore characterized as having a low viscosity. As mafic magma erupts on to the surface through a central vent (see Figure 9.6, and photo in Figure 9.3), the magma (now called lava) will spread out quite easily due to its low viscosity. Mafic lava flows can travel quite far before solidifying completely, and the type of volcano that forms from mafic lava is called a shield volcano (Figure 9.9). Shield volcanoes are very broad at the base, and have relatively gentle slopes. There are several locations around the world where large amounts of mafic basalt flows are found, without forming a shield volcano: places such as the Deccan Traps in India, Siberian Traps in Russia, and here in the U.S. (Columbia River Basalt Group). Such vast outpourings of mafic lava are called flood basalts, and are believed to be caused by mantle plumes (hotspots), which partially melt the mantle beneath the earth's crust. The mafic magma that is generated by the mantle plume reaches the earth's surface through fractures (fissures) instead of one central vent.

Figure 9.9 also shows a much smaller volcano type, called a **composite volcano**. This volcano also forms from eruptions through a central vent, but the smaller size indicates that any lava that is generated from a central vent did not travel far before solidifying completely, which indicates a more viscous magma,

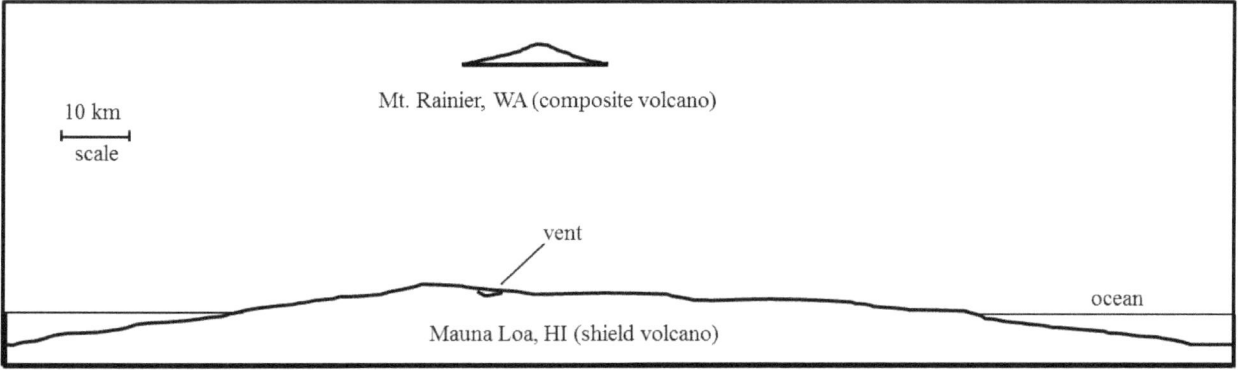

Figure 9.9 | A comparison of a typical shield volcano versus a smaller, steeply sloped composite volcano. The size and shape of these volcano types are due to the different types of magma and associated eruption styles.
Author: Karen Tefend
Source: Original Work
License: CC BY-SA 3.0

such as an intermediate or felsic magma type. The term "composite" comes from the layers of lava flows, and the accumulation of ash and other volcanic material produced during a more explosive type of eruption due to the dissolved gases present in the magma (Figure 9.10, and photos in Figures 9.1 and 9.2). It is common to have a certain amount of dissolved gases within magma, and some gases may escape from the magma while still underground, but the most spectacular release of these gases occurs during eruption. The common gases associated with magma are usually water vapor, carbon dioxide, carbon monoxide, and hydrogen sulfide. Mafic magmas erupting onto the surface may form a lava fountain which spurts the magma (now called lava) into the air, where the height of the lava fountain depends on the gas content. However, gases within felsic magmas are not released as easily due the magma's high viscosity; as viscous magmas move closer to the earth's surface, the dissolved gases are under less pressure and large gas bubbles can form. At some point the gas pressure becomes so great that an explosive eruption occurs, and fragmented volcanic material is released. This material can travel great distances along air currents, but enough deposits around the central vent to build up the composite volcano structure. Once the gases are released and the explosive eruptions end, felsic magma may still be extruded onto the surface inside the volcanic crater. This felsic material, now a viscous lava, will flow with great difficulty, and a large amount of the felsic lava will cool completely in the area around the central vent, and may end up "plugging" the vent and forming what is called a **lava dome** (Figure 9.11).

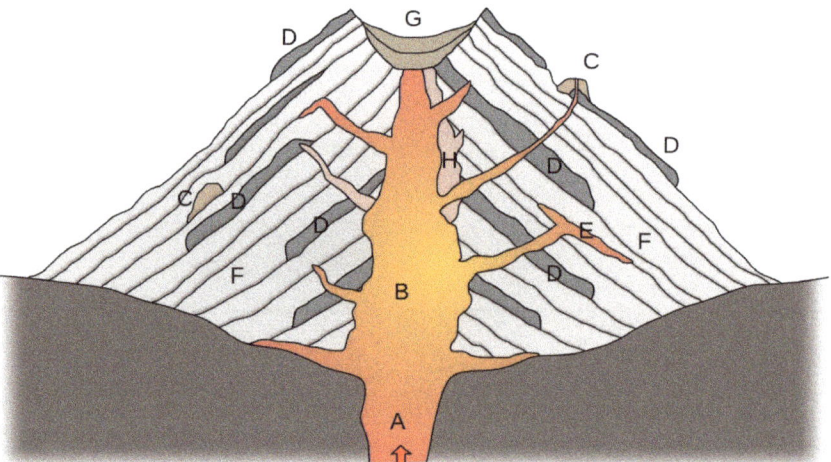

Figure 9.10 | Diagram of a composite volcano. Notice the structure is composed of layered ashflows (F) and viscous lava flows (D). Also of note is the buildup of viscous lava at locations labelled C; these are lava domes.
Author: User "Woudloper"
Source: Wikimedia Commons
License: CC BY-SA 3.0

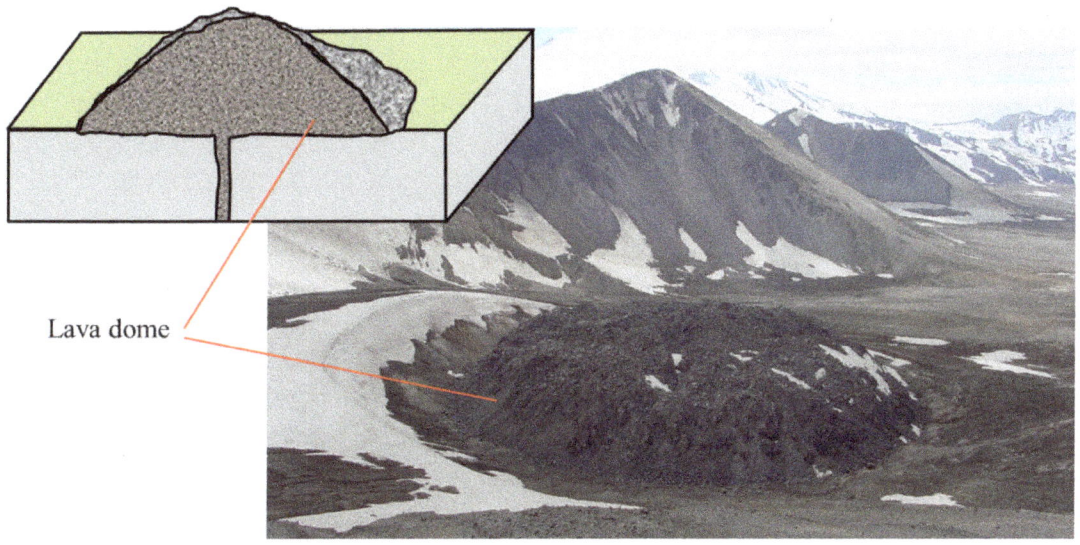

Figure 9.11 | Lava domes are extrusions of viscous lava on the surface of the Earth. The lava does not travel too far from the vent before solidifying completely.
Author: Karen Tefend
Source: Original Work
License: CC BY-SA 3.0

9.9 LAB EXERCISE

Part D - Volcanic Landforms

The composition of a magma or lava may also control what type of volcanic features or landforms are seen on the earth's surface. In this exercise, you will use Google Earth to identify these landforms. Figures 9.9 through 9.11 may help you in this section. For the Google Earth questions, copy and paste the latitude and longitude coordinates into the search bar (or just type them in).

13. Type 43 25 04.18N 113 31 37.38 W in the search bar of Google Earth. Zoom out to an eye elevation of ~90 miles. Based on the size of the area which is dark colored and sparsely vegetated, this region is:

    a. a basaltic dike　　b. a shield volcano　　c. a lava dome　　d. a flood basalt

14. Click on the nearby photo icon (SW of your latitude/longitude coordinate in Question 13) to view a picture of the area as seen from the ground. This is a picture of:

    a. a basalt flow　　b. a felsic dome　　c. a plutonic rock　　d. an ultramafic rock

15. How many miles in length is this dark colored feature? (measure the greatest length)

    a. ~8 miles　　b. ~20 miles　　c. ~40 miles　　d. ~50 miles

16. Type 46 12 07.84N 121 31 02.85W in the search bar of Google Earth. This volcano (Mt. Adams) has a lot of snow cover and small glaciers on it, but you can still see the volcanic rock, especially on the eastern flank (side) of the volcano. Zoom in to an eye altitude of ~9000ft to closely examine the rocks on this eastern side of the volcano; do you see any evidence of layering? This volcano is:

    a. a shield volcano

    b. a lava dome

    c. a composite volcano

    d. a large volcanic dike

17. How tall is Mt. Adams, and what is the length of its base (in the widest, N-S dimension)? You will need to zoom out to an eye altitude of ~17 miles to measure the base.

    a. ~12,290 ft above sea level, and over 7 miles long

    b. ~11, 890 ft above sea level, and over 4 miles long

    c. at least 12, 703 ft above sea level, and over 6 miles long

    d. unable to determine because of the ice

18. Compared to the Big Island of Hawaii (19 53 48.36 N 155 34 58.11 W, or refer to Figure 9.6), which is actually the volcano Mauna Loa, Mt. Adams is:

    a. the same height, but the base is a lot smaller

    b. the same height, but the base is a lot wider

    c. smaller in height, and smaller at the base

    d. smaller in height, but larger at the base

19. Type in 58 15 58.56 N 155 09 35.98 W in the search bar of Google Earth. Examine the shape of this feature by zooming in to an eye altitude of ~3330 ft; use the eye icon in the upper right corner to rotate the view. Now zoom out to an eye altitude of ~7306 ft to see the entire structure and the surrounding area. Based on the size and appearance of this volcanic feature, this is a:

    a. dike   b. shield volcano   c. lava dome   d. flood basalt

9.10 VOLCANIC HAZARDS

When comparing the two volcano types, shield and composite, it is obvious that although the shield volcanos are more massive (see Figure 9.9), they are far less dangerous to the population than the smaller composite volcanoes. Shield volcanoes produce basaltic lavas that may fountain at the vent, due to gases, but end up flowing passively down the flanks of the volcano (Figures 9.3 and 9.12). Other than property damage, anyone living on or near a shield volcano is not likely to perish due to a volcanic eruption. This is not the case for the composite volcanoes; explosive eruptions produce a lot of volcanic fragments, called pyroclastic debris or tephra, that range in size from dust and ash to large blocks (or bombs) of volcanic material (Figure 9.13). Pyroclastic debris at first travels high up into the atmosphere into an eruption column, which then spreads outward along with the prevailing wind direction, but during the explosive phase of the eruption, the central vent widens as the rocks around the vent are also blown during the eruption; this widening of the vent results in less upward momentum of the eruption column. As a result, the pyroclastic material travels down the flanks of the volcano as a **pyroclastic flow**, a very dangerous mixture of hot volcanic material and noxious gases.

Figure 9.12 | Photo of a lava fountain, caused by gas release from a mafic magma. Notice the mafic lava flows down the sides of this structure.
Author: G. E. Ulrich
Source: Wikimedia Commons
License: Public Domain

Although pyroclastic flows are extremely dangerous, most deaths associated with composite volcanoes are from mudflows, called **lahars**. Many composite volcanoes are capped by snow and ice, and even a small eruption can result in meltwater running down the sides of the volcano. This water can easily erode the ash and other volcanic debris on the flanks of the volcano and result in a fast moving slurry of mud and larger material such as trees and boulders. Lahars move swiftly through river channels and can endanger any town or city that is built in the low lying areas downstream from the volcano (Figure 9.13). Lahars can also be generated by large amounts of rainfall in the area. Several ancient mudflow deposits are recognized in the Cascade area of Washington and Oregon. Because of the proximity of major cities and many towns in the vicinity of these dormant volcanoes (and the recently classified "active" Mt. St. Helens), volcano monitoring systems are in place. For example, seismic activity is monitored, as earthquakes are generated by the upward migration of magma beneath the volcanic structure, and GPS technology is used to monitor any changes in slope, as magma may push the sides of the volcano outwards and increase the angle of the volcano's slope. Also monitored are the stream valleys on the volcano to detect lahars that can happen at any time, regardless of the active or dormant status of a composite volcano.

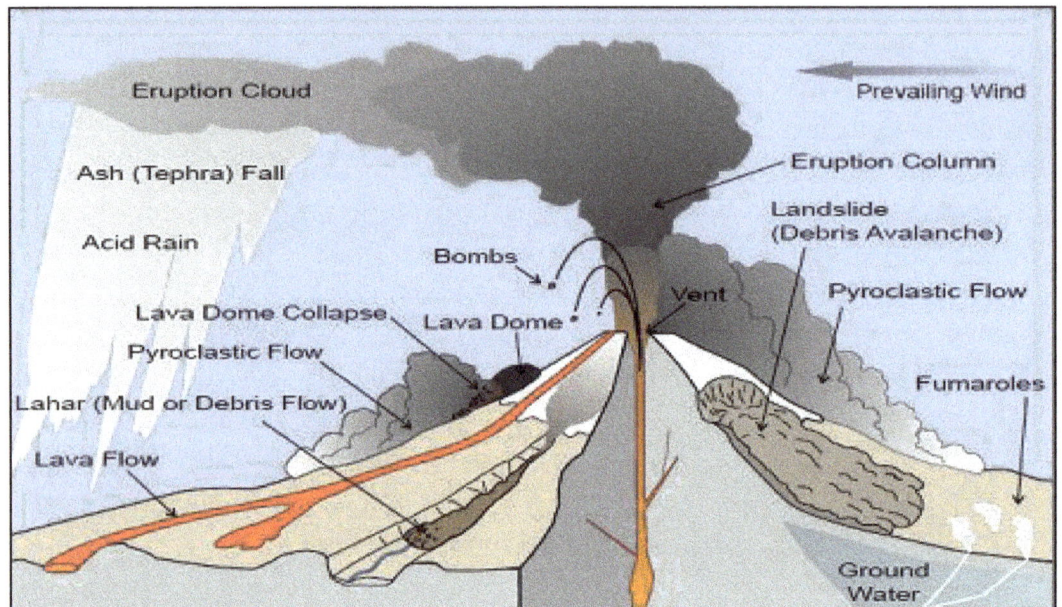

Figure 9.13 | Diagram demonstrating the hazards and other features associated with composite volcanoes.
Author: volcanoes-42325.wikispaces.com
Source: Volcanoes-42325
License: CC BY_SA 3.0

9.11 LAB EXERCISE

Part E - Volcanic Hazards

Potential hazards associated with certain volcanic types can be identified using topographic maps or aerial photographs. In this exercise, we will use Google Earth.

20. Type Mount St. Helens, WA in the search bar in Google Earth and examine the crater at an eye altitude of ~26,000 ft. Based on the appearance of the crater:

    a. a lahar removed the north side of the volcano

    b. a pyroclastic eruption removed the north side of the volcano

    c. a lava dome grew so large that it is higher than the north side of the crater

    d. a glacier has eroded the north side of the crater

21. Type Mount Rainier, WA in the search bar in Google Earth. The Carbon River flows from the north part of Mount Rainier (fed by meltwater from the Carbon Glacier on the flank of the volcano). At an eye altitude of ~20,000ft, follow the course of the Carbon River, past the town of Carbonado and stop at the town of Orting. Based on the locations of these two towns, which do you think is in danger from a lahar?

    a. Both are in danger of lahars

    b. Carbonado, because it is closer to Mount Rainier

    c. Orting, because it is in a low lying area along Carbon River

    d. Neither one is in danger because Mount Rainier is dormant

9.12 STUDENT RESPONSES

1. Type 19 53 48.36 N 155 34 58.11 W in the search bar on Google Earth, and zoom out to an eye altitude of ~615 miles. The magma that resulted in the formation of these islands was generated by what process?

    a. decreased pressure (arrow "b")

    b. the addition of water (arrow "d", which shifts the solidus to the left)

    c. increased temperature (arrow "a")

2. Type 50 16 27.25 N 29 22 05.11 W in the search bar on Google Earth, and zoom out to an eye altitude of ~2663 miles. The magma generated at this location is due to:

    a. decreased pressure (arrow "b")

    b. the addition of water (arrow "d", which shifts the solidus to the left)

    c. increased temperature (arrow "a")

3. Type 36 11 46.85 S 71 09 48.03 W in the search bar on Google Earth, and zoom out to an eye altitude of ~ 2865 miles. The magma responsible for this volcano was generated by what process?

    a. decreased pressure (arrow "b")

    b. the addition of water (arrow "d", which shifts the solidus to the left)

    c. increased temperature (arrow "a")

4. Type "San Andreas Fault" in the search bar on Google Earth; no magma is generated here, because this plate boundary is _____.

    a. a divergent plate boundary, with decompression melting occurs

    b. a convergent plate boundary, where water lowers the melting temperature of rock

    c. a transform plate boundary, where no magma is produced

5. A(n) _____ at the Mid-Ocean Ridge, where oceanic plates are diverging and magma is generated by partial melting of the mantle.

    a. ultramafic magma is produced   b. mafic magma is produced

    c. intermediate magma is produced   d. felsic magma is produced

6. Type 19 28 19.70 N 155 35 31.94 W in the search bar on Google Earth, and zoom out to an eye altitude of ~119 miles. This volcano is composed of:

   a. mafic rocks, because a hotspot partially melted the mantle below the oceanic crust

   b. mafic rocks, because a hotspot partially melted the oceanic crust

   c. ultramafic rocks, because a hotspot partially melted the mantle below the oceanic crust

   d. ultramafic rocks, because a hotspot partially melted the oceanic crust

7. Type 35 35 08.45 S 70 45 08.22 W in the search bar in Google Earth. This volcano formed from an intermediate magma type, because:

   a. subduction of oceanic crust beneath the continental crust occurs here

   b. continental crust is subducting, causing magma to form

   c. a hot spot is partial melting the continental crust

   d. this is the result of a divergent plate boundary

8. Type 36 40 41.62 N 108 50 17.22 W in the search bar in Google Earth. The dark colored rock that forms a straight line on the surface is most likely:

   a. a pluton of ultramafic rock

   b. a dike of ultramafic rock

   c. a sill of mafic rock

   d. a dike of mafic rock

9. Intermediate lavas can flow _____ than mafic lavas, due to the _____ viscosity.

   a. slower, higher   b. slower, lower   c. faster, higher   d. faster, lower

10. Let's relate food items to magmas of different viscosity; if we compare how honey and water flow when poured from a container:

    a. then the honey represents felsic magma, and the water represents mafic magma

    b. then the honey represents mafic magma, and the water represents mafic magma.

11. Imagine putting the honey in the refrigerator overnight; will its viscosity be affected?

    a. yes; the viscosity will increase

    b. yes, the viscosity will decrease

    c. no. there will be no change as the composition stays the same

12. Keep the honey in mind while you answer this question: when it is first erupted, basalt lava typically erupts at 1200°C; after flowing away from the vent, the temperature falls, therefore the viscosity of the basaltic lava will:

    a. increase   b. decrease   c. stay the same

13. Type 43 25 04.18N 113 31 37.38 W in the search bar of Google Earth. Zoom out to an eye elevation of ~90 miles. Based on the size of the area which is dark colored and sparsely vegetated, this region is:

    a. a basaltic dike   b. a shield volcano   c. a lava dome   d. a flood basalt

14. Click on the nearby photo icon (SW of your latitude/longitude coordinate in Question 13) to view a picture of the area as seen from the ground. This is a picture of:

    a. a basalt flow   b. a felsic dome   c. a plutonic rock   d. an ultramafic rock

15. How many miles in length is this dark colored feature? (measure the greatest length)

    a. ~8 miles   b. ~20 miles   c. ~40 miles   d. ~50 miles

16. Type 46 12 07.84N 121 31 02.85W in the search bar of Google Earth. This volcano (Mt. Adams) has a lot of snow cover and small glaciers on it, but you can still see the volcanic rock, especially on the eastern flank (side) of the volcano. Zoom in to an eye altitude of ~9000ft to closely examine the rocks on this eastern side of the volcano; do you see any evidence of layering? This volcano is:

    a. a shield volcano   b. a lava dome

    c. a composite volcano   d. a large volcanic dike

17. How tall is Mt. Adams, and what is the length of its base (in the widest, N-S dimension)? You will need to zoom out to an eye altitude of ~17 miles to measure the base.

   a. ~12,290 ft above sea level, and over 7 miles long

   b. ~11, 890 ft above sea level, and over 4 miles long

   c. at least 12, 703 ft above sea level, and over 6 miles long

   d. unable to determine because of the ice

18. Compared to the Big Island of Hawaii (19 53 48.36 N 155 34 58.11 W, or refer to Figure 9.6), which is actually the volcano Mauna Loa, Mt. Adams is:

   a. the same height, but the base is a lot smaller

   b. the same height, but the base is a lot wider

   c. smaller in height, and smaller at the base

   d. smaller in height, but larger at the base

19. Type in 58 15 58.56 N 155 09 35.98 W in the search bar of Google Earth. Examine the shape of this feature by zooming in to an eye altitude of ~3330 ft; use the eye icon in the upper right corner to rotate the view. Now zoom out to an eye altitude of ~7306 ft to see the entire structure and the surrounding area. Based on the size and appearance of this volcanic feature, this is a:

   a. dike   b. shield volcano   c. lava dome   d. flood basalt

20. Type Mount St. Helens, WA in the search bar in Google Earth and examine the crater at an eye altitude of ~26,000 ft. Based on the appearance of the crater:

   a. a lahar removed the north side of the volcano

   b. a pyroclastic eruption removed the north side of the volcano

   c. a lava dome grew so large that it is higher than the north side of the crater

   d. a glacier has eroded the north side of the crater

21. Type Mount Rainier, WA in the search bar in Google Earth. The Carbon River flows from the north part of Mount Rainier (fed by meltwater from the Carbon Glacier on the flank of the volcano). At an eye altitude of ~20,000ft, follow the course of the Carbon River, past the town of Carbonado and stop at the town of Orting. Based on the locations of these two towns, which do you think is in danger from a lahar?

    a. Both are in danger of lahars

    a. Carbonado, because it is closer to Mount Rainier

    b. Orting, because it is in a low lying area along Carbon River

    c. Neither one is in danger because Mount Rainier is dormant

10 Sedimentary Rocks
Bradley Deline

10.1 INTRODUCTION

We are particularly interested in the history and events that occur on the surface of the Earth both because it is easier to directly observe and test, and has direct relevance to our lives and our own history. Sedimentary rocks are the pages in which history is written, since they contain powerful environmental indicators, traces of life, and chemical signatures that can inform us about a wealth of subjects from the occurrence of ancient catastrophes to the productivity of life.

The identification of sedimentary rocks is more than applying names, since each name is a loaded term that conveys information regarding its history, where it was formed, potentially when it was formed, and the processes that lead to its formation. Each sedimentary rock is a puzzle and by identifying a set of rocks, how they are layered, the fossils within, and patterns in the rocks a geologist can reconstruct an entire environment and ecosystem. Solving these puzzles is both an academic exercise to better understand the world around us as well as a tool for finding the resources that are important to our lives. In particular, fossil fuels as well as many other natural resources are, or are contained within, sedimentary rocks such as coal, natural gas, petroleum, salt, and the materials that go into wallboard or in the making of cement. Therefore, a better understanding of sedimentary rocks and how and where they are formed directly influences your everyday life.

10.1.1 Learning Outcomes

After completing this chapter, you should be able to:
- Describe how erosion and weathering relate to the formation of sedimentary rocks
- Identify sedimentary rocks and their features
- Describe the formation and history of different types of sedimentary rocks

- Describe the formation of different sedimentary structures
- Discuss the distribution of sedimentary rocks, sedimentary structures, and types of weathering in varying sedimentary environments

10.1.2 Key Terms

- Beach Depositional Environments
- Biochemical Sedimentary Rocks
- Chemical Sedimentary Rocks
- Chemical Weathering
- Clastic Sedimentary Rocks
- Continental Environments
- Deep Marine Depositional Environments
- Deltas
- Eolian Depositional Environments
- Erosion
- Fluvial Depositional Environments
- Glacial Depositional Environments
- Lacustrine Depositional Environments
- Marine Environments
- Maturity
- Mechanical Weathering
- Organic Sedimentary Rocks
- Reef Depositional Environments
- Sedimentary Structures
- Shallow Marine Depositional Environments
- Tidal Mudflat Depositional Environments
- Transitional Environments

10.2 WEATHERING AND EROSION

Sedimentary rocks are formed by the weathering, erosion, deposition, and lithification of sediments. Basically, sedimentary rocks are composed of the broken pieces of other rocks. The obvious place to start this chapter is a discussion of how rocks are broken down, which is a process called weathering. There are two basic ways that weathering occurs in nature. First, rocks can be physically broken into smaller pieces (imagine hitting a rock with a hammer), which is called **mechanical weathering**. Alternatively, rocks can be broken down and altered at the atomic level (imagine dissolving salt in a glass of water), which is called **chemical weathering**. There are multiple ways each type of weathering can occur and, therefore, both the rate that rock breaks down and how it breaks down vary dramatically depending on the area and environment.

The most prevalent type of mechanical weathering is the collision, breaking, and grinding of rock by the movement of a fluid, either water or air. The size of the carried sediment depends on the type of fluid and speed of the movement. A fast fluid (like a rapid flowing river) can carry large particles and cause immense amounts of weathering while a slow fluid (like a calm stream) would hardly cause any weathering. The density of the fluid also controls the size of particle that can be transported, for instance denser fluid (like water) can carry larger particles than less dense fluid (like air). Another common method of mechanical weathering is called frost wedging, which occurs when water seeps into a crack in the rock and freezes. Water has a unique property in that it expands when frozen, which puts pressure on the rock and can potentially split boulders. The addition and subtraction of heat or pressure can also cause rock to break, for instance spilling cold water on a hot light bulb will cause it to shatter. This breakage can also occur with rocks when they cool very quickly or immense pressure is released. Finally, plants, animals, and humans can cause significant amounts of weathering. These sediments then undergo **erosion**, which is the transport of sediment from where it is weathered to where it will be deposited and turned into a rock.

Rocks can also be chemically weathered, most commonly by one of three processes. The first, which you are probably familiar with, is called dissolution. In this case, a mineral or rock is completely broken apart in water into individual atoms or molecules. These individual ions can then be transport with the water and then re-deposited as the concentration of ions increases, normally because of evaporation. Chemical weathering can also change the mineralogy and weaken the original material, which again is caused by water. A mineral can undergo hydrolysis, which occurs when a hydrogen atom from the water molecule replaces the cation in a mineral; this normally alters minerals like feldspar into a softer clay mineral. A mineral can also undergo oxidation, which is when oxygen atoms alter the valence state of a cation, this normally occurs on a metal and is commonly known as rusting.

Chemical and mechanical weathering can work together to increase the overall rate of weathering. Chemical weathering weakens rocks making them more prone to breaking physically, while mechanical weathering increases the surface area of the sediment, which increases the surface area that is exposed to chemical weathering. Therefore, environments with multiple types of weathering can erode very quickly. As you go through the following sections (on rocks and environments) think about the types of weathering required to make the sediment that will then make up different types of sedimentary rocks as well as what types of weathering you would expect to occur in different environments.

10.3 IDENTIFYING SEDIMENTARY ROCKS

The classification of sedimentary rocks is largely based on differentiating the processes that lead to their formation. The biggest division in types of sedimentary rocks types is based on the primary type of weathering that leads to the material building the sedimentary rock. If the rock is largely made from broken pieces (called clasts) of rock that have been mechanically weathered the rocks are referred to as Detrital or **Clastic Sedimentary Rocks**. Simply put, these are rocks that are composed of the broken pieces of other rocks. In this case, the mineralogy of the clasts is not important, but instead we need to note the properties of the sediment itself. Alternatively if the rock is largely the product of chemical weathering the classification is then based on the composition of the material as well as the processes involved in the materials precipitation from solution. **Chemical Sedimentary Rocks** form from the inorganic precipitation of minerals from a fluid. If the ions present within a fluid (water) become very concentrated either by the addition of more ions or the removal of water (by freezing or evaporation), then crystals begin to form. In this case the identification of the type of sedimentary rocks is based on the minerals present. If organisms facilitate the precipitation of these minerals from water we refer to the rocks as **Biochemical Sedimentary Rocks.** An example of biochemical precipitation is the formation of skeletal minerals in many organisms: from starfish and clams that grow calcite, to sponges that grow silica-based material, to humans that have bones made of hydroxyapatite. In many cases it is hard to differentiate whether a mineral was formed organically or inorganically, so in the current lab we will mostly group these two types of sedimentary rocks together. Rocks can also be formed from the carbon-based organic material produced by ancient life and are called **Organic Sedimentary Rocks**. Now we can discuss the identification and formation of particular sedimentary rocks.

10.3.1 Clastic Sedimentary Rocks

Weathering and erosion occur normally in areas that are at high elevation, such as mountains, while deposition occurs in lower areas such as valleys, lakes, or the ocean. The sediment is transported from the area of erosion to area of deposition by ice, water, or air. Not surprisingly, the sediment changes during its journey and we can recognize the amount of change and the distance the material has traveled, and the transport mechanism, by looking at its maturity (Figure 10.1). **Maturity** is defined as the texture and composition of a sedimentary rock resulting from varying amounts of erosion or sedimentary transport. Imagine a mountain composed of the igneous rock granite and let us explore how the sediment from this mountain changes as it makes the long distance trek via river to the ocean. The first process is just breaking the rock down into smaller pieces mechanically, which creates sediment that has large and small pieces, the pieces are jagged, and all of the minerals remain. The sizes of clasts in these rocks can range from large boulders, to cobbles, to pebbles, to the smallest particles, clay. As this sediment is transported in the river the pebbles collide with other pebbles and the rocks get smaller and the sharp edges are broken off. Also, as the slope of the land decreases the river slows leaving behind the large boulders and cobbles while carrying away the smaller particles. This results in sediments further from the source to be more uniform in size, which is a process called sorting. Chemical weathering also occurs, altering the feldspars into clay-sized particles. The end result of this process is the granite reduced from boulders and cobbles close to the mountain, to pebbles in the rivers, and finally to pure and uniform quartz sand at the beach and miniscule clay grains on the ocean floor. Therefore, different clastic rocks are found in different areas and have traveled different distances.

In this lab, we will look at three types of clastic rocks (Figure 10.1, Table 10.1), conglomerate, sandstone, and shale. Conglomerate is an immature sedimentary rock (rock that has been transported a short distance) that is a poorly sorted mixture of clay, sand, and rounded pebbles. The mineralogy of the sand and pebbles (also called clasts) can vary depending on its source. These rocks would be found on the continent in a several types of deposits such as ancient landslides or pebble beds in rivers. Sandstone is defined as a clastic sedimentary rock that consists of sand-sized clasts. These clasts can vary from jagged to rounded as well as containing many minerals or just quartz. Therefore, sandstone ranges from being relatively immature to mature which makes sense because we can find layers of sand associated with mountain rivers to pure white quartz beaches. Last we have shale, which is composed of clay particles and has a finely layered or fissile appearance. This extremely mature sedimentary rock is made from the smallest particles that can be carried by wind or barely moving water and can be found thousands of miles away from the original source.

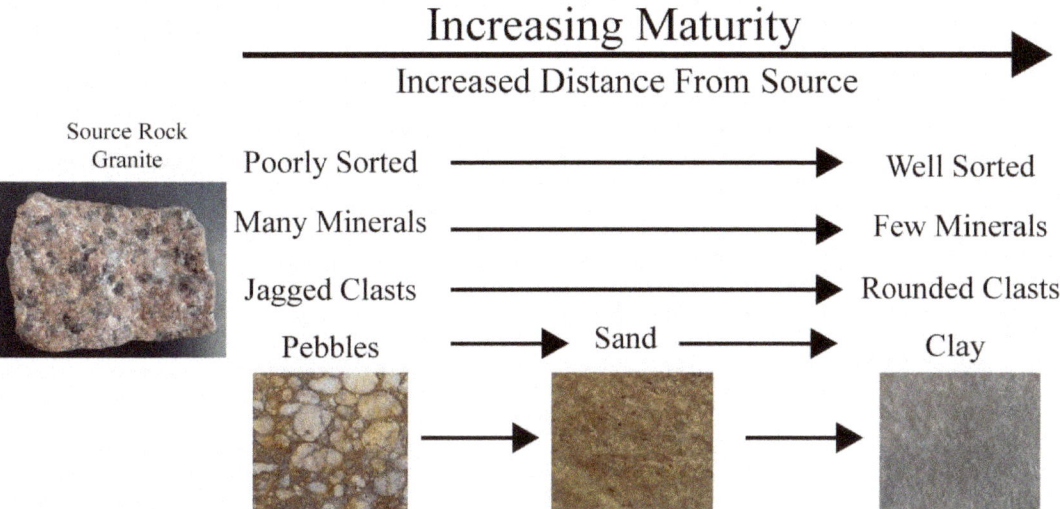

Figure 10.1 | Maturity in clastic sedimentary rocks showing how the sediments change as they are eroded further from their source.
Author: Bradley Deline
Source: Original Work
License: CC BY-SA 3.0

	Composition	Texture and Properties	
Detrital Sedimentary Rocks			
Shale	Rock fragments smaller than 1/16 mm	Clay-sized particles that cannot be differentiated by the naked eye. May be fissile, splits into distinctive layers.	
Sandstone	Rock fragments between 1/16 - 2mm	Composed of sand-sized rock fragments. The fragments can vary in mineralogy including mainly quartz, along with feldspar, and clay	
Conglomerate	Rock fragments ranging in size with the largest >2mm	Poorly sorted mixture of rock fragments including subround or rounded pebbles.	
Chemical and Biochemical Sedimentary Rocks			
Crystalline Limestone	Calcite crystals or microcrystalline calcite	Masses of large interlocking calcite crystals or microscopic crystals not visible with the naked eye. Effervesces in dilute HCl.	
Coquina	Calcareous skeletal fragments of coral or shells	Consisting of gravel-sized shell or coral fragments composed of calcite or aragonite. Effervesces in dilute HCl.	
Dolostone	Microcrystalline dolomite	Masses of large interlocking dolomite crystals or microscopic crystals not visible with the naked eye. Effervesces in dilute HCl.	
Chert	Microcrystalline polymorphs of quartz	Microcrystalline varieties of quartz. Scratches glass and may show conchoidal fracturing.	
Organic Sedimentary Rocks			
Coal	Plant fragments and carbonized organic material	Black brittle rock with possible plant fragments. Light in weight with a sooty or shiny appearance. Dark gray streak.	

Table 10.1 | Classification of Sedimentary rocks.
Author: Bradley Deline
Source: Original Work
License: CC BY-SA 3.0

10.3.2 Biochemical and Chemical Sedimentary Rocks

As mentioned before, biochemical and chemical sedimentary rocks either precipitated directly from water or by organisms. The most recognizable chemical sedimentary rocks are evaporites. These are minerals that are formed by the precipitation of minerals from the evaporation of water. You have already examined multiple examples of these minerals/rocks in a previous lab, such as halite and gypsum. In this current lab, we will focus on siliceous and carbonate biochemical sedimentary rocks. Chert is a rock composed of microcrystalline varieties of quartz, and thus it has properties that are associated with quartz itself, such as conchoidal fracturing and hardness greater than glass. Chert is often formed deep in the ocean from silicious material that is either inorganic (silica clay) or biologic (skeletons of sponges and single-celled organisms) in origin. Carbonates are one of the most important groups of sedimentary rocks and as you have previously learned (Chapter 5), can result in distinctive landscapes (karst) and human hazards (sinkholes). Limestone is a sedimentary rock composed of the carbonate mineral calcite and can vary greatly in its appearance depending on how it is formed, but can easily be identified by its chemical weathering. Limestone composed of the mineral calcite undergoes dissolution in acids, in other words it effervesces dramatically when we apply dilute HCl. As with chert, limestone can be formed inorganically from a supersaturation of calcium and carbonate ions in water in varying environments from caves to tropical beaches. Limestone that consists of crystals of calcite or microcrystalline masses of calcite is called crystalline limestone. Alternatively, limestone can be formed biologically with the most striking example called a coquina, which are rocks made exclusively of fragmented carbonate (calcite or its polymorph aragonite) shells or coral. Lastly, we have dolostones, which are made from crystals (large or microscopic) of the mineral dolomite and are a carbonate that only weakly reacts to dilute HCl; you can scratch and powder dolostone to increase the surface area to see the reaction with acid. Dolostone is formed by the inorganic chemical alteration of limestones, therefore they are classified as chemical rather than biochemical sedimentary rocks.

10.3.3 Organic Sedimentary Rocks

Organic compounds are materials that contain a significant amount of the element carbon and are often associated with life. Organic sedimentary rocks are, therefore, rocks that consist mostly of carbon and are associated with significant biological activity. Other sedimentary rocks such as limestone and shale can contain carbon, but at much lower concentration (though shale can appear black from their carbon content). The most common organic sedimentary rock is coal, which is a very low density (light) black rock that has a dusty (sooty) or shiny appearance. It also produces a dark gray streak that can be seen both on a streak plate or a piece of paper. Coal is formed from the preservation and compaction of abundant plant material often in areas where oxygen is lacking, such as a swamp.

10.4 LAB EXERCISE

Materials

Unpack your HOL Sedimentary Rock bag with rocks labeled S1-S8 following what you see in Figure 10.2. In addition to these samples you will need the following from your HOL Rock Kit: 1) your glass plate, 2) your streak plate, 3) your hand lens, and 4) your bottle of diluted HCl.

Using Table 10.1, start identifying the rocks by separating out the organic sedimentary rocks from the chemical and biological sedimentary rocks from the clastic sedimentary rocks. Make sure to use all of the tools available including the glass plate and the diluted HCl to identify the chemical and biochemical sedimentary rocks (chert will be harder than glass, limestone will strongly react with dilute HCl, and dolostone will weakly react with dilute HCl when powdered). The streak plate can be helpful in identifying coal, which will easily produce a dark gray streak. Finally, use the hand lens to closely examine the size of the grains in the clastic sedimentary rocks. Once you are confident of your identifications, answer the following questions.

Figure 10.2 | The eight rocks (S1- S8) in the Sedimentary Rocks Bag in the HOL kit.
Author: Bradley Deline
Source: Original Work
License: CC BY-SA 3.0

Part A – Identifying Sedimentary Rocks

1. Sample S1 is called _____.

    a. Conglomerate   b. Crystalline Limestone   c. Coal   d. Shale

    e. Coquina   f. Chert   g. Dolostone   h. Sandstone

2. Sample S1 is an example of a _____.

    a. Clastic Sedimentary Rock

    b. Organic Sedimentary Rock

    c. Chemical or Biochemical Sedimentary Rock

3. Sample S1 has the following characteristic:

    a. effervesces in diluted HCl acid

    b. weakly effervesces in diluted HCl acid if powdered

    c. contains fossil shells and effervesces in diluted HCl acid

    d. contains pebbles and finer sediments

    e. a sooty or shiny appearance

4. The formation of Sample S1 includes: _____.

    a. chemical weathering, transport of ions, precipitation of minerals, lithification

    b. mechanical weathering, transport of sediment a short distance, deposition of sediment, lithification

    c. photosynthesis, growth of organic material, deposition of organic materials, lithification

    d. chemical weathering, transport of ions, precipitation of minerals as shells by organisms, deposition, lithification

5. Sample S2 is called _____.

    | a. Conglomerate | b. Crystalline Limestone | c. Coal | d. Shale |
    | e. Coquina | f. Chert | g. Dolostone | h. Sandstone |

6. Sample S2 is composed of _____.

    | a. clastic sediments | b. calcite crystals | c. dolomite crystals |
    | d. organic material | e. calcite shells | |

7. Closely examine the individual grains in Sample S2. Which of the following is true about its maturity?

    a. It is mature because it contains a variety of different minerals.

    b. It is immature because it is poorly sorted.

    c. It is mature because it contains mostly rounded quartz grains.

    d. It is immature because the grains are jagged.

8. The formation of Sample S2 includes: _____.

    a. chemical weathering, transport of ions, precipitation of minerals, lithification

    b. mechanical weathering, transport of sediment a long distance, deposition of sediment, lithification

    c. mechanical weathering, transport of sediment a very short distance, deposition of sediment, lithification

    d. photosynthesis, growth of organic material, deposition of organic materials, lithification

    e. chemical weathering, transport of ions, precipitation of minerals as shells by organisms, deposition, lithification

9. Sample S3 is called _____.

    a. Conglomerate   b. Crystalline Limestone   c. Coal   d. Shale

    e. Coquina   f. Chert   g. Dolostone   h. Sandstone

10. Sample S3 is an example of a _____.

    a. Clastic Sedimentary Rock   b. Organic Sedimentary Rock

    c. Chemical Sedimentary Rock   d. Biochemical Sedimentary Rock

11. Sample S3 has the following characteristic:

    a. effervesces in diluted HCl acid

    b. weakly effervesces in diluted HCl acid if powdered

    c. contains fossil shells and effervesces in diluted HCl acid

    d. contains pebbles and finer sediments

    e. has a sooty or shiny appearance

12. The formation of Sample S3 includes: _____.

    a. chemical weathering, transport of ions, precipitation of minerals, lithification

    b. mechanical weathering, transport of sediment a short distance, deposition of sediment, lithification

    c. chemical weathering, transport of ions, precipitation of minerals, lithification, chemical alteration

    d. photosynthesis, growth of organic material, deposition of organic materials, lithification

    e. chemical weathering, transport of ions, precipitation of minerals as shells by organisms, deposition, lithification

13. Sample S4 is called _____.

    a. Conglomerate   b. Crystalline Limestone   c. Coal   d. Shale

    e. Coquina   f. Chert   g. Dolostone   h. Sandstone

14. Sample S4 is an example of a _____.

    a. Clastic Sedimentary Rock   b. Organic Sedimentary Rock

    c. Chemical Sedimentary Rock   d. Biochemical Sedimentary Rock

15. Closely examine the individual grains in Sample S4. Which of the following is true about its maturity?

    a. It is immature because it is poorly sorted.

    b. It is mature because it is poorly sorted.

    c. It is mature because it contains mostly rounded grains.

    d. It is immature because it contains mostly rounded grains.

    e. It is immature because it has clay-sized particles.

16. The formation of Sample S4 includes: _____.

    a. chemical weathering, transportation of ions, precipitation of minerals, lithification

    b. mechanical weathering, transportation of sediment a short distance, deposition of sediments, lithification

    c. photosynthesis, growth of organic material, deposition of organic materials, lithification

    d. chemical weathering, transportation of ions, precipitation of minerals as shells by organisms, deposition, lithification

17. Sample S5 is called _____.

    a. Conglomerate   b. Crystalline Limestone   c. Coal   d. Shale

    e. Coquina   f. Chert   g. Dolostone h. Sandstone

18. Sample S5 is an example of a _____.

    a. Clastic Sedimentary Rock

    b. Organic Sedimentary Rock

    c. Chemical or Biochemical Sedimentary Rock

19. The history of formation of Sample S5 includes: _____.

    a. chemical weathering, transportation of ions, precipitation of minerals, lithification

    b. mechanical weathering, transportation of sediment a short distance, deposition of sediments, lithification

    c. photosynthesis, growth of organic material, deposition of organic materials, lithification

    d. chemical weathering, transportation of ions, precipitation of minerals as shells by organisms, deposition, lithification

20. Sample S5 has the following characteristic:

    a. effervesces in diluted HCl acid

    b. weakly effervesces in diluted HCl acid if powdered

    c. contains fossil shells and effervesces in diluted HCl acid

    d. contains pebbles and finer sediments

    e. a sooty or shiny appearance

21. Sample S6 is called _____.

   a. Conglomerate   b. Crystalline Limestone   c. Coal   d. Shale

   e. Coquina   f. Chert   g. Dolostone   h. Sandstone

22. The history of formation of Sample S6 includes: _____.

   a. chemical weathering, transportation of ions, precipitation of minerals either biologically or inorganically, lithification

   b. mechanical weathering, transportation of sediments a short distance, deposition of sediments, lithification

   c. photosynthesis, growth of organic material, deposition of organic materials, lithification

   d. chemical weathering, transportation of ions, precipitation of minerals as shells by organisms, deposition, lithification

23. Sample S6 is composed of _____.

   a. clastic sediments   b. microcrystalline calcite crystals

   c. microcrystalline dolomite crystals   d. microcrystalline quartz crystals

   e. organic material

24. Sample S6 can be easily recognized by which of the following properties?

   a. conchoidal fracturing

   b. weakly effervesces in diluted HCl acid if powdered

   c. fissule appearance

   d. a sandpaper texture

   e. a sooty or shiny appearance

25. Sample S7 is called _____.

   a. Conglomerate   b. Crystalline Limestone   c. Coal   d. Shale

   e. Coquina   f. Chert   g. Dolostone   h. Sandstone

26. Compared to Sample S2, how mature is Sample S7?

   a. more mature   b. less mature   c. same level of maturity

27. Sample S7 consists of _____.

   a. fragments of calcite shells   b. clay-sized sediments

   c. sand-sized sediments   d. organic material

   e. dolomite crystals

28. Sample S7 can be easily recognized by which of the following properties?

   a. conchoidal fracturing

   b. weakly effervesces in diluted HCl acid if powdered

   c. fissile appearance

   d. a sandpaper texture

   e. a sooty or shiny appearance

29. Sample S8 is called _____.

   a. Conglomerate   b. Crystalline Limestone   c. Coal   d. Shale

   e. Coquina   f. Chert   g. Dolostone   h. Sandstone

30. Sample S8 is an example of a _____.

   a. Clastic Sedimentary Rock   b. Organic Sedimentary Rock

   c. Chemical Sedimentary Rock   d. Biochemical Sedimentary Rock

31. Sample S8 is most likely to weather by which of the following processes?

   a. Dissolution   b. Frost Wedging

   c. Oxidation   d. Hydrolysis

   e. The addition and subtraction of heat

32. The history of formation of Sample S8 includes: _____.

   a. chemical weathering, transportation of ions, precipitation of minerals, lithification

   b. mechanical weathering, transportation of sediments a long distance, deposition of sediments, lithification

   c. mechanical weathering, transportation of sediments a very short distance, deposition of sediments, lithification

   d. photosynthesis, growth of organic material, deposition of organic materials, lithification

   e. chemical weathering, transportation of ions, precipitation of minerals as shells by organisms, deposition, lithification

10.5 SEDIMENTARY STRUCTURES

Sedimentary rocks often show distinctive patterns that are unrelated to their type of rock, yet reflect events or conditions during deposition and are called **sedimentary structures**. These patterns in the rocks can be very informative to geologists attempting to reconstruct the environment in which a sedimentary rock was formed. Imagine wind blowing steadily along a beach; this wind pushes the sand into dunes that can be preserved in the rock record, informing us about the strength and direction of the wind along with the rock type.

Examples of sedimentary structures are given in Table 10.2, but let us discuss them in more detail. The sedimentary structures that most students are familiar with are ripples and dunes. We are familiar with seeing dunes at the beach or in deserts or smaller ripples in mud puddles. In each case,

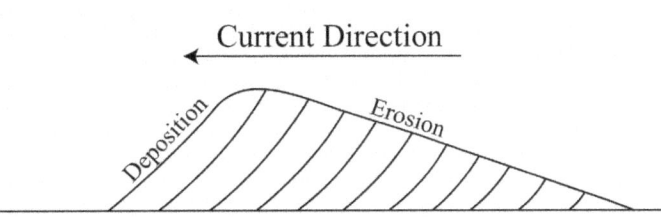

Figure 10.3 | Cross-section of a ripple showing the area of erosion and deposition.
Author: Bradley Deline
Source: Original Work
License: CC BY-SA 3.0

these ripples are formed from either a wind or water current. Ripples form by the current pushing sediment into a pile; on the down-current side the sediment is shadowed and protected from the wind or water current. This means that there is erosion on the up-current side making a shallow slope and deposition on the steeper, down-current side. Therefore, if you cut a ripple in half and look at it in prospect you can see inclined layers of sediment building up on the steep down current side of the ripple. We can often see multiple layers of beds consisting of these inclined layers, which represent multiple generations of migrating ripples or dunes that are called cross-beds. Both ripples and cross-beds can indicate the presence and direction of the current in an environment.

Sedimentary Structure	Description	Required conditions	
Ripple Marks	Either symmetrical or asymetrical ripples present on bedding surfaces.	Water or air currents. Asymetrical ripples indicate unidirectional currents with the steep slope facing down current and the shallow slope facing up current. Symmetrical ripples indicate bidirectional currents.	
Cross-bedding	Layers of inclined beds often altering directions from bed to bed.	Water or air currents.	
Graded Beds	Beds of sedimentary rocks with a change in sediment size through the layer with smaller grains at the top of the bed and larger grains on the bottom.	A turbulent water current carrying sediment loses energy and slows.	
Raindrop impressions	Small indentations on a bedding plane from the impact of raindrops.	Soft muddy substrate that is exposed and subsequently buried with sediment.	
Mudcrack casts	Muddy cast infills of a polygonal pattern of cracks formed in mud as it drys.	Alternating wet and dry conditions create cracks in mud, which are then buried by sediment and preserved.	
Trace fossils	Tracks, burrows, or other traces left by the activity of plants or animals.	The presence of animals or plants. These traces occur where the organism lives and more specific information can be obtained based on the identity of the tracemaker.	

Table 10.2 | Descriptions of common sedimentary structures. Images provided by (from top to bottom) Grand Canyon National Park, Zion National Park, B. Deline, James St. John, Grand Canyon National Park, and B. Deline.
Author: Bradley Deline
Source: Original Work
License: CC BY-SA 3.0

We can also see structures that indicate changes in the strength of a current. Imagine a fast moving current carrying a variety of sediment sizes; if the current slows it will no longer be able to carry the largest particles and they will be deposited first. Then as the current continues to slow progressively smaller particles are deposited on top of the bigger particles, forming a sedimentary deposit called a graded bed. This graded bed is a sedimentary layer with larger clasts on the bottom and smaller clasts on the top. These types of beds can be the result of floods in a river or storms in the ocean. We can also observe features that are pretty self-explanatory such as casts of mud cracks (covered and preserved cracks that are the result of the drying of wet mud), and rain drop impressions (covered and preserved impacts of rain drops in soft mud). In both of these cases, the sedimentary structure tells us about the sediment, the water content, and its exposure at the surface above water level.

Related to sedimentary structures are trace fossils, which are patterns in the rocks that are caused by the activity of organisms. These can occur in many different ways and can indicate many different aspects about the environment depending on the trace and the identity of the tracemaker. Traces can be terrestrial such as footprints, burrows or dens, or the traces of roots, which can inform us about the climate, ecosystem, and the development of soils. Traces can also be found in freshwater and marine environments, such as burrows, borings, footprints, or feeding traces, which tell us about the sediment, chemistry of the poor water, and the life that lived within it.

10.6 DEPOSITIONAL ENVIRONMENTS

A depositional environment is the accumulation of chemical, biological, and physical properties and processes associated with the deposition of sediments that lead to a distinctive suite of sedimentary rocks. Sedimentary environments are interpreted by geologists based on clues within such as rock types, sedimentary structures, trace fossils, and fossils. We can then compare these clues to modern environments to reconstruct ancient environments. We can break up the numerous depositional environments found on earth into common environments we find on land **(continental environments)**, in the ocean **(marine environments)**, and at the interface between the two **(transitional environments)**. As each of the environments is briefly described, think about and fill in Table 3 with the following information: 1) select the image that best depicts the depositional environment in Figure 10.4 and put the corresponding letter (from Figure 10.4) in the first column, 2) the types of sediments and sedimentary rocks most likely to occur in that environment, 3) the maturity of those sediments if clastic, and 4) the types of sedimentary structures that may be present.

10.6.1 Continental Environments

There are many different environments on the continents, but again we are limited to those that are dominated by the deposition rather than the erosion of sediments. Erosion occurs in high altitude areas and although continents are overall topographically elevated compared to the oceans, there are several different areas on the continent where we get distinctive depositional properties. Continental depositional environments are dominated by clastic sedimentary rocks, largely because of their proximity to the source of the sediments.

Glacial depositional environments are controlled mostly by the weathering and erosion by glaciers and glacial meltwater. Glaciers most commonly occur in areas that are both high elevation and/or high latitudes. Glaciers are fairly slow moving (centimeters a day) and normally travel short distances from their source, but they can cause immense mechanical weathering. Glaciers grind and bulldoze rock and create piles of poorly sorted sediment called moraines. Glaciers also produce a significant though fluctuating amount of melt water, which flows through the moraines building a system of braided rivers. These rivers carry the small sediments further from the end of the glaciers into an area called the outwash plain, which consists of poorly sorted sediment.

We have spent a significant amount of time in chapter five discussing rivers and how they erode, transport, and deposit sediment. Sediments that are deposited through the action of rivers are referred to as **fluvial depositional environments**. The gradient and discharge of a river can greatly control the shape of the river, how it flows, and how it deposits sediment. Rivers alter sediment both chemically and physically as well as sort the sediment since there is a limit to the size or particles a river can carry. Within a meandering river we see several different types of sediments, from the pebbles and stones within the river channel to sandy point bars along the outer edge of the meander where the water slows. In addition, we see multiple types of sedimentary structures related to the flow of the river as well as those related to flood events.

We also have sediments deposited within lakes, which are called **lacustrine depositional environments**. Unlike rivers, lakes do not have rapidly flowing water and thus there is significantly less movement of sediment. The sediment that accumulates in lakes can come from several sources including rainwater carrying sediment into the lake from the shores, rivers that flow into the lake, and sediment that is transported by the wind. Once the sediment reaches the lake it remains undisturbed so we see thin layers of fine sediment, with varying amounts of trace fossil.

Lastly, there are **eolian depositional environment**, which are dominated by currents of wind rather than water. Since air is less dense than water, only smaller particles can be transported. In addition, wind is not restrained within distinctive channels like water and, therefore, the features of eolian deposits are more widespread than those of fluvial deposits. Certain areas have predominant wind patterns, such that the wind is fairly consistent in direction and strength, which can generate significant sedimentary structures. When water is present within these environments it is often in the form of seasonal lakes that undergo significant evaporation and sometimes leaving behind chemical sedimentary rocks (evaporite deposits).

10.6.2 Transitional Environments

The interface between the continents and oceans are complicated areas that can be influenced by rivers, ocean currents, winds, waves, and tides. In addition, the sediments that are present in these areas are a mixture of materials derived from the continents (clastic and organic) and those from the ocean (chemical and biochemical). Finally, with the abundant currents (both air and water), sediments influenced by the elements, and abundant life these areas, which results in abundant sedimentary structures and trace fossils. However, we can distinguish transitional areas that are dominated by different forces such as tides, ocean currents, and rivers.

Shorelines that are influenced by strong daily tidal currents are called **tidal mudflat depositional environments**. Tides are currents that are the result of the gravitational forces exerted by the moon and the rotation of the earth. Shorelines that have strong tidal currents as well as seafloors with low gradients can have large areas that are submerged during high tide and exposed to air during low tide. These areas often have smaller particles than a normal shoreline since the tidal currents can pull marine sediments into the area. In addition, the strong bidirectional currents, daily drying out, exposure to the elements, and abundant life create abundant indicators of these environments.

Shorelines that are dominated by ocean currents are called **beach depositional environments**. Shorelines have constant wind on and off shore that are the result of the difference in the way the land and water heat and cool through the day. These winds produce the waves that are iconic at the beach, but as these waves move onto shore they curve, mimicking the shape of the shore and result in a current that runs parallel to the shore itself. This current carries and deposits sand along the beach. In addition, the wind can also produce dunes, which promote a diverse and complicated ecosystem. Within beach depositional environments there are areas where rivers flow into the ocean, which are called **Deltas**. As a river that is carrying material empties into the ocean the water slows and deposits sediment. Most of the sediment is deposited at the mouth of the river with some spilling out into the surrounding areas building a distinctive fan of sediment. Since the sediment is coming from the river, the delta is largely a thick sequence of clastic sediment showing indications of the strong flow of the river.

10.6.3 Marine Environments

Marine depositional environments differ in multiple ways, but the controlling factors in the rocks that are produced is related to the proximity and supply of continental sediment, the water depth, and the community of organisms that live in the area. The further an environment is from the shore the less clastic sediment will be present and the area will have a higher concentration of the chemical and biological sedimentary rocks that are formed within the ocean. In addition, some organisms in the right environmental conditions can produce huge amounts of skeletal material.

Shallow marine depositional environments are areas that are close to shore, but always submerged. These areas have a significant amount of mature clastic sediment along with marine algae (like sea grass) as well as skeletal material from animals like coral, echinoderms (sea urchins and sand dollars), and mollusks (clams and snails). These areas can have a significant difference in their energy level from very shallow areas influenced by waves to deeper areas only influenced by large storms. A better understanding of the relative depth can often be determined based on the sedimentary structures as well as the community of organisms and types of trace fossils, which can be very sensitive to depth.

In warm tropical shallow water area we often find **reef depositional environments**. Reefs are formed through the growth of coral colonies building a large three-dimensional structure built from calcite skeletons. Corals can grow in many different marine environments, but they can only produce reefs when their symbiotic algae that live within their tentacles can photosynthesize effectively, resulting in more energy for the coral to grow faster. Reefs also create a barrier between shallow water environments and ocean currents producing shallow, low-energy environment called lagoons. Lagoons can have thin layers of fine sediment that we would expect in quiet water along with chemical sedimentary deposits that are the result of evaporation.

Most of the ocean consists of **deep marine depositional environments**. These areas are beyond the reach of most clastic sediment other than the dust carried by the wind. Therefore, the sediment is being produced chemically and biologically within the ocean. The largest source of sediment in these deep settings is skeletal material from some of the smallest marine organisms. Multiple types of single-celled organisms can produce shells composed of either silica or calcite. These shells are mostly produced in the surface waters that are bathed in sunlight permitting photosynthesis. When the organism dies, these shells then rain down into deeper water; this slow accumulation of sediment produces fine layers of biochemical sedimentary rocks. In some cases, these shells are dissolved or altered before they reach the bottom (which can be miles away) and are precipitated as chemical sedimentary rocks. Obviously, there is not a clear boundary between shallow and deep-water environments given the gradient of the ocean floor. The deep marine depositional environment is normally thousands of feet deep and beyond the influence of even large storms.

10.6.4 Depositional Environments

Based on the descriptions of the different sedimentary environments along with your understanding of weathering, sedimentary rocks, and sedimentary structures please fill out the following table, which can then be used as a guide for answering the following questions on depositional environments.

Figure 10.4 | Modern examples of the discussed depositional environments.

Photo A
Author: Kelsey Roberts, USGS
Source: Flickr
License: Public Domain

Photo B
Author: Alan Cressler
Source: Flickr
License: CC BY 2.0

Photo C
Author: Kim Keating, USGS
Source: Flickr
License: Public Domain

Photo D
Author: Deborah Bergfeld, USGS
Source: Flickr
License: Public Domain

Photo E
Author: Paul Hutchinson
Source: Wikimedia Commons
License: CC BY-SA 2.0

Photo F
Author: James St. John
Source: Wikimedia Commons
License: CC BY 2.0

Photo G
Author: NOAA
Source: Wikimedia Commons
License: CC BY 2.0

Photo H
Author: Peter Dowley
Source: Wikimedia Commons
License: CC BY 2.0

Photo I
Author: Diueine Monteiro
Source: Wikimedia Commons
License: CC BY-NC-SA 2.0

Example from Figure 4	Depositional Environment	Description	Clastic Rock Maturity, distance from source, and type	Other sedimentary Rocks	Sedimentary Structures
Continental					
	Fluvial				
	Lacustrine				
	Glacial				
	Eolian				
Transitional					
	Beach				
	Mudflats				
Marine					
	Shallow Marine				
	Reefs				
	Deep Marine				

Table 10.3
Author: Bradley Deline
Source: Original Work
License: CC BY-SA 3.0

INTRODUCTORY GEOLOGY   SEDIMENTARY ROCKS

10.7 LAB EXERCISE
Part B – Depositional Environments

The following exercises use Google Earth to explore the depositional environments that are the source for sedimentary rocks. For each locality think about the types of sediments that are accumulating, the types of weathering that would occur, as well as the presence of any sedimentary structures.

33. Search for 39 05 52.46N 84 30 56.16E and zoom out to an eye altitude of ~30,000 ft. The large-scale structures in this sedimentary environment are asymmetrical ripples (known as dunes at this size). Zoom in and out and examine the sedimentary environment. What type of sedimentary environment is this?

    a. Lacustrine   b. Fluvial   c. Eolian   d. Glacial

34. What type of weathering do you think is most prominent in this sedimentary environment?

    a. Mechanical weathering from a current   b. Dissolution

    c. Frost Wedging   d. Hydrolysis

35. Study the large dunes in this image (zoom out to an eye altitude of ~25,000 feet). These structures can indicate the direction that the wind is blowing. What is the predominant wind direction in this area? (Hint: it is easier to see these features if we exaggerate the vertical scale to do this go to Tools, options, and on the 3D view tab change the Elevation Exaggeration to 3. To do this on a Mac go to Google Earth then Preferences)

    a. north to south   b. south to north   c. east to west   d. west to east

36. Search for 64 01 03.61N 16 52 56.63W and zoom out to an eye altitude of ~25,000 ft. What type of sedimentary environment is this?

    a. Lacustrine   b. Fluvial   c. Eolian   d. Glacial

37. What type of weathering do you think is most prominent in this sedimentary environment? (Make sure you evaluate all choices.)

    a. Mechanical weathering from ice   b. Dissolution

    c. Frost Wedging   d. Hydrolysis

    e. Both a and c are correct

38. When the sediment you see today lithifies, what type of sedimentary rock would you expect to be most abundant in this area and how mature is this rock type?

   a. Shale, Mature   b. Shale, Immature   c. Sandstone, Immature

   d. Sandstone, Mature   e. Conglomerate, Immature f. Conglomerate, Mature

39. Search for 44 42 16.98N 1 05 23.88W and zoom out to an eye altitude of ~17 miles. Notice the depth of this shallow bay (Arcachon Bay, France). This bay dries out at times and later greatly expands in size daily. What sedimentary structure would you likely find in this area because of this?

   a. graded beds   b. finely layered beds

   c. cross beds   d. mud cracks

40. What type of sedimentary environment is this?

   a. lacustrine   b. tidal mudflats

   c. deep marine   d. beach

41. Search for 20 20 23.94S 150 38 29.14E and zoom out to an eye altitude of ~25 miles (also zoom far out to notice where you are in the world). What type of sedimentary environment is this?

   a. shallow marine   b. deep marine   c. delta   d. reef

42. Think about the origin of this marine formation and consider the latitude to assess climatic conditions. What is the most abundant type of rock you would expect to form in this sedimentary environment?

   a. Limestone   b. Sandstone   c. Coal   d. Shale

43. If the sea level dropped 1,000 feet and this sedimentary environment stopped being built and began to break down, what type of weathering would be most likely to occur on these rocks?

   a. Fracturing from the addition or subtraction of pressure

   b. Dissolution

   c. Frost Wedging

   d. Hydrolysis

44. Search for 31 00 43.61N 81 25 40.92W and zoom out to an eye altitude of 1,500 ft. What type of sedimentary environment is this?

   a. shallow marine   b. reef   c. fluvial   d. each   e. tidal mudflat

45. What type of sediment would you expect to find in this depositional environment?

   a. Mostly mature clastic sediments with some biochemical and chemical sedimentary sediments

   b. Mostly biochemical and chemical sedimentary sediments with some mature clastic sediments

   c. Exclusively organic sediments

   d. Exclusively biochemical sediments

   e. Exclusively immature clastic sediments

INTRODUCTORY GEOLOGY   SEDIMENTARY ROCKS

10.8 STUDENT RESPONSES

1. Sample S1 is called _____.

    a. Conglomerate   b. Crystalline Limestone   c. Coal   d. Shale

    e. Coquina   f. Chert   g. Dolostone   h. Sandstone

2. Sample S1 is an example of a _____.

    a. Clastic Sedimentary Rock

    b. Organic Sedimentary Rock

    c. Chemical or Biochemical Sedimentary Rock

3. Sample S1 has the following characteristic:

    a. effervesces in diluted HCl acid

    b. weakly effervesces in diluted HCl acid if powdered

    c. contains fossil shells and effervesces in diluted HCl acid

    d. contains pebbles and finer sediments

    e. a sooty or shiny appearance

4. The formation of Sample S1 includes: _____.

    a. chemical weathering, transport of ions, precipitation of minerals, lithification

    b. mechanical weathering, transport of sediment a short distance, deposition of sediment, lithification

    c. photosynthesis, growth of organic material, deposition of organic materials, lithification

    d. chemical weathering, transport of ions, precipitation of minerals as shells by organisms, deposition, lithification

5. Sample S2 is called _____.

    a. Conglomerate   b. Crystalline Limestone   c. Coal   d. Shale

    e. Coquina   f. Chert   g. Dolostone   h. Sandstone

6. Sample S2 is composed of _____.

   a. clastic sediments b. calcite crystals c. dolomite crystals

   d. organic material e. calcite shells

7. Closely examine the individual grains in Sample S2. Which of the following is true about its maturity?

   a. It is mature because it contains a variety of different minerals.

   b. It is immature because it is poorly sorted.

   c. It is mature because it contains mostly rounded quartz grains.

   d. It is immature because the grains are jagged.

8. The formation of Sample S2 includes: _____.

   a. chemical weathering, transport of ions, precipitation of minerals, lithification

   b. mechanical weathering, transport of sediment a long distance, deposition of sediment, lithification

   c. mechanical weathering, transport of sediment a very short distance, deposition of sediment, lithification

   d. photosynthesis, growth of organic material, deposition of organic materials, lithification

   e. chemical weathering, transport of ions, precipitation of minerals as shells by organisms, deposition, lithification

9. Sample S3 is called _____.

   a. Conglomerate b. Crystalline Limestone c. Coal d. Shale

   e. Coquina f. Chert g. Dolostone h. Sandstone

10. Sample S3 is an example of a _____.

    a. Clastic Sedimentary Rock b. Organic Sedimentary Rock

    c. Chemical Sedimentary Rock d. Biochemical Sedimentary Rock

INTRODUCTORY GEOLOGY											SEDIMENTARY ROCKS

11. Sample S3 has the following characteristic:

    a. effervesces in diluted HCl acid

    b. weakly effervesces in diluted HCl acid if powdered

    c. contains fossil shells and effervesces in diluted HCl acid

    d. contains pebbles and finer sediments

    e. has a sooty or shiny appearance

12. The formation of Sample S3 includes: _____.

    a. chemical weathering, transport of ions, precipitation of minerals, lithification

    b. mechanical weathering, transport of sediment a short distance, deposition of sediment, lithification

    c. chemical weathering, transport of ions, precipitation of minerals, lithification, chemical alteration

    d. photosynthesis, growth of organic material, deposition of organic materials, lithification

    e. chemical weathering, transport of ions, precipitation of minerals as shells by organisms, deposition, lithification

13. Sample S4 is called _____.

    a. Conglomerate   b. Crystalline Limestone   c. Coal   d. Shale

    e. Coquina   f. Chert   g. Dolostone   h. Sandstone

14. Sample S4 is an example of a _____.

    a. Clastic Sedimentary Rock   b. Organic Sedimentary Rock

    c. Chemical Sedimentary Rock   d. Biochemical Sedimentary Rock

15. Closely examine the individual grains in Sample S4. Which of the following is true about its maturity?

    a. It is immature because it is poorly sorted.

    b. It is mature because it is poorly sorted.

    c. It is mature because it contains mostly rounded grains.

    d. It is immature because it contains mostly rounded grains.

    e. It is immature because it has clay-sized particles.

16. The formation of Sample S4 includes: _____.

    a. chemical weathering, transportation of ions, precipitation of minerals, lithification

    b. mechanical weathering, transportation of sediment a short distance, deposition of sediments, lithification

    c. photosynthesis, growth of organic material, deposition of organic materials, lithification

    d. chemical weathering, transportation of ions, precipitation of minerals as shells by organisms, deposition, lithification

17. Sample S5 is called _____.

    a. Conglomerate   b. Crystalline Limestone   c. Coal   d. Shale

    e. Coquina   f. Chert   g. Dolostone h. Sandstone

18. Sample S5 is an example of a _____.

    a. Clastic Sedimentary Rock

    b. Organic Sedimentary Rock

    c. Chemical or Biochemical Sedimentary Rock

19. The history of formation of Sample S5 includes: _____.

   a. chemical weathering, transportation of ions, precipitation of minerals, lithification

   b. mechanical weathering, transportation of sediment a short distance, deposition of sediments, lithification

   c. photosynthesis, growth of organic material, deposition of organic materials, lithification

   d. chemical weathering, transportation of ions, precipitation of minerals as shells by organisms, deposition, lithification

20. Sample S5 has the following characteristic:

   a. effervesces in diluted HCl acid

   b. weakly effervesces in diluted HCl acid if powdered

   c. contains fossil shells and effervesces in diluted HCl acid

   d. contains pebbles and finer sediments

   e. a sooty or shiny appearance

21. Sample S6 is called _____.

   a. Conglomerate   b. Crystalline Limestone   c. Coal   d. Shale

   e. Coquina   f. Chert   g. Dolostone   h. Sandstone

22. The history of formation of Sample S6 includes: _____.

   a. chemical weathering, transportation of ions, precipitation of minerals either biologically or inorganically, lithification

   b. mechanical weathering, transportation of sediments a short distance, deposition of sediments, lithification

   c. photosynthesis, growth of organic material, deposition of organic materials, lithification

   d. chemical weathering, transportation of ions, precipitation of minerals as shells by organisms, deposition, lithification

23. Sample S6 is composed of _____.

    a. clastic sediments

    b. microcrystalline calcite crystals

    c. microcrystalline dolomite crystals

    d. microcrystalline quartz crystals

    e. organic material

24. Sample S6 can be easily recognized by which of the following properties?

    a. conchoidal fracturing

    b. weakly effervesces in diluted HCl acid if powdered

    c. fissule appearance

    d. a sandpaper texture

    e. a sooty or shiny appearance

25. Sample S7 is called _____.

    a. Conglomerate   b. Crystalline Limestone   c. Coal   d. Shale

    e. Coquina   f. Chert   g. Dolostone   h. Sandstone

26. Compared to Sample S2, how mature is Sample S7?

    a. more mature   b. less mature   c. same level of maturity

27. Sample S7 consists of _____.

    a. fragments of calcite shells

    b. clay-sized sediments

    c. sand-sized sediments

    d. organic material

    e. dolomite crystals

28. Sample S7 can be easily recognized by which of the following properties?

    a. conchoidal fracturing

    b. weakly effervesces in diluted HCl acid if powdered

    c. fissile appearance

    d. a sandpaper texture

    e. a sooty or shiny appearance

29. Sample S8 is called _____.

   a. Conglomerate   b. Crystalline Limestone   c. Coal   d. Shale

   e. Coquina   f. Chert   g. Dolostone   h. Sandstone

30. Sample S8 is an example of a _____.

   a. Clastic Sedimentary Rock   b. Organic Sedimentary Rock

   c. Chemical Sedimentary Rock   d. Biochemical Sedimentary Rock

31. Sample S8 is most likely to weather by which of the following processes?

   a. Dissolution   b. Frost Wedging

   c. Oxidation   d. Hydrolysis

   e. The addition and subtraction of heat

32. The history of formation of Sample S8 includes: _____.

   a. chemical weathering, transportation of ions, precipitation of minerals, lithification

   b. mechanical weathering, transportation of sediments a long distance, deposition of sediments, lithification

   c. mechanical weathering, transportation of sediments a very short distance, deposition of sediments, lithification

   d. photosynthesis, growth of organic material, deposition of organic materials, lithification

   e. chemical weathering, transportation of ions, precipitation of minerals as shells by organisms, deposition, lithification

33. Search for 39 05 52.46N 84 30 56.16E and zoom out to an eye altitude of ~30,000 ft. The large-scale structures in this sedimentary environment are asymmetrical ripples (known as dunes at this size). Zoom in and out and examine the sedimentary environment. What type of sedimentary environment is this?

   a. Lacustrine   b. Fluvial   c. Eolian   d. Glacial

34. What type of weathering do you think is most prominent in this sedimentary environment?

   a. Mechanical weathering from a current   b. Dissolution

   c. Frost Wedging   d. Hydrolysis

35. Study the large dunes in this image (zoom out to an eye altitude of ~25,000 feet). These structures can indicate the direction that the wind is blowing. What is the predominant wind direction in this area? (Hint: it is easier to see these features if we exaggerate the vertical scale to do this go to Tools, options, and on the 3D view tab change the Elevation Exaggeration to 3. To do this on a Mac go to Google Earth then Preferences)

    a. north to south   b. south to north   c. east to west   d. west to east

36. Search for 64 01 03.61N 16 52 56.63W and zoom out to an eye altitude of ~25,000 ft. What type of sedimentary environment is this?

    a. Lacustrine   b. Fluvial   c. Eolian   d. Glacial

37. What type of weathering do you think is most prominent in this sedimentary environment? (Make sure you evaluate all choices.)

    a. Mechanical weathering from ice   b. Dissolution

    c. Frost Wedging   d. Hydrolysis

    e. Both a and c are correct

38. When the sediment you see today lithifies, what type of sedimentary rock would you expect to be most abundant in this area and how mature is this rock type?

    a. Shale, Mature   b. Shale, Immature   c. Sandstone, Immature

    d. Sandstone, Mature   e. Conglomerate, Immature   f. Conglomerate, Mature

39. Search for 44 42 16.98N 1 05 23.88W and zoom out to an eye altitude of ~17 miles. Notice the depth of this shallow bay (Arcachon Bay, France). This bay dries out at times and later greatly expands in size daily. What sedimentary structure would you likely find in this area because of this?

    a. graded beds   b. finely layered beds

    c. cross beds   d. mud cracks

40. What type of sedimentary environment is this?

    a. lacustrine   b. tidal mudflats

    c. deep marine   d. beach

41. Search for 20 20 23.94S 150 38 29.14E and zoom out to an eye altitude of ~25 miles (also zoom far out to notice where you are in the world). What type of sedimentary environment is this?

   a. shallow marine   b. deep marine   c. delta   d. reef

42. Think about the origin of this marine formation and consider the latitude to assess climatic conditions. What is the most abundant type of rock you would expect to form in this sedimentary environment?

   a. Limestone   b. Sandstone   c. Coal   d. Shale

43. If the sea level dropped 1,000 feet and this sedimentary environment stopped being built and began to break down, what type of weathering would be most likely to occur on these rocks?

   a. Fracturing from the addition or subtraction of pressure

   b. Dissolution

   c. Frost Wedging

   d. Hydrolysis

44. Search for 31 00 43.61N 81 25 40.92W and zoom out to an eye altitude of 1,500 ft. What type of sedimentary environment is this?

   a. shallow marine   b. reef   c. fluvial   d. each   e. tidal mudflat

45. What type of sediment would you expect to find in this depositional environment?

   a. Mostly mature clastic sediments with some biochemical and chemical sedimentary sediments

   b. Mostly biochemical and chemical sedimentary sediments with some mature clastic sediments

   c. Exclusively organic sediments

   d. Exclusively biochemical sediments

   e. Exclusively immature clastic sediments

11 Metamorphic Rocks
Karen Tefend

11.1 INTRODUCTION

In any introductory textbook on physical geology, the reader will find the discussion on metamorphic rocks located after the chapters on igneous and sedimentary rocks, and for very good reason. Metamorphic rocks form by the physical and sometimes chemical alteration of a pre-existing rock, whether it is igneous or sedimentary. In some cases, even metamorphic rocks can be altered into a completely different metamorphic rock. With igneous rocks forming from the melt produced by any rock type and a sedimentary rock forming from the weathered product of any rock type, the alteration of any rock to produce a metamorphic one completes the components of what is known as the rock cycle. Basically, the rocks we encounter today that we classify as either igneous, metamorphic, or sedimentary, could have belonged to a different rock classification in the past, as rocks are recycled throughout geologic time, driven by the motion of the tectonic plates (see Chapter 4). It is easy to see that increasing the temperature of a rock can produce magma, and that rocks on the surface of the earth can break up into sediment that can ultimately lithify into a sedimentary rock. But how can we alter a solid rock into a new rock, without melting it or making it become sediment?

All rocks are formed at certain temperatures and pressures on or more commonly, beneath the earth's surface, and these rocks are the most stable at the conditions under which they form. Therefore, changing the temperature and/or pressure conditions may lead to a different rock, one that changed in order to be stable under new external conditions. This new rock that forms in response to changes in its physical and chemical environment is called a metamorphic rock; the word metamorphism means to change form, and for rocks this means a recrystallization of minerals (crystals) under subsolidus (temperatures too low for melt production) conditions. A metamorphic change can also occur if the rock's composition is altered by hot, chemically reactive fluids, causing a change in the mineral content of the rock. To distinguish between the pre-existing rock and the new metamorphic one, the term **protolith** or parent rock is used to describe the pre-existing rock, and all metamorphic rocks have at least one protolith that has altered during metamorphism. In this chapter you will learn that all metamorphic rocks are identified by the mineral content and texture of the rock; for metamorphic rocks, texture refers to the orientation of the minerals in the rock, although crystal size does convey important information regarding the temperature conditions during metamorphism.

To summarize, metamorphism is the process by which a pre-existing rock (the protolith) is altered by a change in temperature, pressure, or by contact with chemically reactive fluids, or by any combination of these three parameters. The alteration process is a **recrystallization** event, where the initial rock's minerals (crystals) have changed size, shape, and/or composition in response to these new external conditions. The end result is a new (metamorphic) rock that has an altered appearance, sometimes strikingly different from what it used to look like before metamorphism occurred. What metamorphic rock you end up with is strongly dependent on what rock you started with before the metamorphic event. Of secondary importance is the agent of metamorphic change: was recrystallization due to increased temperature, pressure, or both, or were chemically reactive fluids involved? Also, one must consider how high the temperatures and pressures were, as metamorphism can occur in a range of temperature and pressure conditions. However, as mentioned in the preceding paragraph, metamorphism occurs under subsolidus conditions, meaning that elevated temperatures that promote the recrystallization of a rock are not high enough to cause melting.

11.1.1 Learning Outcomes

After completing this chapter, you should be able to:
- Correctly identify the metamorphic rocks and their distinguishing features
- Identify foliated metamorphic rocks, and the type of foliation
- Determine the grade of metamorphism based on foliation and possible index minerals
- Recognize metamorphic environments and types of metamorphism

11.1.2 Key Terms

- Contact metamorphism
- Differential pressure
- Dynamic (fault zone) metamorphism
- Foliation
- Gneissic banding
- Hydrothermal metamorphism
- Index mineral
- Lineation
- Lithostatic pressure
- Metamorphism
- Mylonite
- Protolith
- Recrystallization
- Regional metamorphism
- Schistose foliation

- Regional metamorphism
- Schistose foliation
- Shock (impact) metamorphism
- Slaty cleavage
- Tektite
- Thermodynamic (dynamothermal) metamorphism

11.2 AGENTS OF METAMORPHISM

11.2.1 Pressure

All rocks beneath the surface of the earth experience an increase in pressure due to the weight of the overlying sediment and rock layers, and with increasing depth there is a corresponding increase in pressure. This increased pressure does not necessarily cause a rock to become metamorphic, because this particular pressure is typically equal in all directions and is known as lithostatic pressure. Lithostatic pressure is similar to hydrostatic pressure, such as the pressure on the eardrums a swimmer will experience as he or she dives deep in the water. Lithostatic pressure on rocks below the earth's surface may have a change in overall rock volume, but will not cause a change in the shape. An example of decreasing volume due to lithostatic pressure would be a closer packing of clasts and reduction of pore space within a clastic sedimentary rock. But what if the pressure on a rock is un-

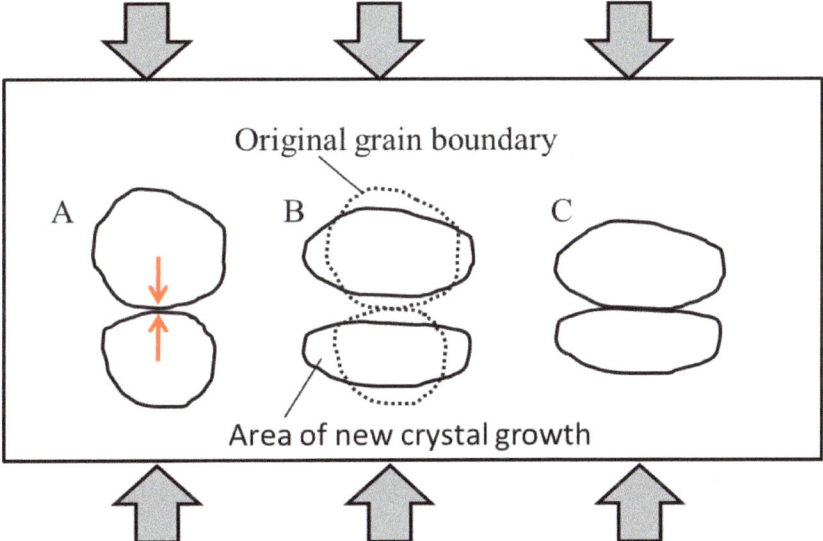

Figure 11.1 | Example of mineral grain deformation due to differential pressure: A) Rounded grains experience the greatest pressure at the contact between the two grains due to the greatest applied pressure as indicated by the large gray arrows. B) The grains change shape by bonds breaking in the high pressure areas, and atoms migrate laterally into the lower pressure areas (the area of new crystal growth). C) The final shape of these two grains. Note that the grains are flattened perpendicular to the direction of greatest pressure.
Author: Karen Tefend
Source: Original Work
License: CC BY-SA 3.0

equal, and the rocks become squeezed in one direction more than another direction? This is known as differential pressure, and it can result in a significant change in the appearance of a rock. Figure 11.1 demonstrates how a mineral can change shape due to differential pressure, in this case with the greatest pressures from the top and bottom (as demonstrated by the large gray arrows). Two initially rounded mineral grains (Figure 11.1A) within a sedimentary rock are experiencing the greatest amount of pressure at the contact between the grains (see red arrows in the figure), and the bonds linking the atoms in this grain will break. The atoms will migrate into the area of lesser pressure and reform a bond with other atoms in the mineral grain (Figure 11.1B). As a result, the grains have a flattened shape that is perpendicular to the direction of greatest pressure (Figure 11.1C).

Figure 11.1 only shows the deformation of two grains; imagine that this is happening to all of the grains in the sedimentary rock, or to all of the phenocrysts (crystals) in an igneous rock. In that case, you will end up with the entire rock having minerals aligned in a certain direction, all by the breaking of bonds between atoms in a mineral, and reforming (recrystallizing) in the lower pressure areas among the grains or crystals in the rock. The end result is a rock with a metamorphic pattern called a **foliation**. Metamorphic foliations are the patterns seen in a rock that has experienced differential pressure; these foliations may be fairly flat, or have a wavy appearance possibly due to more than one direction of greatest pressure. Some rocks may also develop what is called a **lineation**, which can be formed by an elongation of minerals that form a linear feature through the rock. To understand the difference between a foliation and a lineation, let us use some food analogies: a stack of pancakes demonstrates a foliation in your breakfast food, with each pancake layer representing flattened minerals. If you look at the top of the pancakes, you will not see a pattern, but if you view the stack of pancakes from the side, or cut through the stack with your knife in any orientation other than parallel to the pancake layers, you will see the layering or foliation. However, if within that pancake stack, there existed a slice of bacon (yum), the bacon would be the lineation in your breakfast "rock", and you may or may not see it when you cut through the pancake stack. We will discuss foliations, and the different types of foliations, in a later section of this chapter.

11.2.2 Temperature

Probably the most common cause of metamorphism is a change in temperature. Often times metamorphism involves both an increase in temperature along with the pressure changes as described in the above section. Higher temperatures are often associated with metamorphism due to chemically reactive fluids (which we will discuss in the next section). The broad classification for metamorphism into low, medium and high grades of metamorphic change exists mainly due to temperature conditions; this will also be discussed in a later section.

Higher temperatures increase the vibrational energy between the bonds linking atoms in the mineral structure, making it easier for bonds to be broken in order for the recrystallization of the minerals into new crystal shapes and sometimes the development of foliations and lineations as described in the previous section. However, recrystallization can be due to just temperature changes without any differential pressure conditions, and when temperatures are increased, there can be a corresponding increase in mineral sizes as initially small minerals become fused into larger crystals. This fusing of numerous smaller mineral sizes into fewer, yet larger, mineral sizes is known as annealing in metallurgy, but for metamorphic rocks it is still referred to as recrystallization.

In order to understand why increasing temperatures lead to increased grain sizes, we need to again address stability. In general, a mineral grain or crystal is most stable when it has a low surface area to volume ratio, therefore large grains are more stable than small grains, because increasing the grain size results in a greater increase in volume as opposed to a smaller increase in the surface area. Why does stability matter? Because that is why the rocks we are concerned with in this chapter are changing; rocks become unstable when their environment changes, and by a recrystallization processes (metamorphism), they can return to a stable form once again. Figure 11.2 demonstrates the recrystallization process in a sedimentary rock in response to elevated temperature. In this example, the original grains are smaller and rounded, but recrystallization resulted in larger grains that are interlocking; the pore spaces are gone and instead larger crystals exist.

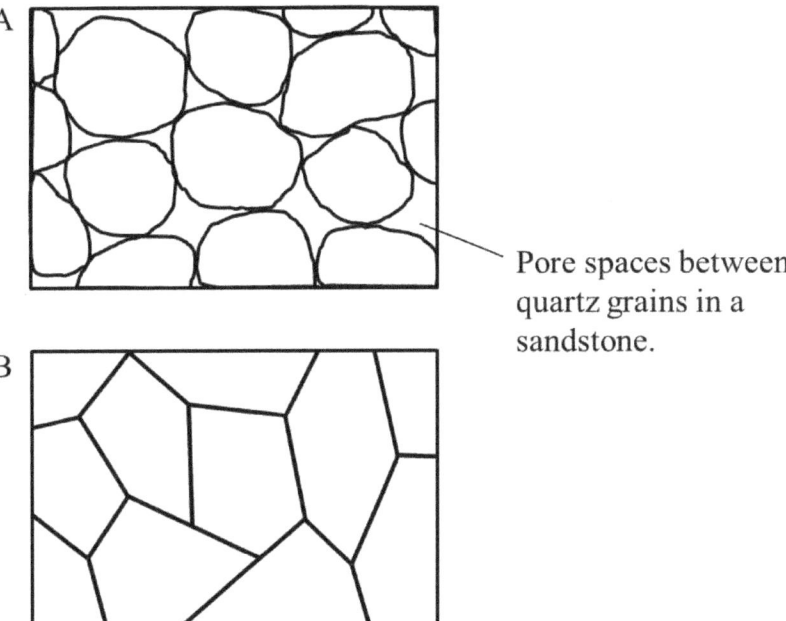

Figure 11.2 | An example of the recrystallization of crystals into larger sizes due to increased temperature: A) a sedimentary rock with rounded quartz grains, and pore spaces between the grains. B) a metamorphic rock with larger, interlocking quartz crystals due to increased temperature conditions.
Author: Karen Tefend
Source: Original Work
License: CC BY-SA 3.0

In addition to increased grain size with increased temperature, occasionally a new mineral forms during metamorphism. These new minerals form at certain temperatures and are called **index minerals**, which can be used to determine the temperature of metamorphism. Index minerals will be covered in more detail in a later section.

11.2.3 Chemically Reactive Fluids

The phrase chemically reactive refers to the dissolved ions in a fluid phase that may react with minerals in a rock; these ions may take the place of some of the atoms in the mineral's structure, which may lead to a significant change in the chemical composition of a rock. Sometimes these fluids are quite hot, especially if they are fluids released from a nearby magma body that is crystallizing while cooling. Metamorphism due to such fluids is known as **hydrothermal metamorphism**.

11.3 METAMORPHIC ROCK NAMES

11.3.1 Foliated Metamorphic Rocks

As mentioned previously, differential pressures can cause a foliation to develop in metamorphosed rocks. There are a few types of foliations that are commonly seen in metamorphic rocks, each foliation type is dependent on the minerals that define the foliation. One type is described as a layering of dark and light colored minerals, so that the foliation is defined as alternating dark and light mineral bands throughout the rock; such a foliation is called **gneissic banding** (Figure 11.3), and the metamorphic rock is called gneiss (pronounced "nice", with a silent g). In Figure 11.3A, the layering in this gneiss is horizontal, and the greatest pressures were at right angles to the gneissic bands. Note that these bands are not always flat, but may be seen contorted as in Figure 11.3B; this rock still is considered to have gneissic banding even though the bands are not horizontal. The typical minerals seen in the dark colored bands are biotite micas and/or amphiboles, whereas the light colored bands are typically quartz or light-colored feldspars. The protoliths for gneiss can be any rock that contains more than one mineral, such as shale with its clay minerals and clay-sized quartz and feldspar, or an igneous rock with both dark-colored ferromagnesian minerals and light-colored non-ferromagnesian minerals (see Chapter 8 for review). In order for gneissic foliation to develop, temperatures

Figure 11.3 | Two examples of the metamorphic rock, gneiss. Each rock exhibits the alternating dark and light mineral bands throughout the rock.
Author: Karen Tefend
Source: Original Work
License: CC BY-SA 3.0

and pressures need to be quite high; for this reason, gneiss rocks represent a high grade of metamorphism.

Sometimes a metamorphic rock is seen with mostly amphibole minerals that can define a foliation pattern as well, and is called an amphibolite (Figure 11.4). Occasionally amphibolites also have layers of light-colored plagioclase minerals present; in that case, the rock may be called an amphibole gneiss. Amphibolites may not always have a foliation, however the rock picture in Figure 11.4 appears to have a parallel alignment of the amphiboles (see red arrows in figure), making this a foliated amphibolite. The protolith for an amphibolite must be a rock type with a large amount of dark ferromagnesian minerals, such as a mafic or ultramafic igneous rock. Amphibolites require higher temperatures and pressures to form as well, and are considered to be high grade metamorphic rocks.

Another type of foliation is defined by the presence of flat or platy minerals, such as muscovite or biotite micas. Metamorphic rocks with a foliation pattern defined by the layering of platy minerals are called schist; the rock name is commonly modified to indicate what mica is present. For example, Figure 11.5 is a photo of a muscovite schist, however it also has garnet present, so the correct name for the rock pictured in Figure 11.5 is a garnet muscovite schist. By convention, when naming a metamorphic rock the mineral in the lowest quantity (garnet, in this case) is mentioned first. Notice that the muscovite micas

Figure 11.4 | Example of a foliated amphibolite. Notice the alignment of the amphibole minerals, best seen where the red arrows are indicating. The lighter colored minerals in this rock are plagioclase feldspars.
Author: Karen Tefend
Source: Original Work
License: CC BY-SA 3.0

Figure 11.5 | Photo of a garnet muscovite schist. A) The muscovite micas are large enough to be seen as very shiny minerals in the top photograph. B) Side view of the schist shows the distinctive wavy foliation made by the platy micas. This metamorphic rock also contains the index mineral, garnet.
Author: Karen Tefend
Source: Original Work
License: CC BY-SA 3.0

define a very wavy foliation in the rock; this textural pattern of wavy micas is called a **schistose foliation** (Figure 11.5B). The sedimentary rock shale is usually the protolith for schist; during metamorphism, the very tiny clay minerals in shale recrystallize into micas that are large enough to see unaided. Temperatures and pressures necessary for schistose foliation are not as high as those for gneiss and amphibolite; therefore schists represent an intermediate grade of metamorphism.

There is one foliation type that is defined by the alignment of minerals that are too small to see, yet the foliation can still be visible. This type of foliation is only seen in the metamorphic rock called slate; slate forms by the low temperature and low pressure alteration of a shale protolith. The clay sized minerals in the shale recrystallize into very tiny micas which are larger than the clay minerals, but still too small to be visible. However, because these tiny micas are aligned, they control how the metamorphic rock (slate) breaks, and the rock tends to break parallel to the mica alignment. Therefore, even though we cannot see the aligned minerals that define the foliation, we can use the alignment of the rock fracture pattern, as the rock is cleaved or split. For this reason, the foliation is called a **slaty cleavage**, and a rock displaying this type of foliation is called a slate. Figure 11.6 is an example of the foliated slate displaying slaty cleavage; notice that this rock has retained its original sedimentary layering (depositional beds), which in this case is quite different from the foliation direction. The only protolith for slate is shale, and the fact that original sedimentary features and even some fossils in shale may be preserved and visible in slate is due to the low temperatures and pressures that barely alter the shale protolith, making slate an example of a low grade metamorphic rock. Slate has great economic value in the construction industry; due to its ability to break into thin layers and impermeability to water, slate is used as roofing tiles and flooring.

Figure 11.6 | Slaty cleavage in the metamorphic rock, slate. The alignment of very tiny micas controls how this rock breaks or cleaves. Note the banded appearance on the top of this rock, which is the depositional layering of the original rock, shale.
Author: Karen Tefend
Source: Original Work
License: CC BY-SA 3.0

11.3.2 Non-Foliated Metamorphic Rocks

The other class of metamorphic rocks is non-foliated; the lack of foliation may be due to a lack of differential pressure involved in the metamorphic process, however this is not necessarily the case for all non-foliated metamorphic rocks. If the protolith rock is monominerallic (composed of one mineral type), such as limestone, dolostone, or a sandstone with only quartz sands, then a foliation will not develop even with differential pressure. Why? The calcite mineral in limestone, the dolomite mineral in dolostone, and the quartz sands in sandstone are neither platy minerals, nor are there different colored minerals in these rocks. These minerals (calcite, dolomite, and quartz) recrystallize into equigranular, coarse crystals (see Figure 11.2B), and the metamorphic rocks that they make are named by their composition, not by foliation type. For example, Figure 11.7 is quartzite, a metamorphosed quartz-rich sandstone. Figure 11.8 shows two examples of marble; note that color can vary for marble, as well as for the quartzite. As a result, quartzites and marbles may be hard to identify based on appearance, therefore you must rely on the properties of the minerals that comprise these rocks; you may recall that quartz is harder than glass, while limestone and dolomite are softer than glass. Also, marble will react (effervesce) to acid, but quartz will not react. If you zoom in for a close view of the marble in Figure 11.8, you will see the calcite crystals are fairly large compared to the quartz crystals in the quartzite in Figure 11.7; this can be attributed to the temperature of metamorphism, as higher temperatures result in larger crystals. These rocks are also of economic importance; marble and quartzite are used for dimension stone in buildings and for countertops in many homes. Furthermore, marble is commonly used for statues and sometimes grave markers.

Figure 11.7 | An example of quartzite; a non-foliated metamorphic rock. Quartzite can appear in a variety of colors, but most are fairly light in color.
Author: Karen Tefend
Source: Original Work
License: CC BY-SA 3.0

Figure 11.8 | Examples of the non-foliated rock, marble. The pink and white colors are common, however some marbles are darker in color, and others may be white with dark gray markings (like the famous Carrara marble from Italy).
Author: Karen Tefend
Source: Original Work
License: CC BY-SA 3.0

One final non-foliated rock type that should be mentioned is anthracite coal (Figure 11.9). As you may recall, coal is a sedimentary rock composed of fossilized plant remains. This sedimentary coal is called bituminous coal; under higher temperatures and pressures bituminous coal can lose more of the volatiles typical of coal (water vapor, for example), but the carbon content is enriched, making metamorphic coal (anthracite coal) a hotter burning coal due to the higher carbon content. Anthracite coal can be distinguished from sedimentary coal by the shinier appearance, and is somewhat harder than bituminous coal, although both coal types are of low density due to their carbon content. Note that this particular metamorphism is not a recrystallization event, per se, as coal is mostly organic remains. However, bonds are still being broken and reforming due to changes in temperature and pressure.

In order to identify and name metamorphic rocks, a logical first step would be to examine the rock for evidence of any pattern or foliation, and if present, identify what mineral or minerals are making the foliation pattern. Non-foliated metamorphic rocks can be identified by the properties defined by their mineral composition. Below is a table summarizing the metamorphic rock types, foliation names, and protolith rock types (Table 11.1).

Figure 11.9 | Anthracite coal. Note the shiny appearance of this metamorphosed coal.
Author: Karen Tefend
Source: Original Work
License: CC BY-SA 3.0

Texture	Characteristics	Protolith	Metamorphic Rock Name
Foliated (Increasing Temperature & Pressure)	Fine-grained rock, tends to split in parallel fragments (known as slaty cleavage)	Shale	**Slate**
Foliated	Contains shiny muscovite (light) or biotite (dark) micas, may see other minerals. Has schistose pattern of foliation	Shale	**Schist**
Foliated	Contains alternating bands of light- and dark-colored minerals (usually biotite or amphibole), called gneissic banding.	Shale or Igneous Rock	**Gneiss**
Possibly Foliated	Most of the minerals in this rock are amphiboles, which may be aligned to form a foliation.	Mafic or Ultramafic Rock	**Amphibolite**
Non-foliated	Equigranular grains of quartz, which has a hardness of 7	Sandstone or Siltstone	**Quartzite**
Non-foliated	Equigranular grains of calcite or dolomite, which has a hardness of 4, and reacts to acid	Limestone or Dolostone	**Marble**
Non-foliated	Contains mostly carbon, is a low density rock, and has a shiny appearance.	Bituminous Coal	**Anthracite Coal**

Table 11.1 | Summary chart of metamorphic rocks discussed in this chapter, including the names of some of the possible sedimentary rock and igneous rock protoliths for each metamorphic rock. A metamorphic rock can also be a protolith: for example, a slate can be a protolith for a schist.
Author: Karen Tefend
Source: Original Work
License: CC BY-SA 3.0

11.4 LAB EXERCISE

Part A - Metamorphic Rock Identification

In order to answer the questions in this lab, you will need the rock samples from your HOL rock kit that are labeled M1 through M7 (located in the metamorphic rock bag). A photo of these samples is given in Figure 11.10. You will also need your hand lens, glass plate, and the HCl bottle.

Figure 11.10 | Metamorphic rocks M1 through M7.
Author: Karen Tefend
Source: Original Work
License: CC BY-SA 3.0

1. Sample M1 has the following texture:

   a. slaty cleavage
   b. schistose foliation
   c. gneissic banding
   d. lineation
   e. non-foliated

2. Sample M1 is called:

   a. marble b. quartzite c. schist d. gneiss
   e. slate f. anthracite coal g. amphibolite

3. The dark minerals in Sample M1 are:

   a. flat biotite micas
   b. tabular amphiboles
   c. flat muscovite micas
   d. tabular feldspars

4. A possible protolith for Sample M1 is:

   a. an ultramafic rock			b. basalt

   c. limestone				d. granite

   e. sandstone

5. Sample M2 has the following texture:

   a. slaty cleavage			b. schistose foliation

   c. gneissic banding			d. lineation

   e. non-foliated

6. Sample M2 is called:

   a. marble	b. quartzite	c. schist	d. gneiss

   e. slate	f. anthracite coal	g. amphibolite

7. Sample M3 has the following texture:

   a. slaty cleavage			b. schistose foliation

   c. gneissic banding			d. lineation

   e. non-foliated

8. Sample M3 is called:

   a. marble	b. quartzite	c. schist	d. gneiss

   e. slate	f. anthracite coal	g. amphibolite

9. A possible protolith for Sample M3 is:

   a. bituminous coal		b. basalt		c. shale

   d. clay			e. rhyolite		f. sandstone

10. Sample M3 is an example of:

    a. high grade of metamorphism

    b. intermediate grade of metamorphism

    c. low grade of metamorphism

11. Sample M3 is used for:

    a. road construction   b. statues   c. roof tiles   d. barbequing

12. Sample M4 has the following texture:

    a. slaty cleavage   b. schistose foliation

    c. gneissic banding   d. lineation

    e. non-foliated

13. Sample M4 is called:

    a. marble   b. quartzite   c. schist   d. gneiss

    e. slate   f. anthracite   g. amphibolite

14. Sample M4 is mainly composed of this mineral:

    a. biotite mica   b. amphibole   c. muscovite mica

    d. calcite   e. quartz

15. Sample M4 is an example of:

    a. high grade of metamorphism

    b. intermediate grade of metamorphism

    c. low grade of metamorphism

16. Sample M5 has the following texture:

    a. slaty cleavage   b. schistose foliation

    c. gneissic banding   d. lineation

    e. non-foliated

17. Sample M5 is called:

   a. marble b. quartzite c. schist d. gneiss

   e. slate f. anthracite g. amphibolite

18. Sample M5 is mainly composed of this mineral:

   a. biotite mica b. amphibole c. muscovite mica

   d. calcite e. quartz

19. A possible protolith for Sample M5 is:

   a. diorite b. quartz c. limestone d. shale e. sandstone

20. Sample M6 is mainly composed of this dark mineral:

   a. biotite mica b. amphibole c. muscovite mica

   d. calcite e. quartz

21. Sample M6 is called:

   a. marble b. quartzite c. schist d. gneiss

   e. slate f. anthracite g. amphibolite

22. A possible protolith for Sample M6 is:

   a. basalt b. granite c. sandstone

   d. limestone e. shale

23. Sample M6 is an example of:

   a. high grade of metamorphism

   b. intermediate grade of metamorphism

   c. low grade of metamorphism

24. Sample M7 has the following texture:

   a. slaty cleavage   b. schistose foliation   c. gneissic banding

   d. lineation   e. non-foliated

25. Sample M7 is called:

   a. marble   b. quartzite   c. schist   d. gneiss

   e. slate   f. anthracite   g. amphibolite

26. A possible protolith for Sample M7 is:

   a. basalt   b. granite   c. sandstone   d. limestone   e. shale

11.5 TYPES OF METAMORPHISM

11.5.1 Regional Metamorphism

Metamorphism that affects entire rock bodies over a broad region is referred to as **regional metamorphism**. There is a wide range of conditions in temperature and pressure that produce a wide range of metamorphic rock types; for example, all of the foliated rocks fall into this metamorphic category and some non-foliated rocks as well. The main point is that a large area is affected by changes in temperature and/or pressure. One type of regional metamorphism occurs at convergent plate boundaries, where rocks are subjected to a variety of pressures and temperatures, resulting in a **thermodynamic metamorphism** ("thermo-" for increased temperature, and "dynamic" for increased pressure). Mountain ranges that form by the converging of tectonic plates are classic examples of thermodynamic metamorphism, such as the Himalayan Mountains located in Asia, or our Sierra Nevada Mountains in North America. These can be viewed if you type the mountain range name in the search bar in Google Earth. Regional metamorphism also occurs along plate boundaries where an oceanic plate descends (subducts) back into the mantle as a result of plate convergence (this was discussed in the plate tectonics chapter); oceanic plates that subduct into the mantle will form a deep ocean trench, such as the trench along the western margin of South America. With increased pressure in these regions metamorphic foliations can get quite pronounced, as we saw in the development of slate to schist to gneiss; with increased temperature we see increases in grain size as well as the introduction of new minerals in the metamorphosed rock. These new minerals are useful in that they define the temperature of metamorphism for both regional and contact metamorphism (next section).

Figure 11.11 is an example of the possible index minerals that form during the metamorphism of a shale protolith. Shale is a sedimentary rock that contains clay minerals and clay-sized quartz and (sometimes) feldspars. With increased temperature, recrystallization involves the formation of new minerals, starting with chlorite at the lowest metamorphic temperature. High temperature metamorphism results in the formation of the mineral sillimanite. The metamorphic sections in this figure are called mineral zones, named for the appearance of each new mineral. For example, the garnet zone begins when garnet first appears in the rock, and the appearance of a new mineral, staurolite, marks the end of the garnet zone and the beginning of the staurolite zone. The study of such index mineral zones have importance in terms of economic exploration, as some of these index minerals have economic importance; for example garnet is used as an abrasive, and sometimes for decorative purposes as a gemstone.

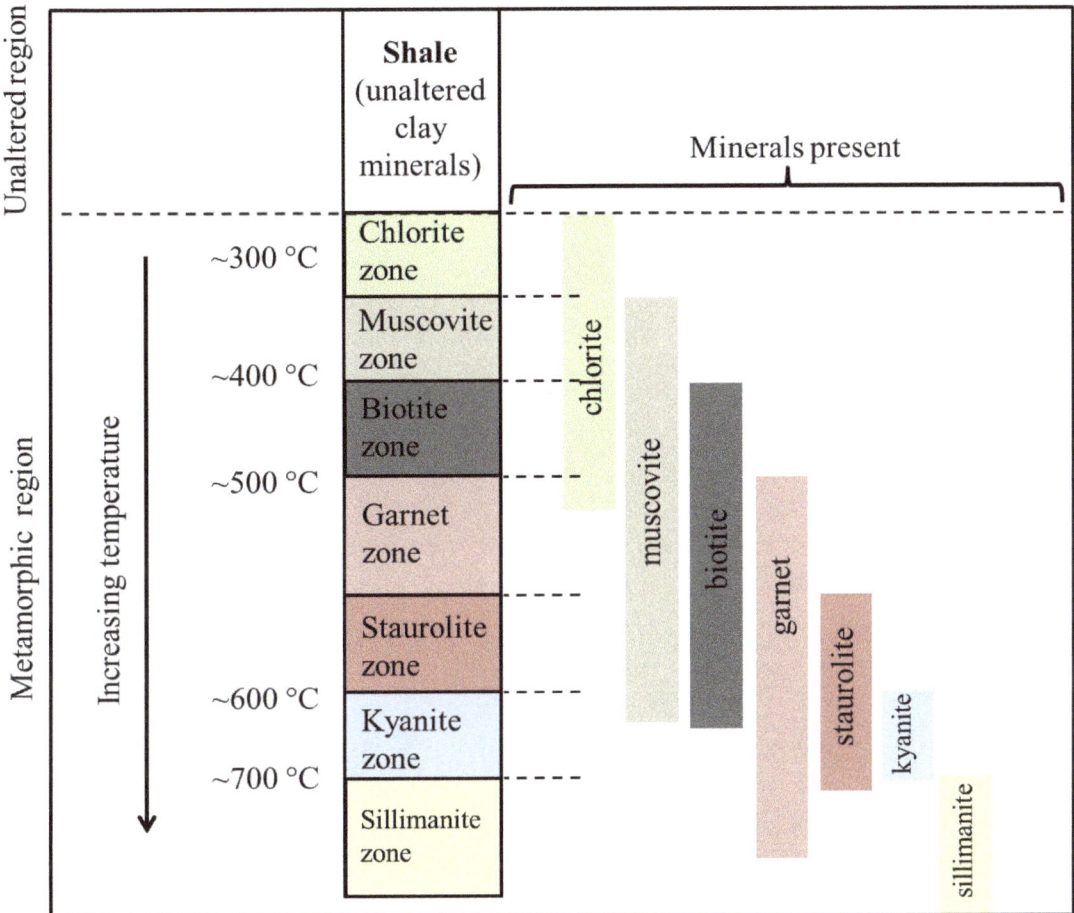

Figure 11.11 | Index mineral chart, showing possible minerals that can form from a shale protolith. The appearance of certain index minerals is also pressure dependent.
Author: Karen Tefend
Source: Original Work
License: CC BY-SA 3.0

11.5.2 Contact Metamorphism

As mentioned previously, recrystallization due to increased temperature results in the formation of larger minerals, or sometimes in the formation of new minerals (Figure 11.11). **Contact metamorphism** describes the type of metamorphism attributed to increased temperature, usually from proximity to a heat source such as an intrusive magma body or a lava flow, and the rock undergoes a **thermal metamorphism**. The rocks that are in closer contact to the magma will form larger crystals due to the higher heat, and may form high temperature index minerals such as garnet, staurolite, kyanite and sillimanite. Rocks that are in contact with hot chemically reactive fluids can also fall within the contact metamorphism category; recrystallization due to this type of contact is called **hydrothermal metamorphism**. Unlike the regional metamorphism described in the previous section, both types of contact metamorphism are confided to smaller areas near the heat source (magma, lava, or hydrothermal fluids). Since differential pressures are not involved, contact metamorphism results in the formation of non-foliated rocks. The non-foliated rock types that were covered previously included marble, quartzite, amphibolite, and anthracite coal. Other non-foliated rocks that have undergone contact metamorphism are broadly classified as hornfels or granofels, to indicate that they are recrystallized rocks that lack foliation; for example, shale can be altered into a hornfels if it is recrystallized, but lacks slaty cleavage due to the lack of any developed foliation.

11.5.3 Shock (Impact) Metamorphism

This type of metamorphism is not as common as regional or contact metamorphism, and it differs from the other types in a spectacular way. Regional and contact metamorphism requires quite a bit of time (up to several hundred thousand years) for rocks to metamorphose; **shock metamorphism** takes place in a matter of seconds. This type of metamorphism occurs at the site of meteorite craters; rocks at the surface and near surface are subjected to extreme stresses upon impact, and sometimes portions of these surface rocks become tiny blobs of melt that cool to form a type of glass called a **tektite**. Age dating of these tektites has helped in the determination and even recognition of ancient meteorite impact sites worldwide. Shock or impact metamorphism is mainly due to the extreme stresses brought on by the impact, and for this reason shock metamorphism is considered to be a **dynamic metamorphism**, as opposed to dynamothermal (regional) metamorphism.

11.5.4 Fault Zone Metamorphism

Rocks along geologic faults are deformed due to pressures associated with shearing, compression, or extensional stresses, with minor changes due to heat. As with shock metamorphism, **fault zone metamorphism** is a type of dynamic metamorphism. Shallow faults (close to the earth's surface) may grind the nearby rocks into smaller, angular fragments called fault breccia, or you may see a fine clay powder called fault gouge; fault gouge forms by the chemical alteration of fault breccia. Fault breccia is a form of brittle deformation of rock; at greater depths, rocks along a fault will undergo a more plastic type of deformation. The rocks that form adjacent to the fault in deeper sections are called **mylonite** (Figure 11.12). The minerals in mylonitic rocks are deformed due to shear stresses associated with movement along the fault.

Figure 11.12 | Example of a mylonite rock; note the deformed light-colored minerals (feldspars) that have developed a tail (red arrow). The direction of shearing can sometimes be determined by examining these minerals.
Author: Randa Harris
Source: Original Work
License: CC BY-SA 3.0

11.6 LAB EXERCISE

Part B - Types of Metamorphism and Metamorphic Regions

For this portion of the lab, you will be using Google Earth, and Figure 11.11 in this chapter. You will also need your rock samples M1 and M7.

27. According to Figure 11.11, what mineral is unstable in the staurolite zone?

    a. chlorite　　　b. muscovite　　　c. biotite　　　d. garnet

28. Refer again to Figure 11.11; which index minerals will not be found together in the same rock?

    a. chlorite and garnet　　　　b. chlorite and kyanite

    c. biotite and staurolite　　　d. biotite and muscovite

INTRODUCTORY GEOLOGY   METAMORPHIC ROCKS

29. Sample M1 from your rock kit is most likely an example of:

   a. hydrothermal metamorphism   b. contact metamorphism

   c. regional metamorphism   d. shock metamorphism

30. Sample M7 from your rock kit is most likely an example of:

   a. fault zone metamorphism   b. contact metamorphism

   c. regional metamorphism   d. shock metamorphism

31. Type 8 56 22.28N 126 55 54.33E in the search bar of Google Earth. Zoom out to an eye altitude of ~1700 miles. What type of metamorphism is likely in this area?

   a. shock metamorphism   b. contact metamorphism

   c. regional metamorphism   d. dynamic metamorphism

32. The agents of metamorphism in this region, are most likely:

   a. chemically reactive fluids   b. temperature

   c. pressure   d. both temperature and pressure

33. Type 35 06 53.42N 119 38 47.59W in the search bar of Google Earth. Zoom out to an eye altitude of ~8500ft. What type of metamorphism is likely in this area?

   a. fault zone metamorphism   b. contact metamorphism

   c. regional metamorphism   d. shock metamorphism

34. Evidence of metamorphism at this location could be from the discovery of:

   a. gneiss   b. quartzite   c. tektites   d. mylonite

35. Type 35 01 38.93N 111 01 21.93W in the search bar of Google Earth. Zoom out to an eye altitude of ~12,000ft. What type of metamorphism is likely in this area?

   a. fault zone metamorphism   b. contact metamorphism

   c. regional metamorphism   d. shock metamorphism

36. Evidence of metamorphism at this location could be from the discovery of:

   a. gneiss   b. quartzite   c. tektites   d. mylonite

11.7 STUDENT RESPONSES

1. Sample M1 has the following texture:

    a. slaty cleavage   b. schistose foliation   c. gneissic banding

    d. lineation   e. non-foliated

2. Sample M1 is called:

    a. marble   b. quartzite   c. schist   d. gneiss

    e. slate   f. anthracite   g. amphibolite

3. The dark minerals in Sample M1 are:

    a. flat biotite micas   b. tabular amphiboles

    c. flat muscovite micas   d. tabular feldspars

4. A possible protolith for Sample M1 is:

    a. an ultramafic rock   b. basalt   c. limestone

    d. granite   e. sandstone

5. Sample M2 has the following texture:

    a. slaty cleavage   b. schistose foliation   c. gneissic banding

    d. lineation   e. non-foliated

6. Sample M2 is called:

    a. marble   b. quartzite   c. schist   d. gneiss

    e. slate   f. anthracite coal   g. amphibolite

7. Sample M3 has the following texture:

    a. slaty cleavage   b. schistose foliation   c. gneissic banding

    d. lineation   e. non-foliated

8. Sample M3 is called:

    a. marble   b. quartzite   c. schist   d. gneiss

    e. slate   f. anthracite   g. amphibolite

9. A possible protolith for Sample M3 is:

    a. bituminous coal   b. basalt   c. shale

    d. clay   e. rhyolite   f. sandstone

10. Sample M3 is an example of:

    a. high grade of metamorphism

    b. intermediate grade of metamorphism

    c. low grade of metamorphism

11. Sample M3 is used for:

    a. road construction   b. statues   c. roof tiles   d. barbequing

12. Sample M4 has the following texture:

    a. slaty cleavage   b. schistose foliation

    c. gneissic banding   d. lineation

    e. non-foliated

13. Sample M4 is called:

    a. marble   b. quartzite   c. schist   d. gneiss

    e. slate   f. anthracite   g. amphibolite

14. Sample M4 is mainly composed of this mineral:

    a. biotite mica   b. amphibole   c. muscovite mica

    d. calcite   e. quartz

15. Sample M4 is an example of:

    a. high grade of metamorphism

    b. intermediate grade of metamorphism

    c. low grade of metamorphism

16. Sample M5 has the following texture:

    a. slaty cleavage b. schistose foliation

    c. gneissic banding d. lineation

    e. non-foliated

17. Sample M5 is called:

    a. marble   b. quartzite   c. schist   d. gneiss

    e. slate   f. anthracite g. amphibolite

18. Sample M5 is mainly composed of this mineral:

    a. biotite mica   b. amphibole   c. muscovite mica

    d. calcite   e. quartz

19. A possible protolith for Sample M5 is:

    a. diorite   b. quartz   c. limestone   d. shale   e. sandstone

20. Sample M6 is mainly composed of this dark mineral:

    a. biotite mica   b. amphibole   c. muscovite mica

    d. calcite   e. quartz

21. Sample M6 is called:

    a. marble   b. quartzite   c. schist   d. gneiss

    e. slate   f. anthracite g. amphibolite

22. A possible protolith for Sample M6 is:

    a. basalt   b. granite   c. sandstone   d. limestone   e. shale

23. Sample M6 is an example of:

    a. high grade of metamorphism

    b. intermediate grade of metamorphism

    c. low grade of metamorphism

24. Sample M7 has the following texture:

    a. slaty cleavage   b. schistose foliation

    c. gneissic banding   d. lineation

    e. non-foliated

25. Sample M7 is called:

    a. marble   b. quartzite   c. schist   d. gneiss

    e. slate   f. anthracite   g. amphibolite

26. A possible protolith for Sample M7 is:

    a. basalt   b. granite   c. sandstone   d. limestone   e. shale

27. According to Figure 11.11, what mineral is unstable in the staurolite zone?

    a. chlorite   b. muscovite   c. biotite   d. garnet

28. Refer again to Figure 11.11; which index minerals will not be found together in the same rock?

    a. chlorite and garnet   b. chlorite and kyanite

    c. biotite and staurolite   d. biotite and muscovite

29. Sample M1 from your rock kit is most likely an example of:

    a. hydrothermal metamorphism   b. contact metamorphism

    c. regional metamorphism   d. shock metamorphism

30. Sample M7 from your rock kit is most likely an example of:

    a. fault zone metamorphism   b. contact metamorphism

    c. regional metamorphism   d. shock metamorphism

31. Type 8 56 22.28N 126 55 54.33E in the search bar of Google Earth. Zoom out to an eye altitude of ~1700 miles. What type of metamorphism is likely in this area?

    a. shock metamorphism   b. contact metamorphism

    c. regional metamorphism   d. dynamic metamorphism

32. The agents of metamorphism in this region, are most likely:

    a. chemically reactive fluids   b. temperature

    c. pressure   d. both temperature and pressure

33. Type 35 06 53.42N 119 38 47.59W in the search bar of Google Earth. Zoom out to an eye altitude of ~8500ft. What type of metamorphism is likely in this area?

    a. fault zone metamorphism   b. contact metamorphism

    c. regional metamorphism   d. shock metamorphism

34. Evidence of metamorphism at this location could be from the discovery of:

    a. gneiss   b. quartzite   c. tektites   d. mylonite

35. Type 35 01 38.93N 111 01 21.93W in the search bar of Google Earth. Zoom out to an eye altitude of ~12,000ft. What type of metamorphism is likely in this area?

    a. fault zone metamorphism   b. contact metamorphism

    c. regional metamorphism   d. shock metamorphism

36. Evidence of metamorphism at this location could be from the discovery of:

    a. gneiss   b. quartzite   c. tektites   d. mylonite

12 Crustal Deformation
Randa Harris and Bradley Deline

12.1 INTRODUCTION

The Earth is an active planet shaped by dynamic forces. Such forces can build mountains and crumple and fold rocks. As rocks respond to these forces, they undergo deformation, which results in changes in shape and/or volume of the rocks. The resulting features are termed geologic structures. This deformation can produce dramatic and beautiful scenery, as evidenced in Figure 12.1, which shows the deformation of originally flat (horizontal) rock layers.

Why is it important to study deformation within the crust? Such studies can provide us with a record of the past and the forces that operated then. The correct interpretation of features created during deformation is critical in the petroleum and mining industry. It is also essential for engineering. Understanding the behavior of deformed rocks is necessary to create and maintain safe engineering struc-

Figure 12.1 | Rocks that have been deformed along the coast of Italy
Author: Randa Harris
Source: Original Work
License: CC BY-SA 3.0

tures. When proper geological planning is not considered in engineering, disasters can strike. For example, the Vajont Dam was constructed at Monte Toc, Italy in the early 1960's. The place was a poor choice for a dam, as the valley was narrow, thorough geological tests were not performed, and the area surrounding the dam was prone to large landslides. The steep canyon walls were composed of limestone with solution cavities, not known for its stability, and shifting and fracturing of rock that occurred during the filling of the reservoir went unheeded. In 1963, a massive landslide in the area displaced much of the water in the dam, causing it to override the top of the dam and flood the many villages downstream, resulting in the deaths of almost 2,000 people (Figure 12.2).

Figure 12.2 | An image of the Vajont reservoir shortly after the massive landslide (you can see the scar from the landslide on the right, and the dam is located in the foreground on the left).
Author: Unknown
Source: Wikimedia Commons
License: Public Domain

12.1.1 Learning Outcomes

After completing this chapter, you should be able to:
- Understand the different types of stress that rocks undergo, and their responses to stress
- Demonstrate an understanding of the concepts of strike and dip
- Recognize the different types of folds and faults, and the forces that create them
- Use block diagrams to display geologic features
- Create a geologic cross-section

12.1.2 Key Terms

- Anticline
- Basin
- Compressional forces
- Contact
- Dip
- Dip-slip fault
- Dome
- Geologic cross-section
- Geologic map
- Horst & graben
- Monocline

- Normal fault
- Reverse fault
- Shear forces
- Strain
- Stress
- Strike
- Strike-slip fault
- Syncline
- Tensional forces
- Thrust fault

12.2 STRESS AND STRAIN

Rocks change as they undergo **stress**, which is just a force applied to a given area. Since stress is a function of area, changing the area to which stress is applied makes a difference. For example, imagine the stress that is created both at the tip of high heeled shoes and the bottom of athletic shoes. In the high heeled shoe, the area is very small, so that stress is concentrated at that point, while the stress is more spread out in an athletic shoe. Rocks are better able to handle stress that is not concentrated in one point. There are three main types of stress: compression, tension, and shear. When **compressional forces** are at work, rocks are pushed together. **Tensional forces** operate when rocks pull away from each other. **Shear forces** are created when rocks move horizontally past each other in opposite directions. Rocks can withstand compressional stress more than tensional stress (see Figure 12.3).

Figure 12.3 | This is a picture of the Roman Forum. Why did the Romans use so many vertical columns to hold up the one horizontal beam? If the horizontal beam spanned a long distance without support, it would buckle under its own weight. This beam is under tensional stress, so it is not as strong.
Author: Randa Harris
Source: Original Work
License: CC BY-SA 3.0

Applying stress creates a deformation of the rock, also known as strain. As rocks are subjected to increased stress and strain, they at first behave in an elastic manner, which means they return to their original shape after deformation (Figure 12.4). This elastic behavior continues until rocks reach their elastic limit (point X on Figure 12.4), at which point plastic deformation commences. The rocks may bend into folds, or behave in a brittle manner by fracturing (brittle behavior can be easily envisioned if you think of a hammer hitting glass), but regardless they do not return to their original shape when the stress is removed in plastic deformation. The resulting deformation from applied stress

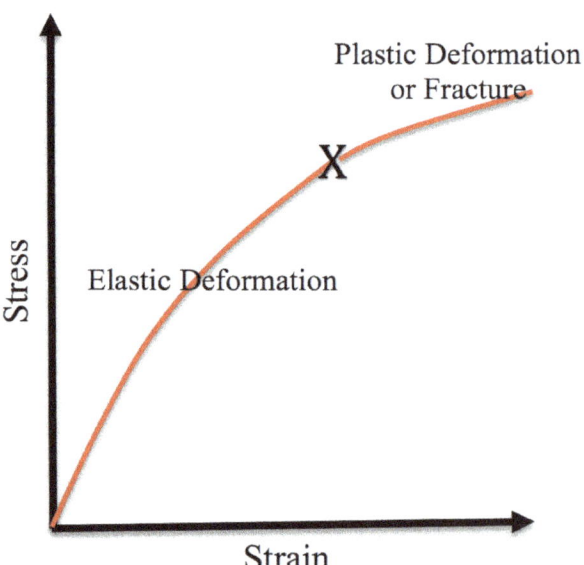

Figure 12.4 | A stress and strain diagram. As stress and strain increase, rocks first experience elastic deformation (and can return to their original shape) until the elastic limit is reached, depicted at point X. After this point, rocks will fracture or experience plastic deformation, so that their original shape is destroyed
Author: Randa Harris
Source: Original Work
License: CC BY-SA 3.0

depends on many factors, including the type of stress, the type of rock, the depth of the rock and pressure and temperature conditions, and the length of time the rock endures the stress. Rocks behave very differently at depth than at the surface. Rocks tend to deform in a more plastic manner at depth, and in a more brittle manner near the Earth's surface.

12.3 STRIKE AND DIP

To learn many of the concepts associated with structural geology, it is useful to use block diagrams. As you examine these blocks, note the different ways that you can view them. If you look at a block from along the side, you are seeing the cross-section view (like what you see along roads that have been cut through the mountains). If you look at the block from directly above it, you are looking at map view (Figure 12.5).

As you think about how rocks have changed through the process of deformation, it is useful to remember how they deposited in the first place. We will briefly review some of the geologic laws that you learned in the Introduction Lab. Sedimentary rocks, under the influence of gravity, will deposit in horizontal layers (Law of Original Horizontality). The oldest rocks will be on the bottom (because they had to be there first for the others to deposit on top of them), and are numbered with the oldest being #1 (Law of Superposition). The wooden block in Figure 12.6 displays how

Figure 12.5 | You are viewing the top block in this image in map view, viewed from directly above the block. The lower block is from the same rock layers, and you are viewing it in cross-section (or from the side). Note that in cross-section, you can see how the rock layers are tilted.
Author: Randa Harris
Source: Original Work
License: CC BY-SA 3.0

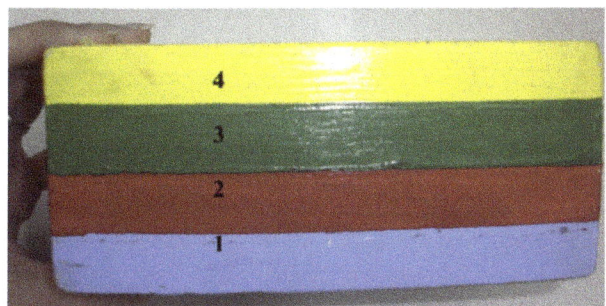

Figure 12.6 | In this image, different rock types are given different colors. The oldest rock, on the bottom, is given a #1. The youngest rock in this image is #4.
Author: Randa Harris
Source: Original Work
License: CC BY-SA 3.0

Figure 12.7 | In this image, beds have been tilted. Which color bed is the oldest? To determine this, it is useful to apply Occam's Razor, which states that the simplest explanation is most likely the best. It is more likely that the gray bed on the left was the bottom bed during deposition, and therefore the oldest.
Author: Randa Harris
Source: Original Work
License: CC BY-SA 3.0

sedimentary rocks originally deposit. Each of the boundaries between the colored rock units represents a geologic **contact**, which is simply the surface between two different rock units. Earth's rock layers are usually not this uncomplicated. Rock layers are often at an angle, not horizontal, indicating that changes have occurred since deposition. Examine Figure 12.7 to see tilted rocks.

In order to measure and describe layers like this, geologists apply the concepts of strike and dip. **Strike** refers to the line formed by the intersection of a horizontal plane and an inclined surface. **Dip** is the angle between that horizontal plane (such as the top of this block) and the tilted surface (the geologic contact between the tilted layers). In Figure 12.8, look at the tilted sedimentary layers. Strike is a line on the horizontal plane created when the dipping green layer intersects the Earth's surface. To better display the horizontal surface, it has been represented by water. The dip angle is measured from the horizontal surface to the dipping bed (outlined in orange on this image).

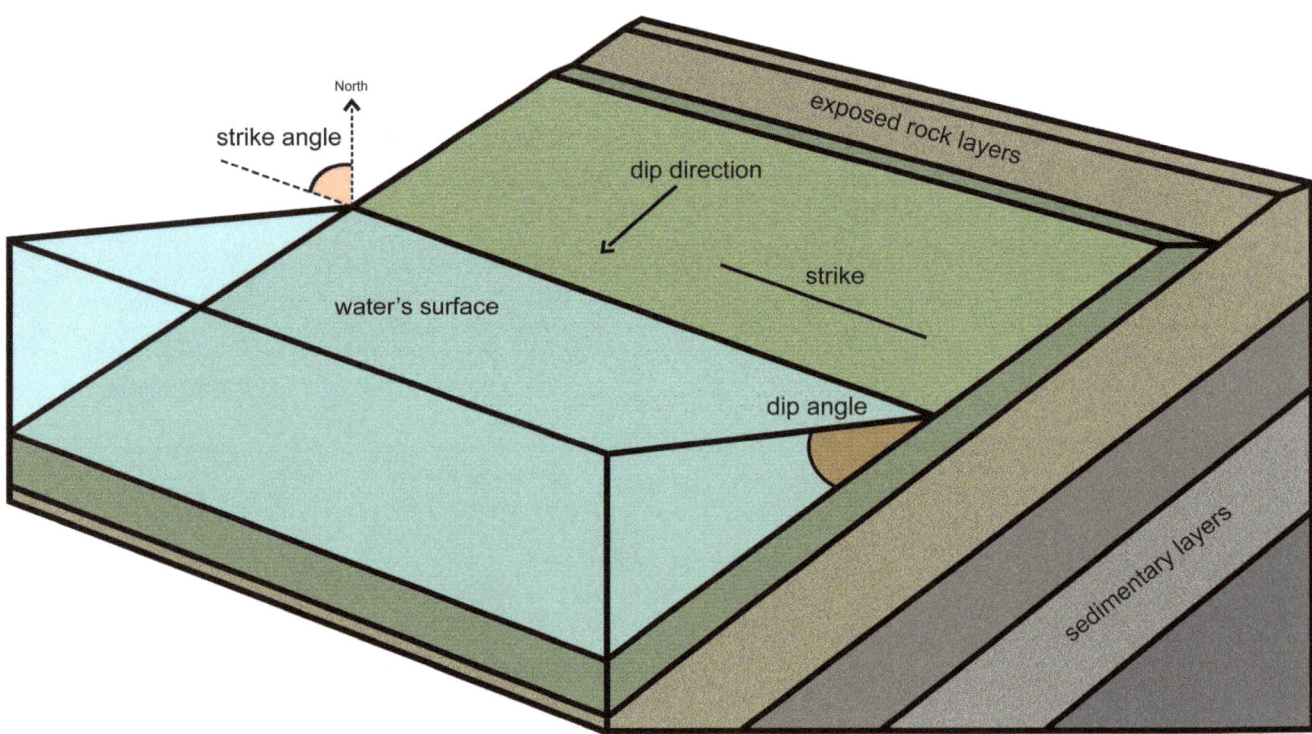

Figure 12.8 | A demonstration of strike and dip. In this example, the beds are dipping to the southwest.
Author: Corey Parson
Source: Original Work
License: CC BY-SA 3.0

Now, let's apply this concept to the block of dipping beds that you just looked at (Figure 12.7). To determine strike, find where the dipping layer intersects the horizontal surface and draw a line parallel to this line of intersection on the top of the block (i.e. our horizontal surface). To determine dip, pretend that there is a drop of water between one bed and the next, for example, along the intersection of the pale blue bed and the red bed. In which direction would the water roll if it followed that contact? That is the direction of dip—towards the right in this case. The symbol for strike and dip is given along the top of the block. Note that the dip symbol (shorter line) is perpendicular to the strike symbol (Figure 12.9).

Useful in map interpretation is the application of the Rule of V's to determine dip direction. As a stream crosses tilted strata, it will cut a V shape into the rock layers. The point of the V is in the direction of dip. In vertical strata, no V shapes are created (Figure 12.10).

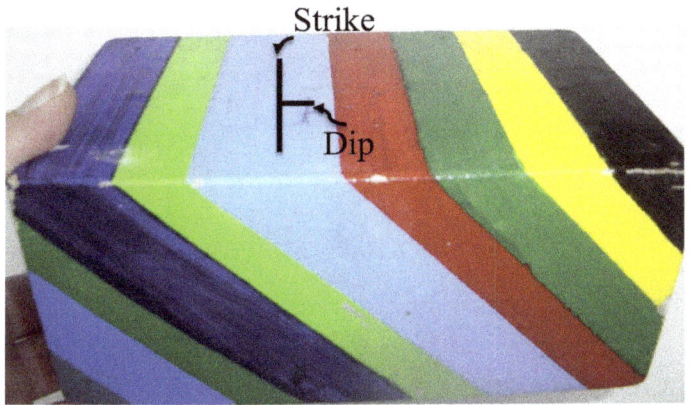

Figure 12.9 | This image depicts the strike and dip for the pale blue bed. As all the beds are oriented in the same direction, they would all have the same strike and dip.
Author: Randa Harris
Source: Original Work
License: CC BY-SA 3.0

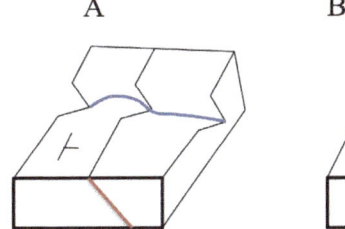

Figure 12.10 | In image A, the stream is encountering tilted strata, which results in a V shape that points towards the dip direction (in this case, to the right). In image B, the strata are vertical, so the stream cannot create a V shape. For reference on common geologic map symbols, refer to Figure 12.31 at the end of this lesson.
Author: Randa Harris
Source: Original Work
License: CC BY-SA 3.0

12.4 LAB EXERCISE

Part A (6 points)

1. For the following diagram, determine the correct map symbol that would go in the oval box.

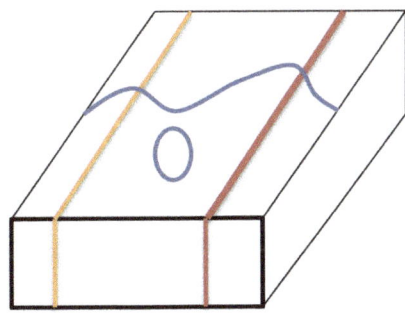

a. ⊢   b. ⊣   c. +   d. ⊕

2. For the following diagram, determine the correct map symbol that would go in the oval box.

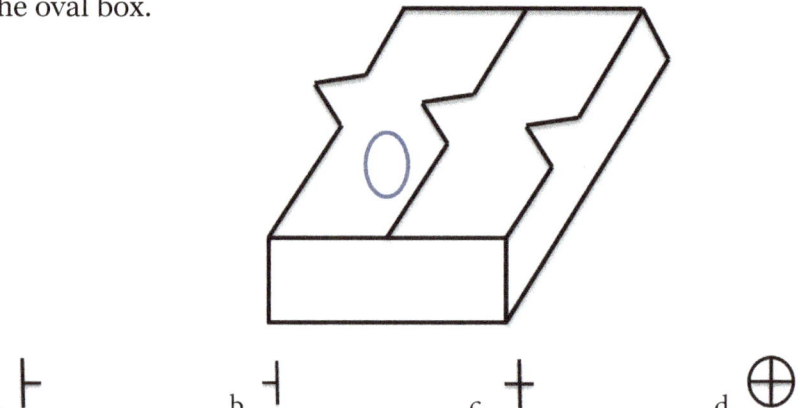

a. ⊢   b. ⊣   c. ┼   d. ⊕

3. Using Google Earth, search for the following area in Pakistan: **27 50 35.00N 67 10 03.70E**. Zoom out to an eye altitude of approximately 25,000 feet. The inclined layers in these folded rocks can be easily seen. As you view them, in which compass direction are the beds dipping?

   a. Northeast (NE)   b. Southeast (SE)   c. Northwest (NW)   d. Southwest (SW)

12.5 GEOLOGIC STRUCTURES CREATED BY PLASTIC AND BRITTLE DEFORMATION

12.5.1 Folds

Folds are geologic structures created by plastic deformation of the Earth's crust. To understand how folds are generated, take a piece of paper and hold it up with a hand on each end. Apply compressional forces (push the ends towards each other). You have just created a fold (bent rock layers). Depending on how your paper moved, you created one of the three main fold types (Figure 12.11).

Figure 12.11 | The three main fold types, from left to right, are monocline, anticline, and syncline (the anticline and syncline are both displayed in one block).
Author: Randa Harris
Source: Original Work
License: CC BY-SA 3.0

A **monocline** is a simple fold structure that consists of a bend in otherwise horizontal rock layers. More commonly found are anticlines and synclines. An **anticline** fold is convex up and one in which the layered strata are inclined down and away from the center of the fold (if you drew a line across it, the anticline would resemble a capital letter A). A **syncline** is a concave upward fold in which the layered strata are inclined up (it resembles a smile). Parts of a fold include the axis (the hinge line), the axial plane, and limbs on either side of the axis (Figure 12.12). It is important to note that anticlines do not always represent mountains or high areas and synclines do not always represent basins or low areas. They are simply folded rock layers, and do not necessarily indicate topographic high and low points (see Figure 12.13).

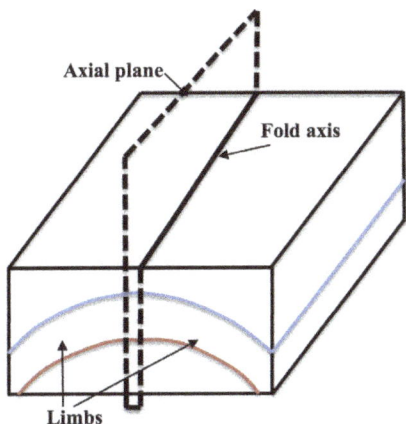

Figure 12.12 | The axial plane and fold axis, along the center of the fold, and corresponding limbs on either side.
Author: Randa Harris
Source: Original Work
License: CC BY-SA 3.0

Figure 12.13 | This topographic high, along Interstate 68 and US 40 in Maryland, is a syncline.
Author: User "Acroterion"
Source: Original Work
License: CC BY-SA 3.0

Folds observed in cross-section look much different from map view. In map view, rather than seeing folds, you will only encounter beds that look like Figure 12.14. To help determine what type of fold you have, it is useful to determine the strike and dip of the beds you encounter. On Figure 12.14, determine the strike and dip for each location marked by an oval. Check yourself against Figure 12.15.

Once rocks are folded and exposed at the Earth's surface, they are subjected to erosion, creating regular patterns. The erosion exposes the interiors of the folds, such that parallel bands of dipping strata can be observed along the fold axis. In an anticline, the oldest rocks are exposed along the fold axis, while it is the youngest rocks exposed at the fold axis in a syncline (Figure 12.16).

Figure 12.14 | A block diagram of an anticline and syncline. Looking from above you see map view, and from the side you observe cross-section view. Determine the strike and dip symbols that should go in the ovals.
Author: Randa Harris
Source: Original Work
License: CC BY-SA 3.0

Figure 12.15 | In this block, the strike and dip symbols, along with the symbols for anticline and syncline, have been drawn in for you. Note that on the anticline, the beds dip AWAY from the axis, and the anticline symbol is drawn along the axis, with arrows pointing away from each other. In the syncline, the beds dip TOWARDS each other, with the syncline symbol having arrows that point inwards.
Author: Randa Harris
Source: Original Work
License: CC BY-SA 3.0

Figure 12.16 | In the top block (A), you view a typical anticline and syncline. Look at the center of the folds. Are the beds that you see in the center of the anticline older or younger than the beds on either side of it? What about for the syncline? In the lower block (B), you see the top portion of the block removed, done so to simulate erosion of rock layers. Note the pale blue bed in the center of the anticline. It is older than the red bed on either side of it (since it is lower in side view—remember older beds are on the bottom). In the syncline, the black bed in the middle is younger than the yellow beds on either side of it.
Author: Randa Harris
Source: Original Work
License: CC BY-SA 3.0

Table 12.1

Fold Type	Direction of dip of layers	Age of beds at axis
Anticline	Away from axis	Oldest
Syncline	Towards the axis	Youngest

Imagine you take that same sheet of paper that you originally created a fold with. Create a fold again, but rather than holding it horizontally, plunge one end of your paper down into your desk surface. You have now created a plunging fold, which is essentially a tilted fold that creates a V shaped pattern on the surface (Figures 12.17, 12.18, and 12.19). In an anticline, the oldest strata can be found at the center of the V, and the V points in the direction of the plunge of the fold axis. In a syncline, the youngest strata are found at the center of the V, and the V points in the opposite direction of the plunge of the fold axis.

Similar to anticlines and synclines are domes and basins, which are basically the circular (or elliptical) equivalent of those folds. A **dome** is an upfold similar to an anticline, and a **basin** is an area where

Figure 12.17 | Both blocks are plunging folds. Note the V shape created in map view in a plunging fold. The fold on the left block is a plunging anticline (observe the end of the block to determine this) and the fold on the right block is a plunging syncline.
Author: Randa Harris
Source: Original Work
License: CC BY-SA 3.0

Figure 12.18 | Note this side view of a plunging anticline, to better see how the rock layers plunge into the Earth.
Author: Randa Harris
Source: Original Work
License: CC BY-SA 3.0

Figure 12.19 | These blocks depict map view for both plunging and non-plunging folds. The top block is a plunging fold, with the characteristic V-shape. The lower block shows map view of non-plunging folds.
Author: Randa Harris
Source: Original Work
License: CC BY-SA 3.0

the rocks are inclined downward towards the center, similar to a syncline (Figure 12.20). The key to identifying these structures is similar to identifying the folds. In a dome, the oldest rocks are exposed at the center, and rocks dip away from this central point. In a basin, the youngest rocks are in the center, and the rocks dip inward towards the center.

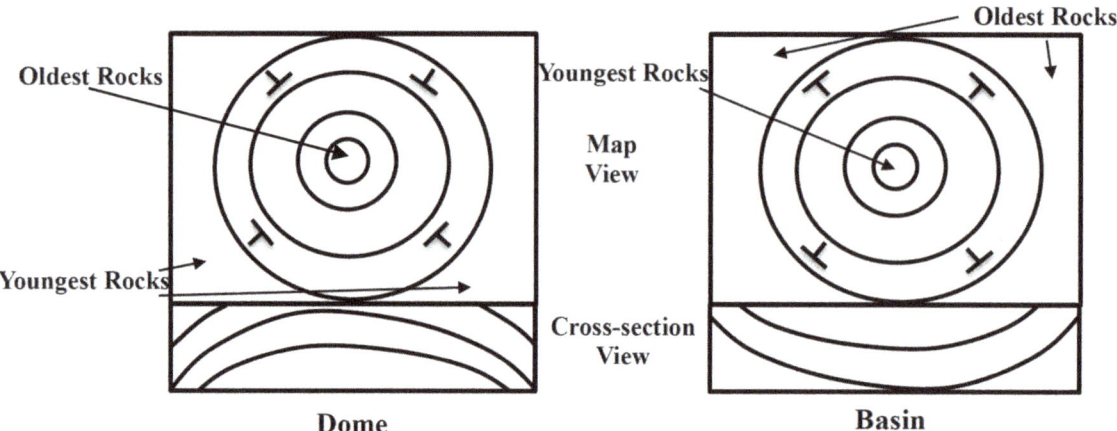

Figure 12.20 | The dome has older rocks in the center, with rocks dipping away from this point, while the basin has rocks dipping inwards and the youngest rocks in the center.
Author: Randa Harris
Source: Original Work
License: CC BY-SA 3.0

12.5.2 Faults

As rocks undergo brittle deformation, they may produce cracks in the rocks. If no appreciable displacement has occurred along these cracks, they are called joints. If appreciable displacement does occur, they are referred to as faults. We will first examine **dip-slip faults**, in which movement along the fault is either up or down. The two masses of rock that are cut by a fault are termed the fault blocks (Figure 12.21). The type of fault is determined by the direction that the fault blocks have moved.

Fault block movement is described based on the movement of the hanging wall, the fault block located above the fault plane. The other fault block, located beneath the fault plane, is called the foot wall. The term hanging wall comes from the idea that if a miner was climbing along the fault plane, she would hang her lantern above her head, along the hanging wall. Alternately, you can draw a stick figure straight up and down across the fault plane. Its head will be on the hanging wall and its feet will be on the foot wall (Figure 12.22).

Figure 12.21 | Two fault blocks. The fault is the break in the block that separates the two fault blocks.
Author: Randa Harris
Source: Original Work
License: CC BY-SA 3.0

When extensional forces are applied to the fault blocks, the hanging wall will move down, creating what is called a **normal fault** (an easy way to remember this is the phrase "It's normal to fall down"). As this happens, crust is stretched out and lengthened (Figure 12.23).

When compressional forces are applied to the fault blocks, the hanging wall will move up, creating a **reverse fault**. This causes the crust to shorten in the area (Figure 12.24). A special type of reverse fault is a **thrust fault**. It is a low angle reverse fault (with a dip angle of less than 45o), and has a much thinner hanging wall.

Figure 12.22 | In this image, the head of the stick figure is on the hanging wall (in mauve) and the feet of the stick figure are on the foot wall (in blue).
Author: Randa Harris
Source: Original Work
License: CC BY-SA 3.0

Figure 12.23 | The hanging wall, on the right, has moved down relative to the footwall, resulting in a normal fault. Notice how close together the cross symbols were in Figure 12.21, compared to this figure, evidence for the lengthening of the block (along red line).
Author: Randa Harris
Source: Original Work
License: CC BY-SA 3.0

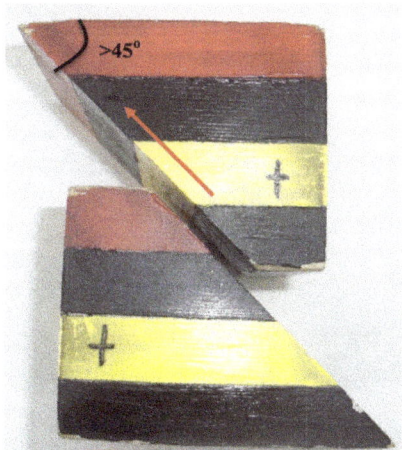

Figure 12.24 | The hanging wall, on the right, has moved up relative to the footwall, resulting in a reverse fault. Note how much closer the cross symbols have become when compared to figure 21, evidence for the shortening of the block.
Author: Randa Harris
Source: Original Work
License: CC BY-SA 3.0

Figure 12.25 | An example of a normal fault near Somerset, UK. Note the fault line in red, and the hanging wall sediments (on the left side) that have moved downward.
Author: Ashley Dace
Source: Wkimedia Commons
License: CC BY-SA 2.0

Table 12.2

Fault Type	Type of Force	Direction HW moved	Length of Block
Normal	Extensional	Down	Lengthened
Reverse	Compressional	Up	Shortened

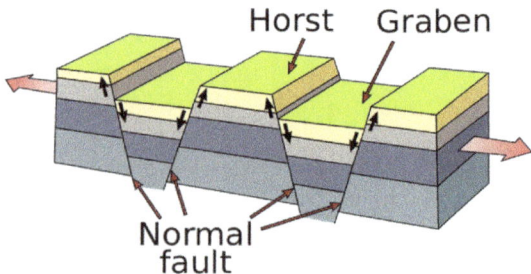

Tensional forces acting over a region can produce normal faults that result in landforms known as horst and grabens. In **horst and grabens**, the graben is the crustal block that downdrops, and is surrounded by two horsts, the relatively uplifted crustal blocks (Figure 12.26). This terrain is typical of the Basin and Range of the western United States.

The previous faults we observed were dip-slip faults, where movement occurred parallel with the fault's dip. In a **strike-slip fault**, horizontal motion occurs (in the direction of strike, hence the name), with blocks on opposite sides of a fault sliding past each other due to shear forces. The classic example of a strike-slip fault is the San Andreas Fault in California (Figure 12.27). These faults can be furthered classified as right-lateral or left-lateral. To determine this, an observer would stand along one side of the fault, looking across at the opposite fault block. If that fault block appears to have moved right, it is right-lateral; if it has moved left, it is left-lateral. Figure 12.28 provides all three fault types for your review.

Figure 12.26 | This figure depicts an area that has been stretched by tensional forces, resulting in numerous normal faults and horst and graben landforms.
Author: User "Gregors
Source: Wikimedia Commons
License: Public Domain

Figure 12.27 | A block diagram of the San Andreas Fault, a right-lateral strike-slip fault.
Author: Randa Harris
Source: Original Work
License: CC BY-SA 3.0

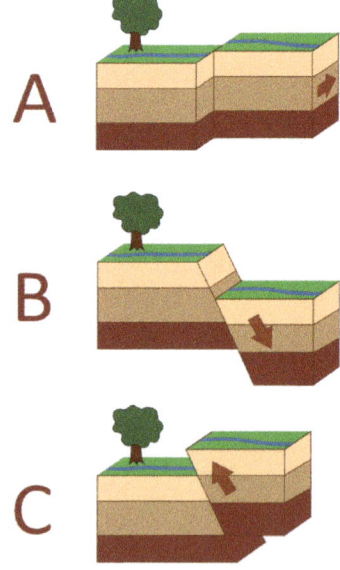

Figure 12.28 | The three types of faults discussed in this lab. **A** is a strike-slip fault, **B** is a normal fault, and **C** is a reverse fault.
Author: User "Karta24"
Source: Wikimedia Commons
License: Public Domain

12.6 GEOLOGIC MAPS AND CROSS-SECTIONS

A **geologic map** uses lines, symbols, and colors, to include information about the nature and distribution of rock units within an area. It includes a base map, over which information about geologic contacts and strikes and dips are included. Geologists make these maps by careful field observations at numerous outcrops (exposed rocks at the Earth's surface) throughout the mapping area. At each outcrop, geologists record information such as rock type, strike and dip of the rock layers, and relative age data. Geologic maps take practice to understand, as three-dimensional features (such as folds) are displayed on a two-dimensional surface. Remember that a geologic map will be seen in map view, as we learned about earlier with the block diagrams. Geologists use information about rocks that are exposed to visualize the unseen rocks beneath the surface, enabling them to complete the cross-sectional views we observed in the blocks. You will examine a geologic map for an area in Georgia and construct a **geologic cross-section**. A geologic cross-section shows geologic features from the side view. They are similar to the topographic profiles that you have already created, but include additional information about the rocks present. In order to construct a geologic cross-section, obey the following steps:

1. Observe the geologic map given. Pay close attention to any strike and dip symbols, geologic contacts, and ages of the rock types.

2. Take a scratch sheet of paper. Line it up along the line provided across the cross-section.

3. At each geologic contact, make a mark on the scratch paper. Position the marks in the direction you believe the rocks are dipping (To determine this, use strike and dip symbols. If they are not provided, use the Rule of V's or the ages given to help determine the geologic structure).

4. Transfer the marks from your paper to a provided diagram.

5. Sketch in the structure, paying careful attention to dip angles (if provided). Structures may be drawn in with a dotted line above the Earth's surface to indicate rocks that were formerly present but that have since been eroded.

There are some helpful hints to remember when constructing a cross-section:

1. Anticlines – these folds have the oldest beds in the middle, with beds dipping away from the axis. Plunging anticlines plunge towards the closed end of the V.

2. Synclines – these folds have the youngest beds in the middle, with beds dipping towards the axis. Plunging synclines plunge towards the open end of the V.

3. As streams intersect dipping beds of rock, they will cut V shapes in the direction of dip.

Figure 12.29 provides an example of a simple geologic cross-section, based off ages of the rock units.

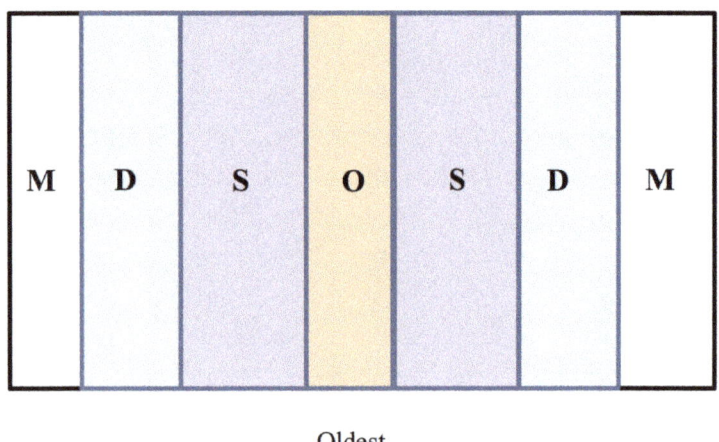
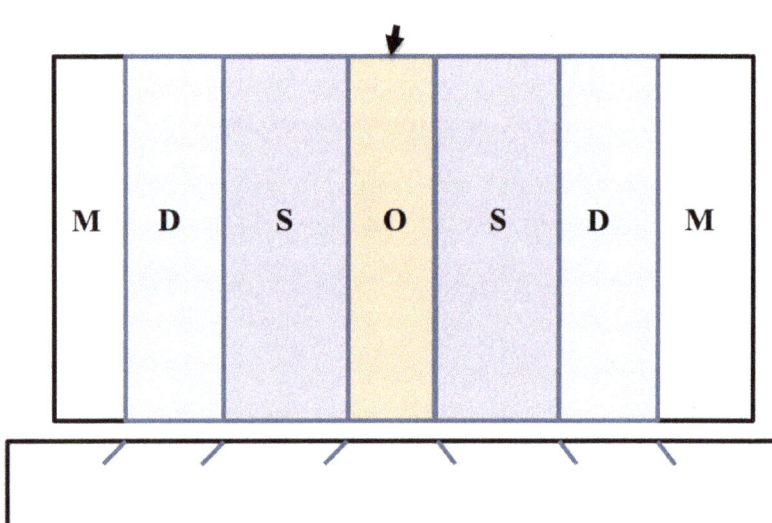
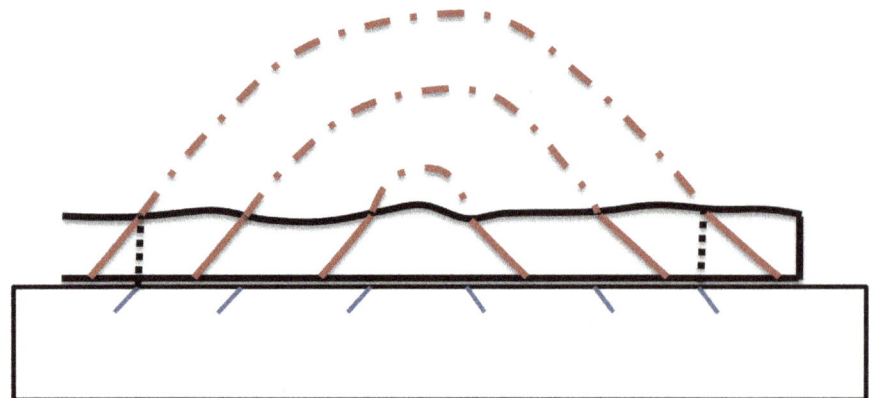

Figure 12.29 | The top figure displays the geologic map, with different rock units of different colors, and age information given (see Figure 12.30). In the second figure, the scratch sheet of paper is stretched along the bottom of the cross-section, and a mark is made at each geologic contact. Since the oldest bed is in the middle, this indicates that this structure is an anticline, so beds have been drawn so that they dip away from each other. In the bottom image, the scratch paper has been superimposed and contacts have been drawn in (as indicated by the dashed lines on each end mark). Beneath the surface, contacts have been drawn in using a solid line. The beds above the surface that have since eroded have been drawn in with a dashed line.
Author: Randa Harris
Source: Original Work
License: CC BY-SA 3.0

Age of Rocks	Geologic Age Symbol	Geologic Time Period
Youngest	Q	Quaternary
↑	T	Tertiary
	K	Cretaceous
	J	Jurassic
	Tr	Triassic
	P	Permian
	lP	Pennsylvanian
	M	Mississippian
	D	Devonian
	S	Silurian
	O	Ordovician
	-C	Cambrian
Oldest	p-C	PreCambrian

Figure 12.30 | Guide to rock ages.
Author: Randa Harris
Source: Original Work
License: CC BY-SA 3.0

Map Symbol	Explanation
 ⊢	Strike & Dip
 +	Vertical strata
 ⊕	Horizontal strata
 +↔+	Anticline axis
 →∣←	Syncline axis
 ⤢	Plunging anticline axis
 ⤢	Plunging syncline axis
 ↯	Strike-slip fault

Figure 12.31 | Guide to common map symbols.
Author: Randa Harris
Source: Original Work
License: CC BY-SA 3.0

12.7 PRACTICE WITH BLOCK DIAGRAMS

For practice, complete the questions below about block diagrams. A key is provided after this section to check your work.

For each diagram, draw in the geological contacts on each side of the block. Add strike and dip symbols, and other symbols to document geologic features (like direction of movement on faults). Also state the name of the geology feature in the diagram.

1.

2.

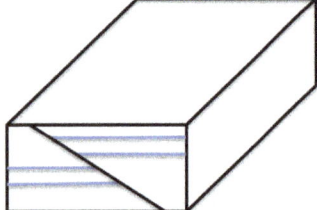

3.

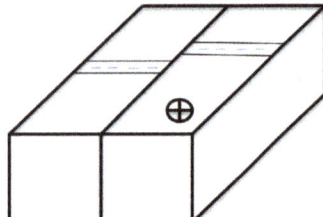

4.

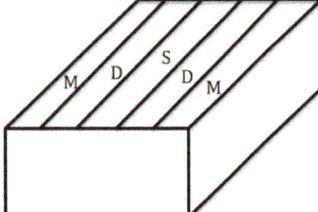

5.

12.8 ANSWERS TO PRACTICE WITH BLOCK DIAGRAMS

1. NORMAL FAULT

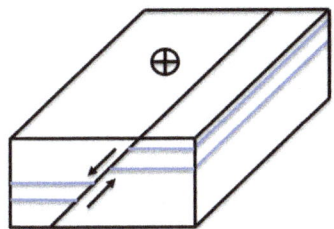

2. REVERSE FAULT

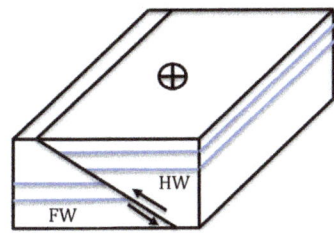

3. STRIKE-SLIP FAULT

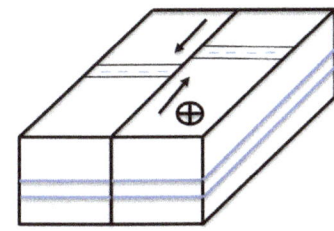

4. ANTICLINE

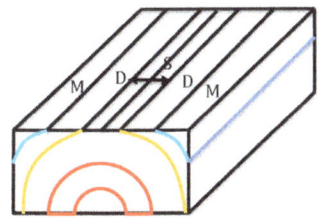

5. PLUNGING SYNCLINE

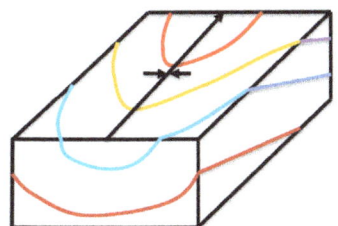

INTRODUCTORY GEOLOGY CRUSTAL DEFORMATION

12.8 LAB EXERCISE
Student Responses   Name_____

This Lab Assignment must be mailed to your Instructor. There is no online assessment for the Crustal Deformation Lab.

Complete the entire assignment and mail to your instructor postmarked by the assessment deadline. You should make an extra copy to practice on and mail in a clean and neat version for grading. Make sure to include your name and staple all of the pages together. It is a good idea to make a copy of what you mail in, just in case it gets lost in the mail. **For several parts of this lab there are multiple interpretations! Ask if you have questions.**

Part A (6 pts) Circle the correct answer to the following questions.

1. For the following diagram, determine the correct map symbol that would go in the oval box.

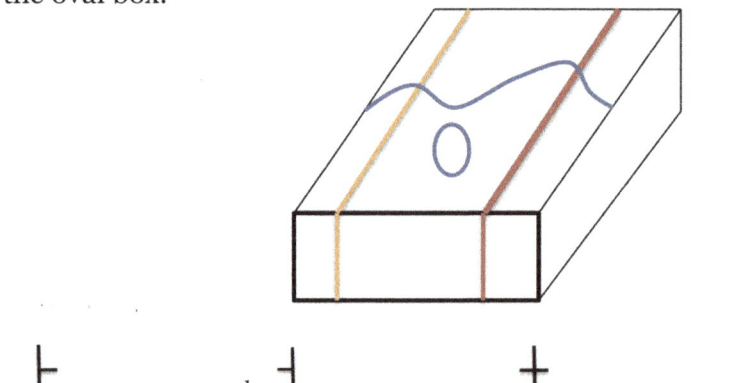

2. For the following diagram, determine the correct map symbol that would go in the oval box.

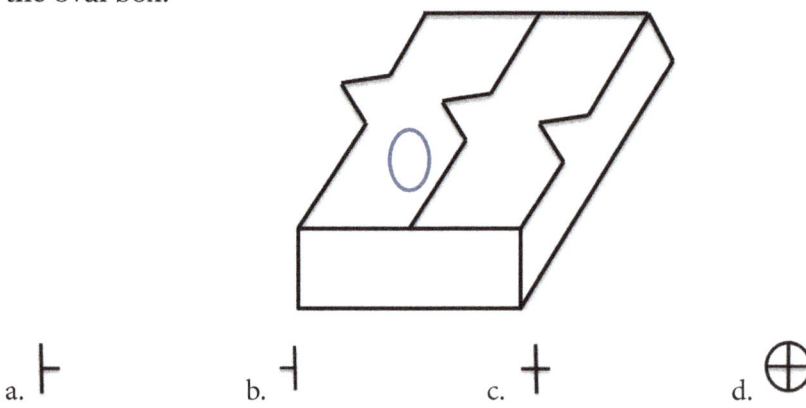

INTRODUCTORY GEOLOGY   CRUSTAL DEFORMATION

3. Using Google Earth, search for the following area in Pakistan: **27 50 35.00N 67 10 03.70E**. Zoom out to an eye altitude of approximately 25,000 feet. The inclined layers in these folded rocks can be easily seen. As you view them, in which compass direction are the beds dipping?

a. Northeast (NE)   b. Southeast (SE)   c. Northwest (NW)   d. Southwest (SW)

Part B (24 pts)
For each of the following block diagrams complete the following: 1- Complete the diagram drawing in geological contacts on each side of the block; 2- Add symbols indicating the strike and dip of each geological layer as well as symbols documenting any other geological features (include the direction of movement for any faults); 3- In the space provided under the block diagram write the specific name of the geology feature in the block diagram. Please note that in several of the blocks the ages of the layers are provided.

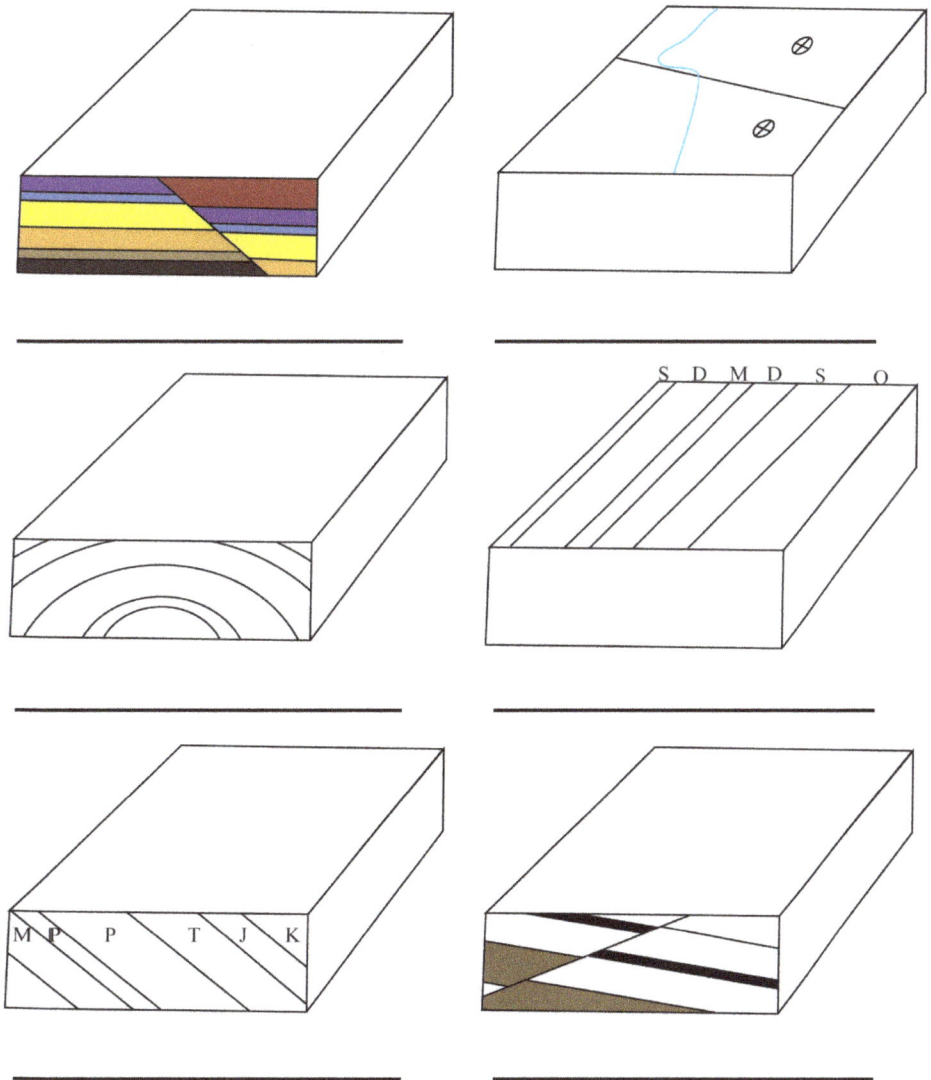

Page | 305

INTRODUCTORY GEOLOGY   CRUSTAL DEFORMATION

Part C (20 pts)

At the end of the lab there are two full-page block diagrams. Cut along the dashed lines and fold along the solid lines to examine the block in three dimensions. For each of the block diagrams complete the following: 1- Complete the diagram drawing in geological contacts on each side of the block; 2- Add symbols indicating the strike and dip of each geological layer as well as symbols documenting any other geological features; 3- Identify the geologic structure presented on the block and write the name of the structure on the top surface of the block. Please unfold the blocks flat to mail in to your instructor.

Part D (24 pts)

The geological map on the following page is from the Paleozoic rocks in the Northwest Georgia Mountains. Please refer to the key to rocks and their ages present on the map in order to answer the questions. Note that the rock key is in chronological order with the oldest rocks on the bottom (Conasauga shale) and the youngest rocks on top (Pennsylvanian Fm).

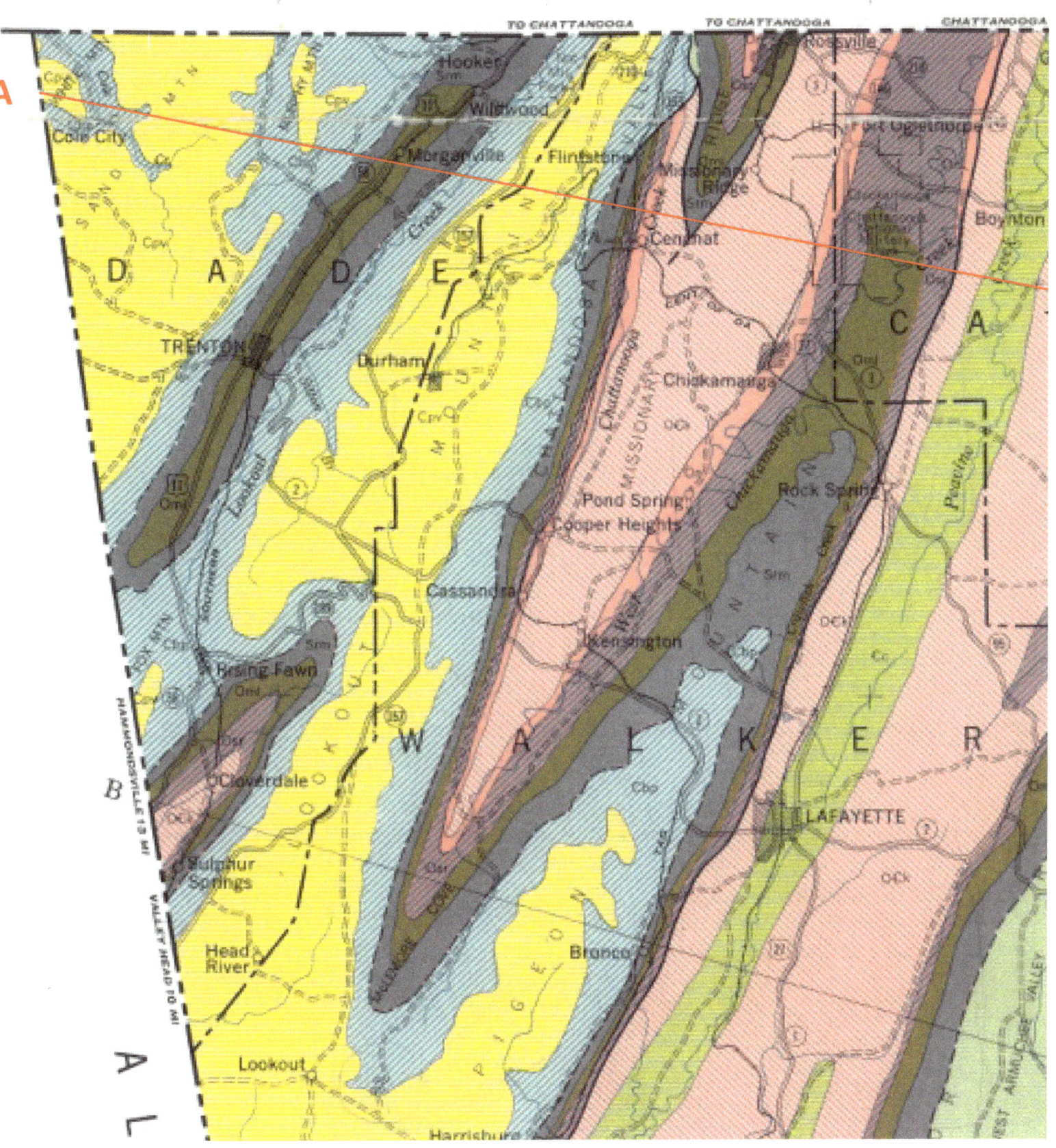

4. (6 pts) Examine the map and the ages of the rock layers. Based on your interpretation, label the structures and add appropriate symbols to the map to show the axis of the folds. What kind of folds are present in this area? Explain how you came to this conclusion.

5. (6 pts) Along the red line starting at point 'A' there are four faults indicated by bold black lines (they are also marked on the cross section in Part C). Examine the age of the rocks and the faults themselves. What type of faults are these? Explain how you came to this conclusion.

6. (12 pts) Complete the following geological cross section based on your understanding of the geological structures present in the area.

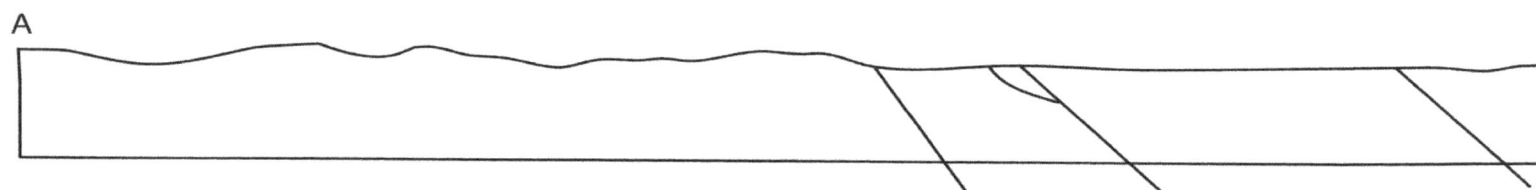

Part E - Google Earth. (26 pts)

7. Search for **38 04 36.00N 109 55 26.61 W** and zoom to an eye altitude of 20,000 ft, make sure to note your geographic location. (hint: re-read the section in the course content unit on Faults and in this lab manual)

    a. You are looking at the surface expression of many fault lines (fault scarp) with the hanging wall occupying the valleys and the foot wall representing the plateaus, what kind of faults are they?

b. What would we call these **paired** features?

c. What type of stress created these features?

d. The amount of displacement (extension or shortening) in the area can be calculated by the following equation based on the angle of the fault surface:

Amount of horizontal displacement = Vertical displacement along the fault surface (ft)/5.67

Measure the change in elevation of one of these features to determine the amount of extension in the area (Show your work, but feel free to use a calculator).

e. How many of these faults would you need to make this area (Utah) a mile longer? Remember there are 5,280 feet in a mile.

8. Search for **22 48 45.63S 117 20 12.85E** and zoom to an eye altitude of 25 miles. This feature is caused by the folding of rocks.

   a. Imagine you are a geologist trying to determine what is going on in the area. How could you tell if this large bullseye structure is a dome or a basin (State whether this is a dome or basin, and give two different ways to tell them apart)?

   b. Examine the feature ~10 miles southwest of the previous feature. Is this fold horizontal or plunging, how can you tell?

(Part C-1)

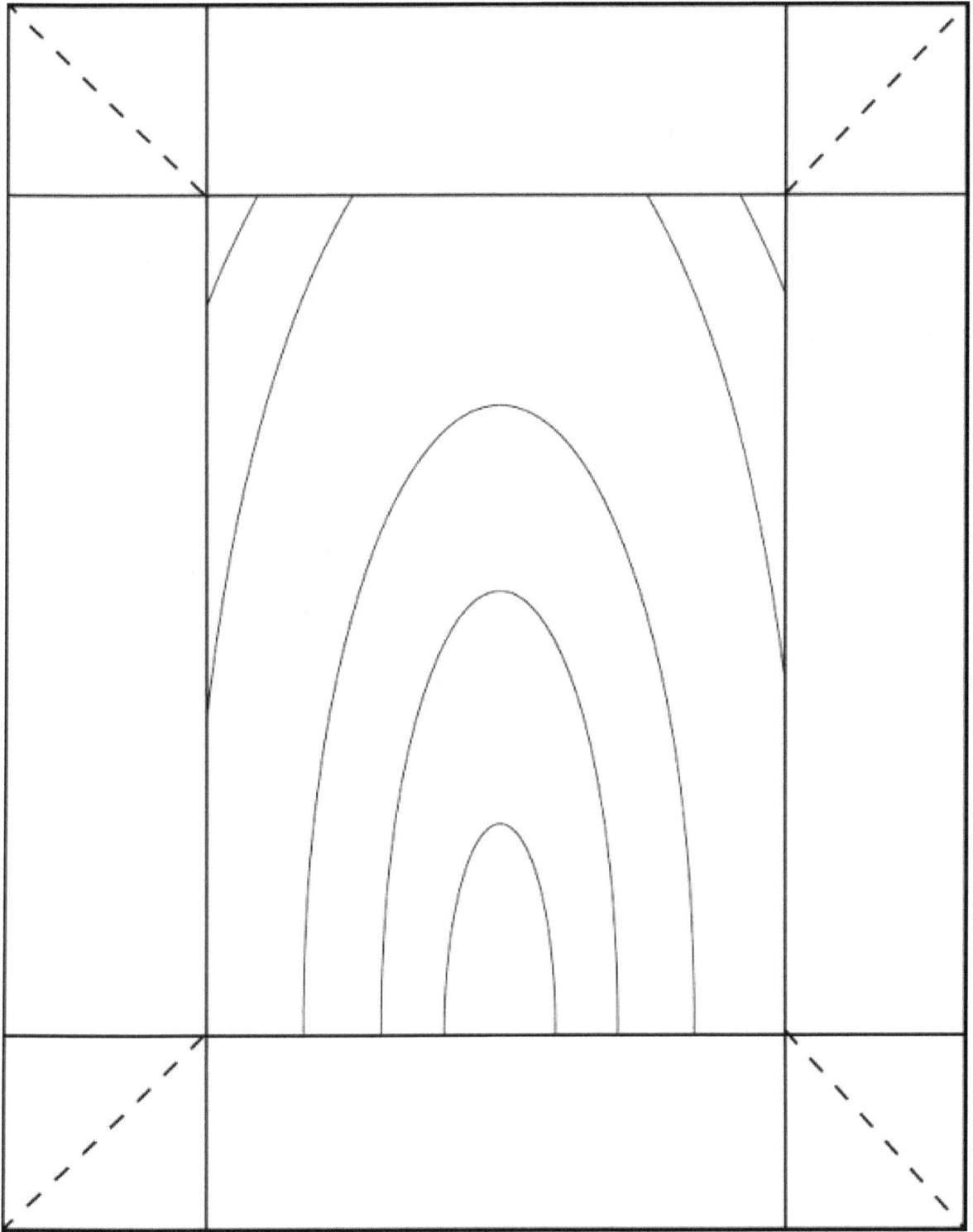

(Part C-2)

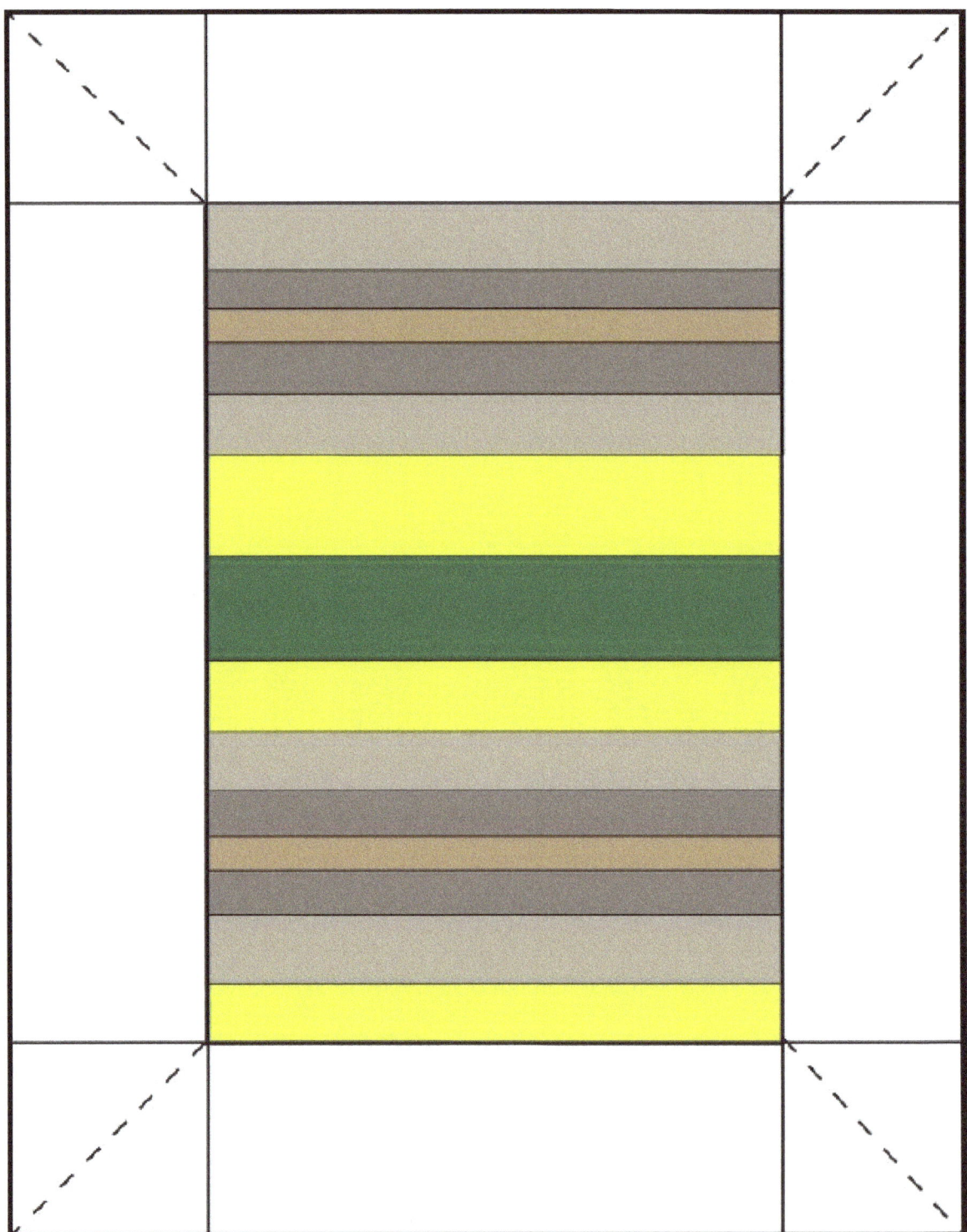

13

Earthquakes
Randa Harris

13.1 INTRODUCTION

It was the deadliest day in the history of Mt. Everest. On April 25, 2015, a 7.8 magnitude earthquake hit Nepal. This triggered an avalanche that killed 19 climbers on Mt. Everest. In Nepal, over 8,800 people died, and many more were injured and made homeless. Hundreds of aftershocks (smaller earthquakes that follow a larger earthquake) have occurred since (Figure 13.1).

Earthquakes are not new to this region. A similar death toll was experienced in a 1934 earthquake, and many other smaller earthquakes have occurred within historical times. An 1833 quake of similar magnitude resulted in less than 500

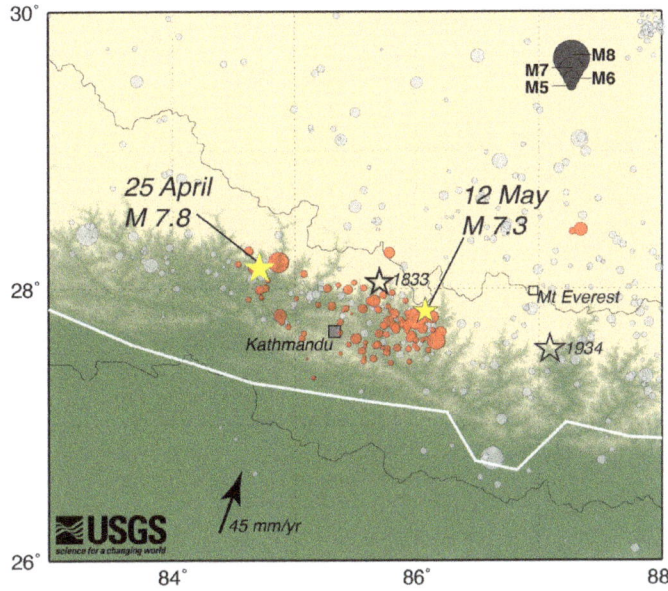

Figure 13.1 | A map of the main earthquake to hit Nepal on April 25, 2015, along with a major aftershock on May 12, and numerous (>100) other aftershocks (in red – note the magnitude scale in the upper right).
Author: USGS
Source: USGS
License: Public Domain

deaths, though this was most likely due to two very large foreshocks (smaller earthquakes that precede the main earthquake) that sent most residents out of doors in alarm, which was safer for them. Worldwide, there have been much deadlier and stronger earthquakes just in this century (Haiti, 2010 – 316,000 dead; Sumatra, 2004 – 227,000 dead, both with deaths related to ground shaking and the other hazards that were created by the earthquake). Earthquakes give geologists valuable information about the Earth, both the interior, as we learned about in the Earth's Interior chapter, and about conditions at the Earth's surface (most earthquakes occur at plate boundaries, as we learned in the Plate Tectonics chapter, Figure 4.8).

13.1.1 Learning Outcomes

After completing this chapter, you should be able to:
- Compare and contrast the different types of seismic waves
- Understand the different scales used to measure earthquakes, and apply them to the amount of devastation
- Understand how different geologic materials behave during an earthquake, and the resulting impact on structures
- Explain how an earthquake epicenter is located
- Explore the relationship between the fracking industry and seismicity

13.1.2 Key Terms

- Benioff Zones
- Body Waves
- Epicenter
- Focus
- Induced Seismicity
- Intensity
- Liquefaction
- Love Waves
- Magnitude
- P Waves
- Rayleigh Waves
- S Waves
- Seismogram
- Seismograph
- Seismology
- Surface Waves

13.2 THE EPICENTER, FOCUS, AND WAVES

An earthquake is a like a telegram from the Earth. It sends a message about the conditions beneath the Earth's surface. The shaking or trembling experienced during an earthquake is the result of a rapid release of energy within the Earth, usually as a result of movement along geologic faults. Think back to the strike-slip fault from the Crustal Deformation chapter. Rocks on either side of the fault are sliding past each other. As they move in opposite directions, the rocks become deformed, as they will bend slightly and build up pressure. Eventually they will reach a breaking point. Once the strength of the rock has been exceeded, the rocks will snap back to their normal shape, releasing all that stored energy as an earthquake. The more energy that has been stored, the larger the earth-

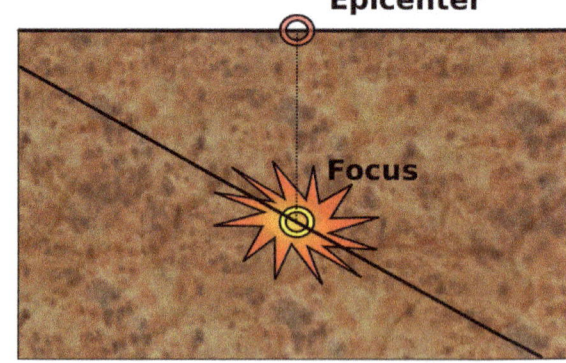

Figure 13.2 | An illustration depicting the focus, where the earthquake originates, and the epicenter, the point on the ground's surface directly above the focus.
Author: Unknown
Source: Wikimedia Commons
License: GNU Free Documentation

quake is. Remember the stress-strain diagram from Crustal Deformation. When rocks are under too great of a stress, they undergo brittle failure (the earthquake). The strength of the rock has been exceeded at this point.

Earthquakes originate at a point called the **focus** (plural foci). From this point, energy travels outward in different types of waves. The place on the Earth's surface directly above the focus is called the **epicenter** (Figure 13.2). Earthquake foci may be shallow (less than 70 km from Earth's surface) to deep (greater than 300 km deep), though shallow to intermediate depths are much more common. Earthquake frequency and depth are related to plate boundaries. The vast majority (95%) of earthquakes occur along a plate boundary, with shallow focus earthquakes tending to occur at divergent and transform plate boundaries, and shallow to intermediate to deep focus earthquakes occurring at convergent boundaries (along the subducting plate). The earthquakes associated with convergent boundaries occur along Wadati-Benioff zones, or simply **Benioff zones**, areas of dipping seismicity along the subducting plate (Figure 13.3).

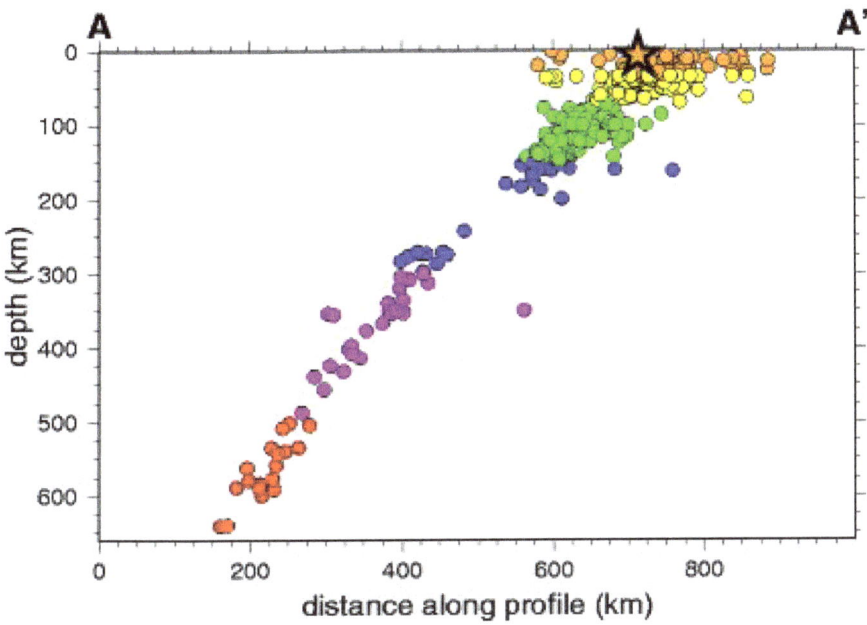

Figure 13.3 | This is a seismicity cross-section, taken along the subducting plate in an ocean-ocean convergent boundary at the Kuril Islands, located just northeast of Japan. Foci are located in the descending slab. Only brittle substances (like the lithosphere) can generate earthquakes, so this must be the subducting slab. The star represents the location of an 8.3 magnitude earthquake that occurred on 11/15/06.
Author: USGS
Source: Wikimedia Commons
License: Public Domain

As an earthquake occurs, two different types of waves are produced; **body waves**, so termed because they travel through the body of the Earth, and **surface waves** that travel along the Earth's surface (Figure 13.4). There are two types of body waves. **P-waves**, or primary waves, are compressional waves that move back and forth, similar to the action of an accordion. As the wave passes, the atoms in the material it is travelling through are being compressed and stretched. Movement is compressional parallel to the direction of wave propagation, which makes P-waves the fastest of the seismic waves. These waves can travel through solids, liquids, and gases, because all materials can be compressed to some degree. **S-waves**, or secondary waves, are shear waves that move material in a direction perpendicular to the direction of travel. S-waves can only travel through solids, and are slower than P-waves. A similar motion to S-wave motion can be created by two people holding a rope, with one snapping the rope quickly. Alternately, you can also think of this wave movement similar to the wave created by fans in a stadium that stand up and sit down. Body waves are responsible for the jerking and shaking motions felt during an earthquake.

Surface waves are slower than body waves, and tend to produce more rolling sensations to those experiencing an earthquake, similar to being in a boat on the sea. Because surface waves are located at the ground's surfacewhere humans (and their structures) are located, and because they move so slowly, which bunches them up and increases their amplitude, they are the most damaging of seismic waves. **Love waves** are the faster surface waves, and they move material back and forth in a horizontal plane that is perpendicular to the direction of wave travel (see Figure 13.4). Buildings do not handle this type of movement well, and Love waves may be responsible for considerable damage to structures. **Rayleigh waves** make the Earth's surface move in an elliptical motion, similar to the movement in a sea wave. This results in ground movement that is up and down and side-to-side.

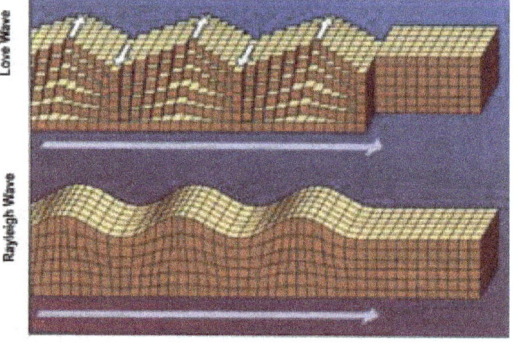

Figure 13.4 | The different types of seismic waves. Body waves, in the upper portion of the figure, consist of P-waves and S-waves. The P-wave motion is compressional. The hammer on the left starts the wave moving. The arrow on the right shows the general direction of the wave. In the S-wave, the motion is undulating. Surface waves are depicted in the lower portion of the chart. Love waves move similarly to S-waves, resulting in horizontal shifting of the Earth's surface. Rayleigh waves are surface waves that travel much like a wave along the water's surface.
Author: USGS
Source: Wikimedia Commons
License: Public Domain

13.3 SEISMOLOGY

Earthquakes have been experienced by humans as long as humans have roamed the Earth, though most ancient cultures developed myths to explain them (including envisioning large creatures within the Earth that were moving to create the quake). The study of earthquakes, called **seismology**, began to take off with the development of instruments that can detect earthquakes; this instrument, called a **seismograph**, can measure the slightest of Earth's vibrations (Figure 13.5). A typical seismograph consists of a mass suspended on a string from a frame that moves as the Earth's surface moves. A rotating drum is attached to the frame, and a pen attached to the mass, so that the relative motion is recorded in a **seismogram**. It is the frame (attached to the ground) that moves during an earthquake—the suspended mass generally stays still due to inertia (the tendency of a body to stay at rest and resist movement).

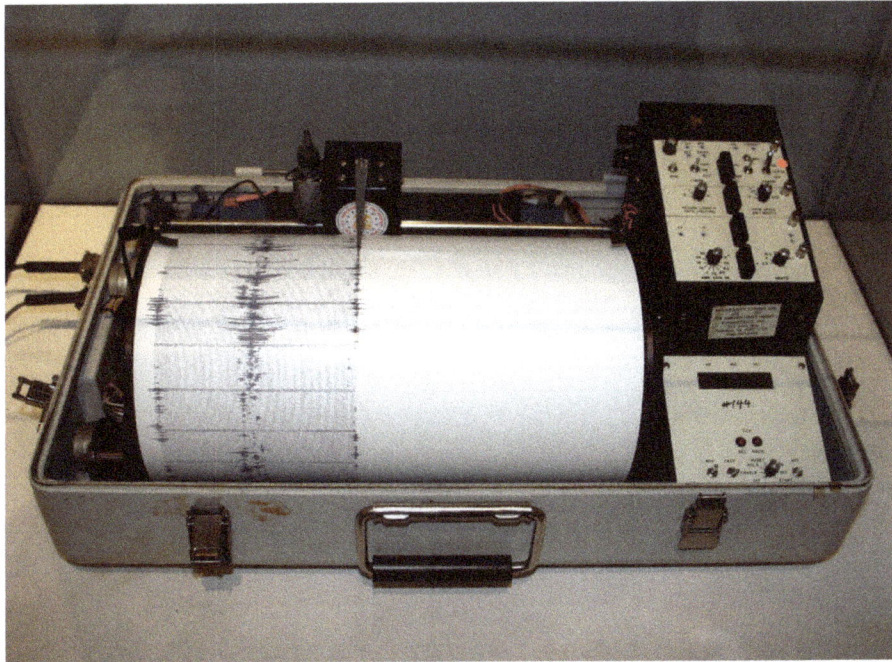

Figure 13.5 | A seismograph and the seismogram it produces.
Author: User "Yamaguchi"
Source: Wikimedia Commons
License: CC BY-SA 3.0

13.3.1 How Are Earthquakes Measured?

The tragic consequences of earthquakes can be measured in many ways, like death tolls or force of ground shaking. Two measures in particular are commonly used. One is a qualitative measure of the damage inflicted by the earthquake, and it is referred to as **intensity**. The second is a quantitative measure of the energy released by the earthquake, termed **magnitude**. Both measures provide meaningful data.

13.3.2 Earthquake Intensity

Intensity measurements take into account both the damage incurred due to the quake and the way that people respond to it. The Modified Mercalli Intensity Scale (Figure 13.6) is the most widely used scale to measure earthquake intensities. This scale has values that range from Roman numerals I to XII which characterize the damage observed and people's reactions to it. Data for this scale is often collected right after an earthquake by having the local population answer questions about the damage they see and what happened during the quake. This information can then be pooled to create an intensity map, which creates colored zones based on the information collected (Figure 13.7). These maps are frequently used by the insurance industry.

Intensity	Characteristics
I	Shaking not felt under normal circumstances.
II	Shaking felt only by those at rest, mostly along upper floors in buildings.
III	Weak shaking felt noticeably by people indoors. Many do not recognize this as an earthquake. Vibrations similar to a large vehicle passing by.
IV	Light shaking felt indoors by many, outside by few. At night, some were awakened. Dishes, doors, and windows disturbed; walls cracked. Sensation like heavy truck hitting a building. Cars rock noticeably.
V	Moderate shaking felt by most; many awakened. Some dishes and windows broken. Unstable objects overturned.
VI	Strong shaking felt by all, with many frightened. Heavy furniture may move, and plaster breaks. Damage is slight.
VII	Very strong shaking sends all outdoors. Well-designed buildings sustain minimal damage; slight-moderate damage in ordinary buildings; considerable damage in poorly built structures.
VIII	Severe shaking. Well-designed buildings sustain slight damage; considerable damage in ordinary buildings; great damage in poorly built structures.
IX	Violent shaking. Well-designed buildings sustain considerable damage; buildings are shifted off foundations, with some partial collapse. Underground pipes are broken.
X	Extreme shaking. Some well-built wooden structures are destroyed; most masonry and frame structures are destroyed. Landslides considerable.
XI	Few structures are left standing. Bridges are destroyed, and large cracks open in the ground.
XII	Total damage. Objects thrown upward in the air.

Figure 13.6 | (Above) An abbreviated table of the Modified Mercalli Intensity Scale. Intensity for a particular earthquake is determined by the maximum damage incurred.
Author: Randa Harris
Source: Original Work
License: CC BY-SA 3.0

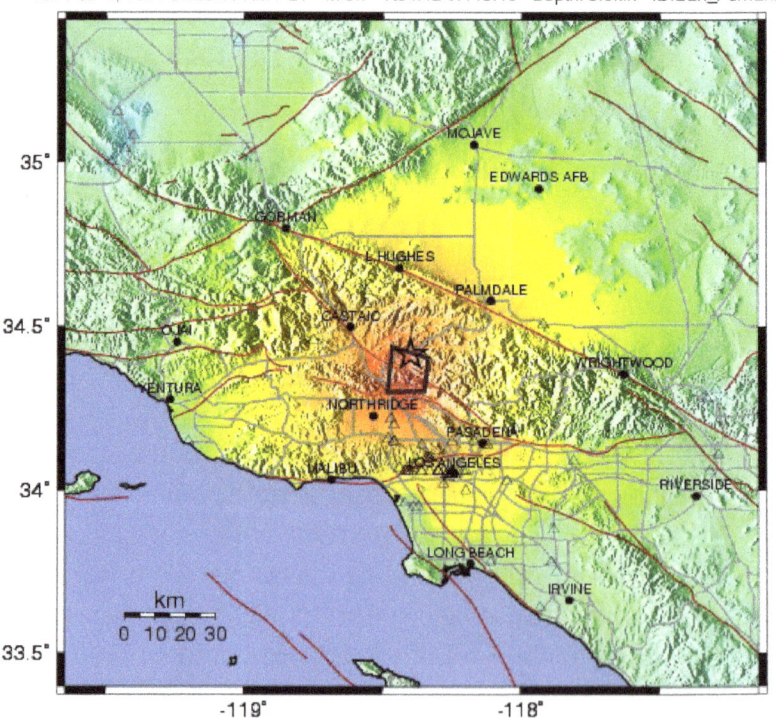

Figure 13.7 | (Right) An intensity map for the San Fernando earthquake in southern California on 2/9/76. Notice that near the epicenter (marked by a star), the intensity was extreme.
Author: USGS
Source: Wikimedia Commons
License: Public Domain

13.3.3 Earthquake Magnitude

Another way to classify an earthquake is by the energy released during the event; this is referred to as the magnitude of the earthquake. While magnitude has been measured using the Richter scale, as the frequency of earthquake measurements around the world increased, it was realized that the Richter magnitude scale was not valid for all earthquakes (it is not accurate for large magnitude earthquakes). A new scale called the Moment Magnitude Intensity Scale was developed, which maintains a similar scale to the Richter scale. This scale estimates the total energy released by an earthquake and can be used to characterize earthquakes of all sizes throughout the world. The magnitude is based on the seismic moment (estimated based on ground motions recorded on a seismogram), which is a product of the distance a fault moved and the force required to move it. This scale works particularly well with larger earthquakes and has been adopted by the United States Geological Survey. Magnitude is based on a logarithmic scale, which means for each whole number that you increase, the amplitude of the ground motion recorded by a seismograph increases by 10 and the energy released increases by 101.5, rather than one (so that a 3 magnitude quake results in ten times the ground shaking as a 2 magnitude quake; a magnitude 4 quake has 10^2 or 100 times the level of ground shaking as a 2 magnitude quake (releasing 10^3 or 1000 times as much energy). For a rough comparison of magnitude scale to intensity, see Figure 13.8. Why is it necessary to have more than one type of scale? The magnitude scale allows for worldwide characterization of any earthquake event, while the intensity scale does not. With an intensity scale, a IV in one location could be ranked a II or III in another location, based off of building construction (ex. poorly constructed buildings will suffer more damage in the same magnitude earthquake as those built with stronger construction).

Magnitude	Typical Maximum Modified Mercalli Intensity
1.0 – 2.9	I
3.0 – 3.9	II – III
4.0 – 4.9	IV – V
5.0 – 5.9	VI – VII
6.0 – 6.9	VII – IX
7.0 and above	VIII or above

Figure 13.8 | A comparison of magnitude versus intensity scales for earthquakes.
Author: Randa Harris
Source: Original Work
License: CC BY-SA 3.0

13.4 LOCATING AN EARTHQUAKE EPICENTER

During an earthquake, seismic waves are sent all over the globe. Though they may weaken with distance, seismographs are sensitive enough to still detect these waves. In order to determine the location of an earthquake epicenter, seismographs from at least three different places are needed for a particular event. In Figure 13.9, there is an example seismogram from a station that includes a minor earthquake.

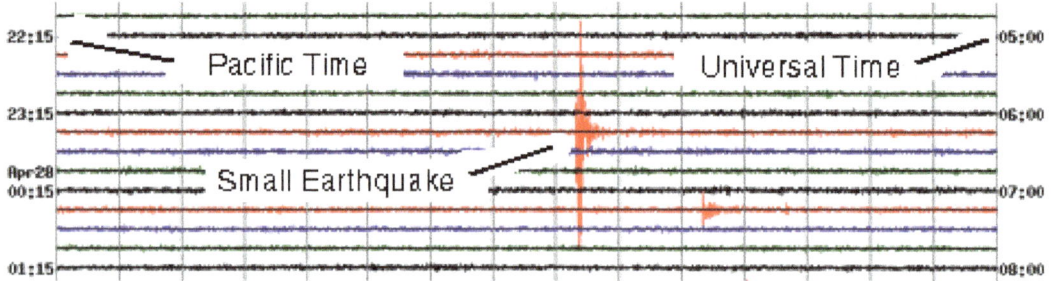

Figure 13.9 | This seismogram is read from left to right and top to bottom. Note the small earthquake that is marked, and the resulting change in wave amplitude at that point.
Author: USGS
Source: USGS
License: Public Domain

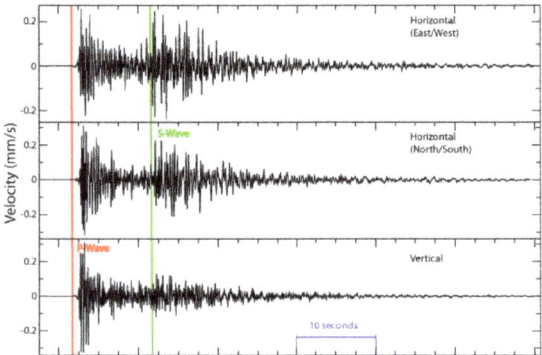

Figure 13.10 | An example seismogram with the arrival of P and S waves included. Note how the arrival of waves is marked by an increase in the wave height (known as amplitude) and by more tightly packed waves. This example does not include time along the bottom, but those in the lab exercise will.
Author: User "Pekachu"
Source: Wikimedia Commons
License: Public Domain

Once three seismographs have been located, find the time interval between the arrival of the P-wave and the arrival of the S-wave. First determine the P-wave arrival, and read down to the bottom of the seismogram to note at what time (usually marked in seconds) that the P-wave arrived. Then do the same for the S-wave. The arrival of seismic waves will be recognized by an increase in amplitude – look for a pattern change as lines get taller and more closely spaced (ex. Figure 13.10).

By looking at the time between the arrivals of the P- and S-waves, one can determine the distance to the earthquake from that station, with longer time intervals indicating longer distance. These distances are determined using a travel-time curve, which is a graph of P- and S-wave arrival times (see Figure 13.11).

Though distance to the epicenter can be determined using a travel-time graph, direction cannot be told. A circle with a radius of the distance to the quake can be drawn. The earthquake occurred somewhere along that circle. Triangulation is required to determine exactly where it happened. Three seismographs are needed. A circle is drawn from each of the three different seismograph locations, where the radius of each circle is equal to the distance from that station to the epicenter. The spot where those three circles intersect is the epicenter (Figure 13.12).

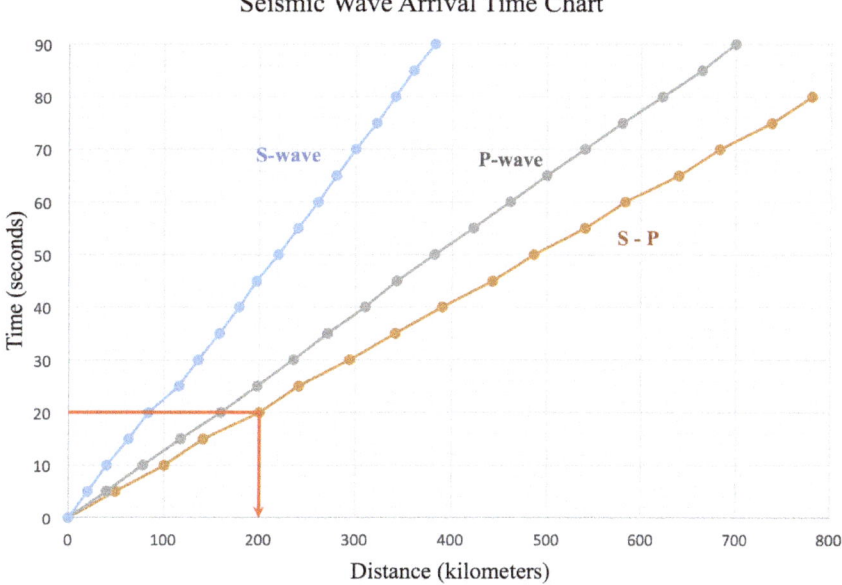

Figure 13.11 | A travel-time graph that includes the arrival of P-waves and S-waves. Note that these curves plot distance versus time, and are calculated based on the fact that the Earth is a sphere. Curves vary with the depth of earthquake because waves behave differently (i.e. their velocities change) with depth and change in material. This particular curve is used for shallow earthquakes (<20 km deep) with stations within 800 km. The S-P curve refers to the difference in time between the arrival of the P-wave and S-wave. If you noted on your seismogram that the P-wave arrived at 10 seconds, and the S-wave arrived at 30 seconds, the difference between arrival times would be 20 seconds. You would read the 20 seconds off the y-axis above to the S-P line, then drop down to determine the distance to the epicenter. In this case, it would be approximately 200 kilometers.
Author: Randa Harris
Source: Original Work
License: CC BY-SA 3.0

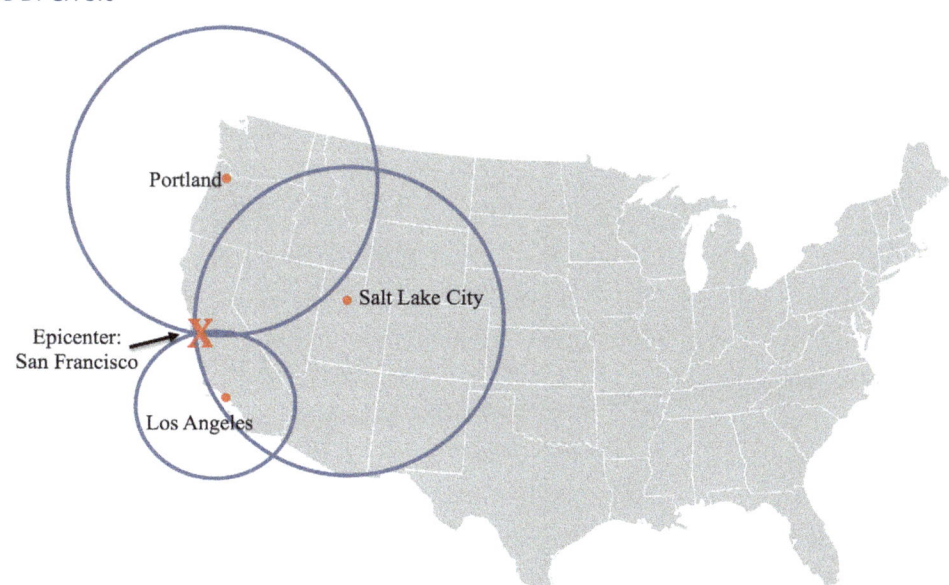

Figure 13.12 | In order to locate this earthquake epicenter, seismograms from Portland, Salt Lake City, and Los Angeles were used. The time between P and S wave arrivals was calculated, and travel time tables gave a distance. Circles with each distance for its radii were drawn from each station. The one resulting overlap, at San Francisco, was the earthquake epicenter.
Author: Randa Harris
Source: Original Work
License: CC BY-SA 3.0

INTRODUCTORY GEOLOGY   EARTHQUAKES

13.5 LAB EXERCISE

Part A – Locating an Epicenter

You will determine the location of an earthquake epicenter using seismograms from Carrier, Oklahoma, Smith Ranch in Marlow, Oklahoma, and Bolivar Missouri available at the end of this chapter. These are actual seismograms that you will be reading, from an actual event. For each, three different readouts are given, as the seismograph measured in three different axes. You may focus on any of the three readouts for each station, as all will have the same arrival times for each wave. First, determine when the P and S waves arrived, and note these times (remember to look for a pattern change as lines get taller and more closely spaced). Mark both the arrival of the P-wave and S-wave, then using the time scale in seconds, note the time difference between the P and S wave arrivals. Add this to the table below for each of the three seismograms.

Table 13.1

Station	P-wave arrival time (sec)	S-wave arrival time (sec)	Difference between P and S travel times (sec)	Distance to Epicenter from Station (km)
Carrier, OK				
Marlow, OK				
Bolivar, MO				

The difference between the P and S wave arrivals will be used to determine the distances to the epicenter from each station using Figure 13.11. Make sure that you use the curve for S-P Difference – find the seconds on the y-axis, read over to the S-P curve, then draw a line down to the x- axis for distance. Add the values to the table above. Now you need to create the circles from each station using Figure 13.13, a map with the three stations on it. This map includes a legend in kilometers. For each station, note the distance to the epicenter. Using a drafting compass (or alternately, tie a string to a pencil, cut the string the length of the distance to the epicenter, pin it at the station, and draw a circle, with the pencil stretched out the full distance of the string), you will create the circle. First, measure the scale on the map in Figure 13.13 in centimeters, and use that to convert your distances in kilometers to centimeters (ex. the map's scale of 100 km = 2.1 cm on your ruler, so if you had a measured distance from one station of 400 km, that would equal 8.4 cm on your ruler). For this fictional example, starting at the station, use the drafting compass to make a circle that is 8.4 cm in radius. Create a circle for each of the three stations, using their different distances to the epicenter. They should overlap (or nearly overlap) in one location. The location where they overlap is the approximate epicenter of the earthquake. Once done, answer the questions below.

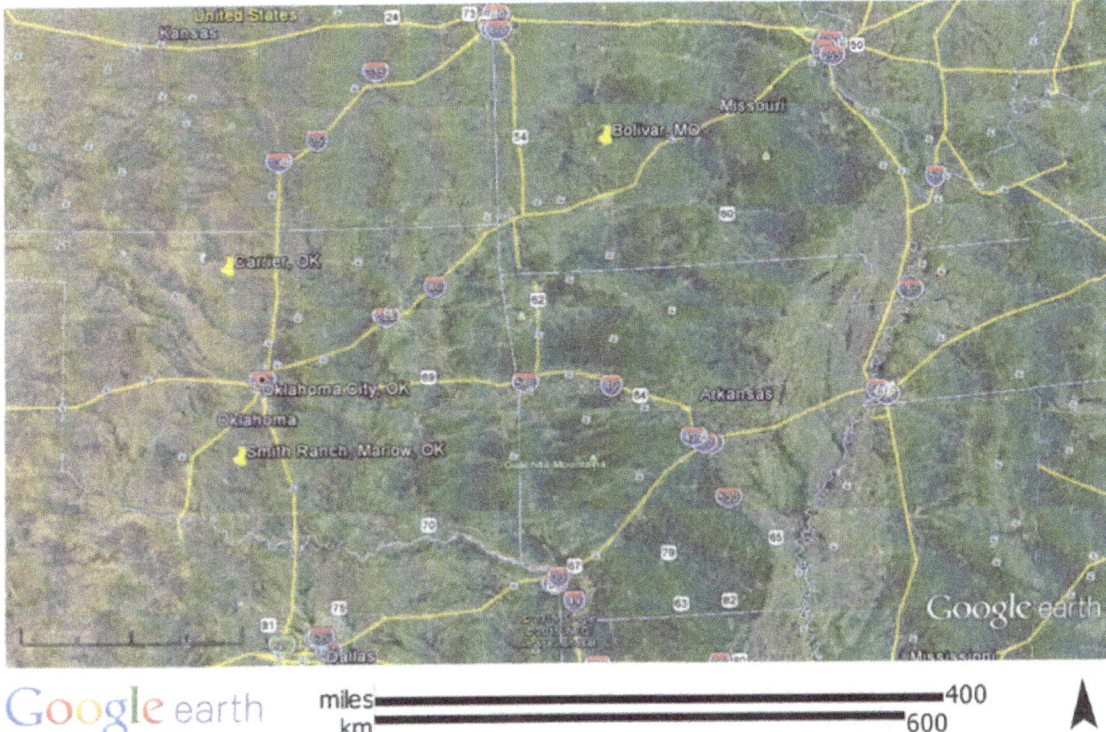

Figure 13.13
Author: Google Earth
Source: Google Earth
License: Used with attribution per Google's Permissions Guidelines

1. For Carrier, Oklahoma, what is the approximate time of the arrival of the first P-wave?

   a. 10 seconds b. 15 seconds c. 21 seconds d. 30 seconds

2. For Marlow, Oklahoma, what is the approximate time of the arrival of the first S-wave?

   a. 19 seconds b. 22 seconds c. 35 seconds d. 42 seconds

3. For Bolivar, Missouri, what is the difference between the P and S wave arrival times?

   a. 10 seconds b. 20 seconds c. 40 seconds d. 55 seconds

4. What is the approximate distance to the epicenter from Carrier, Oklahoma?

   a. 70 km b. 130 km c. 240 km d. 390 km

5. What is the approximate distance to the epicenter from Marlow, Oklahoma?

   a. 70 km b. 130 km c. 240 km d. 390 km

6. What is the approximate distance to the epicenter from Bolivar, Missouri?

   a. 70 km   b. 130 km   c. 240 km   d. 390 km

7. Look at the location that you determined was the earthquake epicenter. Compare its location to Oklahoma City. Which direction is the epicenter located from Oklahoma City?

   a. southeast   b. northwest   c. northeast   d. southwest

On January 12, 2010, a devastating magnitude 7.0 earthquake hit 16 miles west of Port-au-Prince, the capital of Haiti. At the following website, images are given of areas in Port-au-Prince both before the earthquake and soon after the earthquake, with a slider bar so that you can compare them. Access these images at: http://www.nytimes.com/interactive/2010/01/14/world/20100114-haiti-imagery.html (or alternately at http://elearningexamples.com/the-destruction-in-port-au-prince-2/) and note the changes in many areas due to damage from the earthquake.

8. Examine the before and after image of the National Cathedral. Based on the changes seen within the structure, decide where this earthquake would most likely fall on the Modified Mercalli Intensity Scale. Based off this image, the most likely intensity of this earthquake would be:

   a. <IV   b. V-VI   c. VII   d. VIII or greater

9. Residents in Port-au-Prince complained of extreme shaking during the earthquake, while residents of Santo Domingo, the capital of the Dominican Republic that sits 150 miles east of Port-au-Prince, assumed the shaking was caused by the passing of a large truck. Based on the Modified Mercalli Intensity Scale, the residents of Port-au-Prince mostly like experienced an intensity of ____, while the residents of Santo Domingo experienced an intensity of ____.

   a. VII, II   b. VIII, III   c. X, III   d. X, IV

13.6 HAZARDS FROM EARTHQUAKES

Earthquakes are among nature's most destructive phenomena, and there are numerous hazards associated with them. Ground shaking itself leads to falling structures, making it the most dangerous hazard. The intensity of ground shaking depends on several factors, including the size of the earthquake, the duration of shaking, the distance from the epicenter, and the material the ground is made of. Solid bedrock will not shake much during a quake, rendering it safer than other ground materials. Artificial fill refers to areas that have been filled in for construction and/or waste disposal (think of a hill that gets cleared for a shopping mall – the soil that was removed is dumped somewhere else as artificial fill). Sediment is not compacted in areas of artificial fill, but compaction will occur during the shaking of an earthquake, leading to structure collapse. Artificial fill sediments behave similarly to water-saturated sediments. As they shake, they may experience **liquefaction**, in which the sediments behave like a fluid. Normally, friction between grains holds them together. Once an earthquake occurs, water surrounds every grain, eliminating the friction between them and causing them to liquefy (Figure 13.14). This can be very dangerous. Seismic waves will amplify as they come in contact with these weaker materials, leading to even more damage.

Figure 13.14 | A diagram depicting liquefaction. In the water-saturated sediment on the left, the pore (open) spaces between the grains are filled with water, but friction holds the grains together. In liquefaction, on the right, water surrounds the grains so that they no longer have contact with each other, leading them to behave as a liquid.
Author: Randa Harris
Source: Original Work
License: CC BY-SA 3.0

Other hazards associated with earthquakes include fire (as gas lines rupture), which may be difficult to combat as water lines may also be ruptured. The vast majority of damage during the 1906 San Francisco earthquake was due to fire. Earthquakes can trigger tsunamis, large sea waves created by the displacement of a large volume of water during fault movement. The Sumatra-Andaman earthquake in 2004 triggered a tsunami in the Indian Ocean that resulted in 230,000 deaths. Earthquakes can trigger landslides in mountainous areas, and initiate secondary hazards such as fires, dam breaks, chemical spills, or even nuclear disasters like the one at Fukushima Daiichi Nuclear Power Plant in Japan. Earthquake-prone areas can take steps to minimize destruction, such as implementing strong building codes, responding to the tsunami warning system, addressing poverty and social vulnerability, retrofitting existing buildings, and limiting development in hazardous zones.

13.7 LAB EXERCISE
Part B - Liquefaction

Download the kml file from the USGS for Google Earth found here: http://earthquake.usgs.gov/regional/nca/bayarea/kml/liquefaction.kmz (Alternately the file can be downloaded from this site: http://earthquake.usgs.gov/regional/nca/bayarea/liquefaction.php). Note that this file adds a layer of liquefaction susceptibility, with areas more likely to experience liquefaction in yellow, orange, or red. Once in Google Earth, type in San Francisco, CA. Zoom in to less than 25 miles to see the layers added and note where liquefaction is most likely, then answer the following questions. When necessary, type locations into the Search box to locate them.

10. A significant earthquake hits San Mateo, California while you are there. During the shaking you are caught indoors. Would you rather be at the US Social Security Administration Building (located at South Claremont Street, San Mateo) or with the San Mateo Park Rangers (located at J Hart Clinton Drive, San Mateo)?

    a. the US Social Security Administration Building   b. the San Mateo Park Rangers

11. While visiting California, you become violently ill and must visit a hospital. Based off of your fears of a possible earthquake occurring, would you rather go to Highland Hospital in Oakland or Alameda Hospital in Alameda?

    a. Highland Hospital, Oakland, CA   b. Alameda Hospital, Alameda, CA

13.8 INDUCED SEISMICITY

The number of significant earthquakes within the central and eastern United States has climbed sharply in recent years. During the thirty-six year period between 1973 and 2008, only 21 earthquakes with a magnitude of 3.0 or greater occurred. During the 5 year period of 2009-2013, 99 earthquakes of that size occurred within the same area, with 659 earthquakes in 2014 alone and well over 800 earthquakes in 2015 just in Oklahoma (see the blue and red line on the graph in Figure 13.15).

Human intervention is apparently the cause, resulting in **induced seismicity** (earthquakes caused by human activities). Humans have induced earthquakes in the past (for example, impounding reservoirs has led to earthquakes in Georgia), but this rapid increase in induced seismicity has led to much current research into the problem. Evidence points to several contributing factors, all related to types of fluid injection used by the oil industry. Hydraulic fracturing, also referred to as fracking, has been used for decades by oil and gas companies to improve well production. Fluid (usually water, though other fluids are often present) is injected at high pressure into low-permeability rocks in an effort to fracture the rock. As more fractures open up within the rock, fluid flow is enhanced and more distant fluids can be accessed, increasing the production of a well. In the past, this practice was utilized in vertical wells. With the recent advent of horizontal drilling technology, the fracking industry has really taken off. Drillers can now access thin horizontal oil and gas reservoirs over long distances, highly increasing well production in rocks that formerly were not exploited, creating a boom in US gas and oil production. While there have been many reports in the public that blame fracking for all of the increased seismicity rates, this is not entirely the case. Fracking mainly produces very minor earthquakes (less than magnitude 3), though it has been shown to produce significant earthquakes on occasion. The majority of induced earthquakes are caused by injection of wastewater deep underground. This wastewater is the byproduct of fracking, so ultimately the industry is to blame.

As wells are developed (by fracking or other processes), large amounts of waste fluid, which may contain potentially hazardous chemicals, are created. When the fluids cannot be recycled or stored in retention ponds above ground, they are injected deep underground, theoretically deep enough to not come into contact with oil reservoirs or water supplies. These wastewater wells are quite common and are considered a safe option for wastewater disposal. By injecting this water in areas that contain faults, the stress conditions on the faults change as friction is reduced, which can result in movement along faults (resulting in earthquakes).

For our lab exercise, we will focus on the state of Oklahoma, and the increased seismicity there (Figure 13.16).

The USGS has focused some research on the seismicity in Oklahoma and determined that the main seismic hazard within the state is the disposal of wastewater from the oil and gas industry by deep injection, though some smaller quakes (magnitude 0.6 to 2.9) have been shown

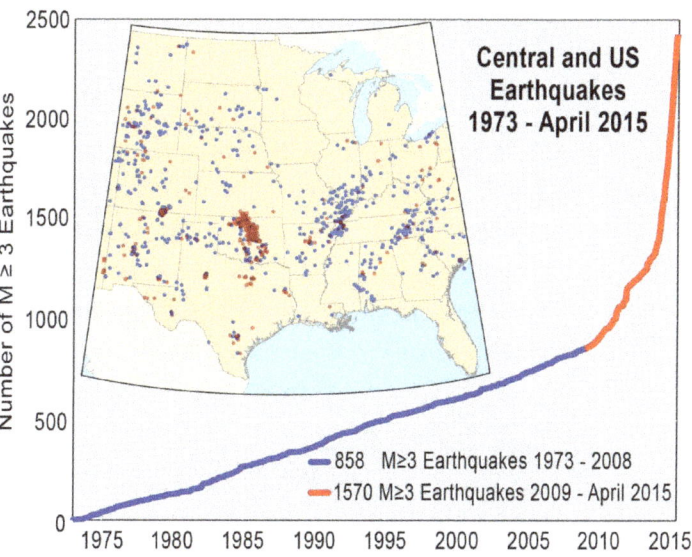

Figure 13.15 | Chart of increased seismicity of magnitude 3 or greater earthquakes within the central and eastern U.S. from 1973-2015. The spatial distribution of the earthquakes is shown on the map, with blue dots representing quakes from 1973-2008, and red dots representing quakes from 2009-2015.
Author: USGS
Source: USGS
License: Public Domain

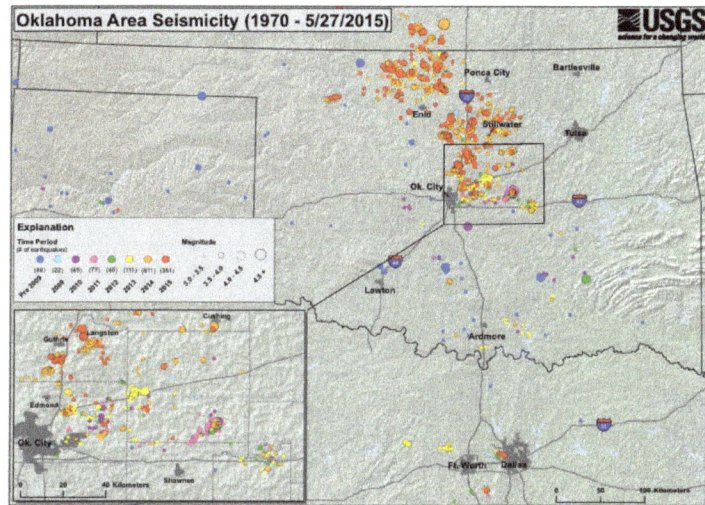

Figure 13.16 | Earthquakes that occurred within Oklahoma from 1970 – 5/27/15 are depicted above. Please note that the colors indicate year and the size indicates magnitude (see legend on image). The inset image is a close-up view of the outlined box.
Author: USGS
Source: USGS
License: Public Domain

to correlate directly to fracking. A 50% increase in earthquake rate has occurred within the state since 2013. One large earthquake of 5.7 magnitude struck in November, 2011, and has been linked to an active wastewater injection site ~200 meters away. A 4.7 magnitude earthquake struck in November 2015, too.

INTRODUCTORY GEOLOGY   EARTHQUAKES

13.9 LAB EXERCISE

Part C – Induced Seismicity

The table below contains data regarding the number of fracking wells within the state of Oklahoma and the number of significant earthquakes (magnitude 3 or greater) that have occurred since 2000. Before answering the questions for this lab exercise, plot the information in the table below on the graph that is provided; note that the graph has two y-axes, one for the number of fracking wells and the other for the number of earthquakes.

Table 13.2

Year	# of Fracking Wells in Oklahoma	# of Earthquakes greater than M 3
2000	0	0
2001	0	0
2002	0	3
2003	0	0
2004	0	2
2005	0	1
2006	0	2
2007	0	1
2008	1	2
2009	4	20
2010	1	43
2011	637	63
2012	1,568	34
2013	1,939	109
2014	3,296	585
2015	1,749	850

(From: http://www.oudaily.com/news/oklahoma-reports-surge-in-earthquakes-during/article_79a364da-a1d4-11e5-894a-5ba84c8399c1.html)

Note: Information on number of fracked wells was obtained by SkyTruth through accessing FracFocus. Oklahomans are required to report all fracked wells, but the site was only created in 2011, so some wells may have not been retroactively added pre-2011. Seismic data was obtained through the USGS.

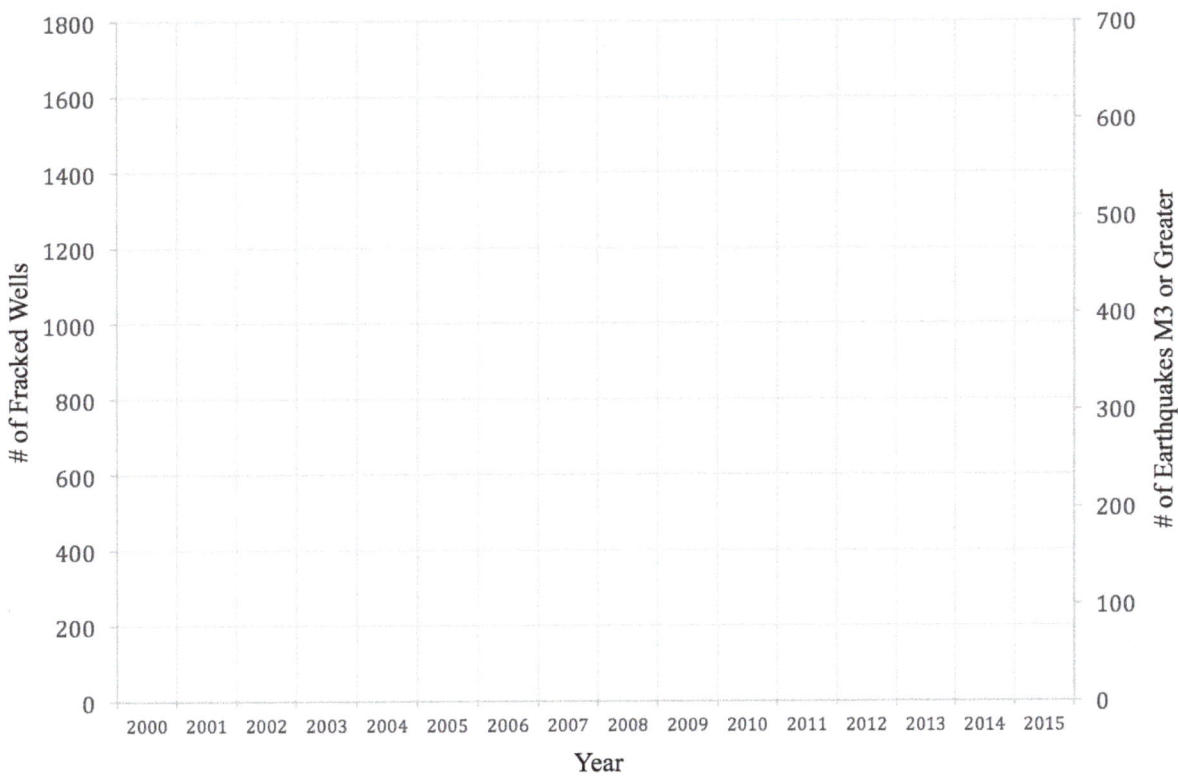

Seismicity Vs Fracked Wells in Oklahoma

12. After what year does the number of magnitude 3 or greater earthquakes begin to rise significantly?

    a. 2007   b. 2009   c. 2011   d. 2015

13. After what year does the number of fracking wells begin to rise significantly?

    a. 2007   b. 2009   c. 2011   d. 2015

14. Based on the graph that you constructed, do significant earthquakes and the number of fracking wells appear to be related?

    a. Yes   b. No

The exercises that follow use Google Earth. Let's start by examining the 1906 earthquake that hit Northern California. Access the following website: http://earthquake.usgs.gov/regional/nca/virtualtour/

There are several links of interest here. Spend some time familiarizing yourself with the site. Scroll down to the section entitled "The Northern California Earthquake, April 18, 1906" and open the link. The San Andreas Fault is ~800 miles long, located in California. In 1906, a major earthquake occurred along a portion of the fault. Scroll down and check out the Rupture Length and Slip.

15. How long was the rupture length (the length of the fault that was affected)?

    a. 25 miles   b. 74 miles   c. 198 miles   d. 296 miles   e. 408 miles

Horizontal slip, or relative movement along the fault, ranged from 2-32 feet. To envision this, imagine that you are facing an object directly across from you. Suddenly, it moves up to 32 feet to your right! Horizontal slip is shown along the rupture as a histogram. Check out all the measurements along the fault by clicking the Rupture Length and Slip on the Google Earth link.

16. Locate the epicenter of the 1906 quake. Does the amount of horizontal slip decrease faster along the northern end or the southern end of the rupture?

    a. northern end of the rupture   b. southern end of the rupture

Go back to the "The Northern California Earthquake, April 18, 1906" page and scroll down to check out the Shaking Intensity. If your map begins to get difficult to read, remember that by clicking on a checked box in the Places folder, you can remove prior data. Use the search box to display the desired location.

17. What was the shaking intensity like in Sacramento?

    a. light   b. strong   c. severe   d. violent   e. extreme

18. What was the shaking intensity like in Sebastopol?

    a. light   b. strong   c. severe   d. violent   e. extreme

Navigate back to the main page and select "Earthquake Hazards of the Bay Area Today." Check out the Earthquake Probabilities for the Bay Area.

19. Based on the map, would you be more likely to experience an earthquake of magnitude >6.7 by 2031 if living in the northwest Bay Area or southeast Bay Area?

    a. northwest   b. southeast

Go back to the "Earthquake Hazards of the Bay Area Today" page and check out the Liquefaction Susceptibility in San Francisco. Look at the overall trend in the areas affected by liquefaction.

20. Based on the liquefaction map, are areas more dangerous inland or along the coast?

    a. inland   b. along the coast

13.10 STUDENT RESPONSES

1. For Carrier, Oklahoma, what is the approximate time of the arrival of the first P-wave?

   a. 10 seconds   b. 15 seconds   c. 21 seconds   d. 30 seconds

2. For Marlow, Oklahoma, what is the approximate time of the arrival of the first S-wave?

   a. 19 seconds   b. 22 seconds   c. 35 seconds   d. 42 seconds

3. For Bolivar, Missouri, what is the difference between the P and S wave arrival times?

   a. 10 seconds   b. 20 seconds   c. 40 seconds   d. 55 seconds

4. What is the approximate distance to the epicenter from Carrier, Oklahoma?

   a. 70 km   b. 130 km   c. 240 km   d. 390 km

5. What is the approximate distance to the epicenter from Marlow, Oklahoma?

   a. 70 km   b. 130 km   c. 240 km   d. 390 km

6. What is the approximate distance to the epicenter from Bolivar, Missouri?

   a. 70 km   b. 130 km   c. 240 km   d. 390 km

7. Look at the location that you determined was the earthquake epicenter. Compare its location to Oklahoma City. Which direction is the epicenter located from Oklahoma City?

   a. southeast   b. northwest   c. northeast   d. southwest

8. Examine the before and after image of the National Cathedral. Based on the changes seen within the structure, decide where this earthquake would most likely fall on the Modified Mercalli Intensity Scale. Based off this image, the most likely intensity of this earthquake would be:

   a. <IV   b. V-VI   c. VII   d. VIII or greater

9. Residents in Port-au-Prince complained of extreme shaking during the earthquake, while residents of Santo Domingo, the capital of the Dominican Republic that sits 150 miles east of Port-au-Prince, assumed the shaking was caused by the passing of a large truck. Based on the Modified Mercalli Intensity Scale, the residents of Port-au-Prince mostly like experienced an intensity of ____, while the residents of Santo Domingo experienced an intensity of ____.

   a. VII, II   b. VIII, III   c. X, III   d. X, IV

10. A significant earthquake hits San Mateo, California while you are there. During the shaking you are caught indoors. Would you rather be at the US Social Security Administration Building (located at South Claremont Street, San Mateo) or with the San Mateo Park Rangers (located at J Hart Clinton Drive, San Mateo)?

    a. the US Social Security Administration Building   b. the San Mateo Park Rangers

11. While visiting California, you become violently ill and must visit a hospital. Based off of your fears of a possible earthquake occurring, would you rather go to Highland Hospital in Oakland or Alameda Hospital in Alameda?

    a. Highland Hospital, Oakland, CA   b. Alameda Hospital, Alameda, CA

12. After what year does the number of magnitude 3 or greater earthquakes begin to rise significantly?

    a. 2007   b. 2009   c. 2011   d. 2015

13. After what year does the number of fracking wells begin to rise significantly?

    a. 2007   b. 2009   c. 2011   d. 2015

14. Based on the graph that you constructed, do significant earthquakes and the number of fracking wells appear to be related?

    a. Yes   b. No

15. How long was the rupture length (the length of the fault that was affected)?

    a. 25 miles   b. 74 miles   c. 198 miles   d. 296 miles   e. 408 miles

INTRODUCTORY GEOLOGY   EARTHQUAKES

16. Locate the epicenter of the 1906 quake. Does the amount of horizontal slip decrease faster along the northern end or the southern end of the rupture?

    a. northern end of the rupture   b. southern end of the rupture

17. What was the shaking intensity like in Sacramento?

    a. light   b. strong   c. severe   d. violent   e. extreme

18. What was the shaking intensity like in Sebastopol?

    a. light   b. strong   c. severe   d. violent   e. extreme

19. Based on the map, would you be more likely to experience an earthquake of magnitude >6.7 by 2031 if living in the northwest Bay Area or southeast Bay Area?

    a. northwest   b. southeast

20. Based on the liquefaction map, are areas more dangerous inland or along the coast?

    a. inland   b. along the coast

14 Physiographic Provinces
Bradley Deline

14.1 INTRODUCTION

If you took a road trip across the continental United States of America you would see significant changes in the landscape in terms of the topography, rocks, soils, geological structures, and plant life that are evident even through the car window on the highway. Regions vary in their geologic history, from the rocky coastline of New England, to the flat plains of the Midwest, to the sharp peaks of the Rocky Mountains, to the volcanoes of the Pacific Northwest, these (and the many other) observable differences across the United States can be broken into **physiographic provinces**. Physiographic provinces are identifiable by their distinctive landforms, geologic features, and suites of rocks.

You can observe such diverse geologic characteristics across the state of Georgia from the foothills of the Appalachian Mountains in the North, to the iconic red clay of Middle Georgia, to the flat Coastal Plain. As we discuss the physiographic provinces of Georgia, we will also explore regional geologic resources that benefit the state as well as the major river systems that provide water for our state. Lastly, the tools and knowledge you have gained in this lab manual will aid your reconstruction of the various physiographic province geologies.

14.1.1 Learning Outcomes

After completing this chapter, you should be able to:
- Distinguish the different physiographic provinces of the United States based on their topography, geology, and other features
- Identify the physiographic features of the different geological provinces of Georgia
- Describe the major natural resources within Georgia including minerals, building rock, and water

14.1.2 Key Terms

- Adirondack Mountains Province
- Appalachian Plateau Province
- Basin and Range Province
- Blue Ridge Province
- Cascade Range
- Coastal Plain
- Colorado Plateau
- Columbia Plateau
- Fall Line
- Great Plains Province
- Interior Lowlands Province
- New England Province
- Physiographic Province
- Piedmont Province
- Rocky Mountains
- Sierra Nevada Range
- Valley and Ridge Province

14.2 PHYSIOGRAPHIC PROVINCES OF THE UNITED STATES OF AMERICA

The physiographic provinces of the United States of America can be broken into three different broad areas: Western, Central, and Eastern regions (Figure 14.1). Adjacent provinces will share features or will at least be affected by the geologic events that define the nearby region. The Western Provinces are shaped by relatively young events (Post-Paleozoic), which are mostly the result of an active plate tectonic margin (the edge of the continent is also the edge of a tectonic plate).

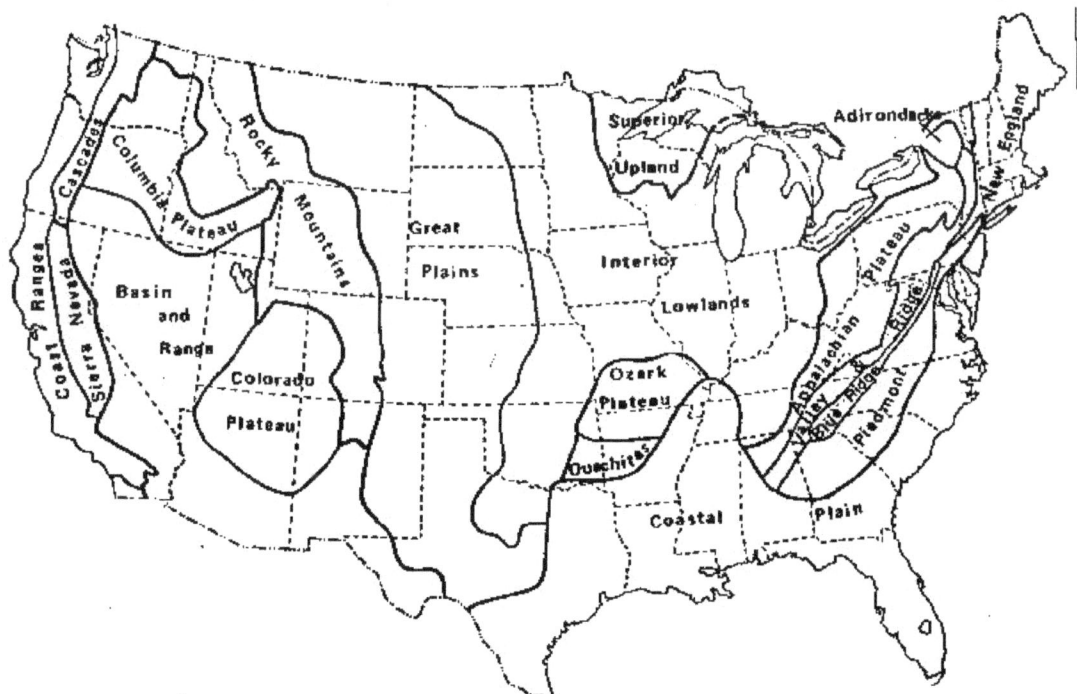

Figure 14.1 | Physiographic provinces of the continental United States of America.
Author: User "Kbh3rd"
Source: Wikimedia Commons
License: CC BY 3.0

The Eastern Provinces are a mix between young and old geologic events. Most of the individual provinces within this region result from the deformation of ancient mountain building as well as the recent passive sedimentary buildup observable today. Since the Central Provinces of the country has been largely shielded from tectonic activity, this area is flatter and less deformed than the United States' coasts. Before the individual provinces are discussed in depth, it will be helpful to review the geologic time scale that was presented in Chapter 1.

14.2.1 Western U.S.

The geologic provinces in the Western United States occupy roughly a third of the country and stretch from the Rocky Mountains to the Pacific Ocean. We can group this tectonically active area into four types of provinces: the Rocky Mountains, the Coastal Pacific Mountain System, Interior Plateaus, and the extensional Basin and Range.

The **Rocky Mountains** are an immense range that stretches from New Mexico to Northern Canada. The Rocky Mountains first started to form during the Late Paleozoic, but their main growth occurred during the Laramide Orogeny during the Late Mesozoic and Early Cenozoic. An orogeny is a large-scale deformational event that is the result of the interaction between tectonic plates, in short it is mountain building. The formation of this range differs from the Continent-Continent convergent boundaries, like the Himalayas you studied in Chapter 4. The Rocky Mountains are the result of low-angle subduction, which caused deformation in the overlying plate. This striking mountain range is no longer active nor growing and has since been glaciated and eroded, which has rounded its highest peaks.

Figure 14.2 | Half Dome at Yosemite National Park, which is part of the Sierra Nevada Range.
Author: Arian Zwegers
Source: Wikimedia Commons
License: CC BY 2.0

The Pacific Coast Mountains stretch from Mexico to the Arctic Circle in Alaska. Not only do the Pacific Coast Mountains differ significantly from the Rocky Mountains, but the mountain ranges within the Coastal Pacific Mountain System as also differ amongst themselves. The **Sierra Nevada Range** runs 400 miles across the middle of California and contains iconic features such Yosemite's Half Dome, which is a mecca for mountain climbers (Figure 14.2). As with the Rocky Mountains, the Sierra Nevada Range is the byproduct of convergence and subduction. During the same time as the growth of the Rocky Mountains (Late Mesozoic), the subduction led to the formation of volcanoes and the intrusion of massive granite batholiths. The volcanoes have long since eroded away exposing the batholiths at the surface. The Sierra Nevada topography is relatively recent (within the last 5 million years), as these rocks have been deformed by extensional stress, creating large fault blocks that build the steep face of the range. The **Cascade Range** is the northern portion of the Pacific Coast Mountain Range that spans an area from Northern California to Canada. The Cascade Range is the result of modern subduction of the Juan de Fuca and Gorda Plates under the North American Plate. Whereas the Sierra Nevada range is the eroded exposed core of a Mesozoic chain of volcanoes, the Cascade Range is modern and active. This chain encompasses thousands of volcanoes including well-known landmarks such as Mount Rainier and Mount Saint Helens.

Between the Rocky Mountains and the Pacific Coast Mountain Range is a large broad area containing a great diversity of geologic features. The Northernmost portion between the Cascades and the Rocky Mountains is the **Columbia Plateau**. The Columbia Plateau is a large igneous province dominated by Cenozoic volcanic rocks that are a product of the Yellowstone Hotspot. This hotspot produced several massive, though infrequent, eruptions; the largest eruption covering over 50,000 square miles. The Yellowstone Hotspot is still active and supplies the heat that feeds the classic geysers and hot springs throughout Yellowstone National Park (Figure 14.3). Southeast of this province is the **Colorado Plateau**, which is a broad flat expanse that considering its neighboring provinces, is remarkably un-deformed. The Colorado Plateau is mostly composed of horizontal Paleozoic and Mesozoic sedimentary rocks that contain abundant fossils, including the massive bone beds

Figure 14.3 | Hot Spring in Yellowstone National Park within the Columbia Plateau Province.
Author: User "daveynin"
Source: Flickr
License: CC BY 2.0

of Dinosaur National Monument. During the Laramide Orogeny this area was dramatically uplifted, allowing rivers to erode downward thus producing immense canyons, such as Grand Canyon National Park. The southeastern portion of the expanse between the Mountain Ranges is called the **Basin and Range Province.** This area can easily be recognized by the abundant Horsts and Grabens. Which you may remember are series of elevated plateaus and low basins produced from the extensional deformation of abundant paired and mirrored normal faults. There are multiple hypotheses for the source of the extension, which range from heating from the mantle to movement along the San Andres Fault.

14.2.2 Eastern U.S.

The physiographic provinces that occupy the eastern third of the country range from New England to Texas, wrapping around the Gulf of Mexico. These provinces can easily be split into three regions, such as the provinces associated with the building of the Appalachian Mountains, provinces built by much older orogenies in the Northeast, and provinces created from more recent passive build-up of sediments along the coast.

The Appalachian Mountains are the product of multiple collisions with small island chains during the Paleozoic, culminating with a collision with the supercontinent of Gondwana during the formation of Pangaea. These ancient mountains have been significantly eroded down to rolling hills that span most of the Southeastern United States as seen in Great Smoky Mountains National Park (Figure 14.4). The Appalachian Mountains can be further divided into four distinctive physiographic provinces based on their topography and geology. The center of the Appalachian

Figure 14.4 | Appalachian Mountains within Great Smoky Mountains National Park, which contains multiple physiographic provinces.
Author: Ernest Duffo
Source: Flickr
License: CC BY 2.0

Mountains is called the **Valley and Ridge Province**, which is composed of highly folded and faulted sedimentary rock. Fossils within these rocks indicate that they are Paleozoic in age with thick and resistant Pennsylvanian-age sandstones forming its ridges while weaker Devonian and Cambrian shales forming its valleys. To the east of the Valley and Ridge is the **Blue Ridge Province,** which was uplifted along with the Valley and Ridge but is composed of much older igneous and metamorphic rock. Absolute dating of these rocks indicates that the Blue Ridge was formed during the Proterozoic and later deformed during the Middle Paleozoic. The topography between these two provinces differs with the more uniform rocks in the Blue Ridge producing random peaks, valleys, and ridges as opposed to the resistant and non-resistant rocks of the Valley and Ridge, which produce more uniformly parallel ridges.

Figure 14.5 | Stone Mountain within the Piedmont Province.
Author: User "kschlot1"
Source: Flickr
License: CC BY 2.0

Southeast of the Blue Ridge is the Piedmont Province, which is typified by much lower, rolling hills along with small isolated mountains such as Stone Mountain in Georgia (Figure 14.5). The Piedmont is composed of igneous and metamorphic rock from the cores of long eroded mountain chains ranging from the Proterozoic to the end of the Paleozoic period. On the adjacent west side of the Valley and Ridge province is the **Appalachian Plateau**, which is an uplifted and largely un-deformed region analogous to the Colorado Plateau. The rocks in this province are similar in age and lithology to those in the Valley and Ridge, but the preservation of fossils is enhanced because of the absence of extensive folding and faulting. In addition, the Appalachian Plateau contains abundant geological resources, including coal in West Virginia and iron in Alabama.

The physiographic provinces in the northeastern United States are very similar to the Piedmont and Blue Ridge Provinces. The **New England Province** is similar to the Piedmont in that it is composed of Late Proterozoic and Paleozoic intrusive igneous and metamorphic rocks. The major difference between these two provinces is in their history following exposure, with the New England Province showing extensive weathering and erosion from glaciers that is absent from its southeastern counterpart. The **Adirondack Mountain Province** located in upstate New York is similar to the Blue Ridge in that they are both composed of igneous and metamorphic rock. However, the Adirondacks are fairly unique considering that they are a circular rather than linear range of mountains. The rocks themselves are billion year old remnants of the building of a Proterozoic supercontinent called Rodinia. The actual mountains in the range are quite young and represent Late Cenozoic uplift, which exposed these ancient rocks.

The last eastern physiographic province is the **Coastal Plain** that spans a vast area along the Atlantic Ocean and Gulf of Mexico from New England to Texas. This province is composed of the sediment that has accumulated since the rifting of Pangaea when the eastern edge of the continent became tectonically inactive. The sediment was derived from the continent and was deposited in shallow marine sedimentary environments with abundant, mature sediment and marine fossils.

14.2.3 Central U.S.

The central third of the country is mostly flat, un-deformed, and dominated by sequences of sedimentary rocks. We separate this broad expanse into two physiographic provinces based largely on the source of the sediment. The **Interior Lowlands Province** covers the Midwestern states and consists largely of un-deformed Paleozoic marine rocks (limestones and shales) that have been since carved and shaped by the activity of glaciers and rivers. This description may not seem as geologically exciting as those

Figure 14.6 | Mammoth Cave National Park in Central Kentucky.
Author: Gary Tindale
Source: Flickr
License: CC BY 2.0

to the east or west, but spectacular cave developments can occur in these marine limestones, such as those from Mammoth Cave, Kentucky (Figure 14.6). To the west of the Interior Lowlands Province is the **Great Plains Province**, which is composed of sediments that eroded from the Rocky Mountains and are, therefore, substantially younger (Mesozoic and Cenozoic) with a higher proportion of clastic material. As with Interior Lowlands, the Great Plains have since been shaped by glaciers and rivers.

14.3 LAB EXERCISE

Materials

We will explore the various physiographic provinces of the United States of America by looking closely at the National Parks that showcase iconic geologic and topological features within each region. For each park, examine the area using Google Earth. To get a better view of the features, making sure to zoom in and out and also click on multiple photographs posted (make sure that photos are checked in the layers box). More information about these and other parks can be found at www.nps.gov.

This lab is also cumulative in that you may need to review the material presented in previous chapters to answer the following questions.

Part A – National Parks

Crater Lake National Park

Figure 14.7 | Crater Lake National Park, Southern Oregon.
Author: Ray Bouknight
Source: Flickr
License: CC BY 2.0

Crater Lake National Park (Figure 14.7) is located in southern Oregon and was established as a National Park in 1902. The main attraction at this park is Crater Lake, which at almost 2,000 feet is one of the deepest lakes on Earth. Search for 42 56 33.15N 122 06 14.89W and zoom out to an eye altitude of 15 miles.

1. What type of volcano is Crater Lake?

    a. Composite Volcano

    b. Shield Volcano

    c. Caldera

    d. Cinder Cone

2. What is the origin of this volcano?

    a. Subduction of an oceanic plate at a Convergent Boundary

    b. Subduction of a continental plate at a Convergent Boundary

    c. Development of a Hotspot

    d. Continental Rifting

3. Crater Lake National Park is located in which of the following physiographic provinces?

   a. Rocky Mountains

   b. Sierra Nevada

   c. Cascades

   d. Columbia Plateau

   e. Colorado Plateau

   f. Basin and Range

Theodore Roosevelt National Park

Theodore Roosevelt National Park (Figure 14.8) is located in North Dakota and was established as a National Park 1978. This park was named after Theodore Roosevelt in honor of his conservation policies that led to the establishment of the National Park System. Following the death of his wife, Roosevelt spent several years in the area that would ultimately become the park, during which time he wrote extensively about the lonely beauty of the surrounding landscape. Search for 46 58 52.55N 103 32 13.91W and zoom out to an eye altitude of 30,000 feet.

Figure 14.8 | Theodore Roosevelt National Park, North Dakota.
Author: User "stereogab"
Source: Flickr
License: CC BY-SA 2.0

4. What is the prominent drainage pattern in this park?

   a. Radial

   b. Trellis

   c. Rectangular

   d. Dendritic

5. One of the main attractions at this park is a forest of large petrified trees. These trees were preserved by being replaced with silica from ash layers within the Triassic (Mesozoic) sandstones. Based on the geologic history and features of the area, which of the following statements about these sedimentary rocks is TRUE?

   a. The clastic sand that forms these rocks was weathered and eroded from the Appalachian Mountains.

   b. The ash was produced from volcanoes associated with the Yellowstone Hotspot.

   c. The clastic sand that forms these rocks was weathered and eroded from the Rocky Mountains.

   d. The clastic sand that forms these rocks was weathered and eroded from the Sierra Nevada Mountains.

6. Theodore Roosevelt National Park is located in which of the following physiographic provinces?

   a. Rocky Mountains b. Colorado Plateau

   c. Interior Lowlands d. Great Plains

   e. Basin and Range f. Columbia Plateau

Acadia National Park

Acadia National Park (Figure 14.9) is located in Southern Maine and was preserved as a National Park in 1916. This was the first park established east of the Mississippi River and helped Maine gain the nickname "Vacationland". Search for 44 21 09.94N 68 13 23.22W and zoom out to an eye altitude of 3,000 feet.

Figure 14.9 | Acadia National Park in Southern Maine.
Author: Ken Lund
Source: Flickr
License: CC BY-SA 2.0

7. Based on the history of the region containing Acadia National Park and the color of the rocks, what type of igneous rock occur in this area?

   a. Gabbro b. Granite c. Rhyolite

   d. Andesite e. Diorite

8. What type of igneous body does this structure represent?

   a. Stock b. Dike c. Sill

9. Acadia National Park is located in which of the following physiographic provinces?

   a. Interior Lowlands b. Adirondacks

   c. Great Plains d. Appalachian Plateau

   e. New England f. Piedmont

Congaree National Park

Congaree National Park (Figure 14.10) is located in South Carolina and was recently established as a National Park in 2003. Unlike most National Parks, this area was preserved for reasons other than geology. Congaree contains the oldest tract of old growth hardwood forest left in the United States. This lush ecosystem contains a diverse assemblage of animals, fungi, and plants. Search for 33 47 57.63N 80 47 49.79W and zoom out to an eye altitude of 35,000 feet.

Figure 14.10 | Congaree National Park in South Carolina.
Author: Miguel Vieira
Source: Flickr
License: CC BY 2.0

10. What is the maturity of the river in this area?

    a. Youthful   b. Mature   c. Old Age

11. Zoom out to an eye altitude of 500 miles to see the source of the sediment that is accumulating in this area. This sediment is _____ and if lithified would be called _____.

    a. Immature, Conglomerate   b. Intermediate, Sandstone   c. Mature, Shale

12. Congaree National Park is located in which of the following physiographic provinces?

    a. Interior Lowlands   b. Blue Ridge   c. Piedmont

    d. Coastal Plain   e. New England

Shenandoah National Park

Shenandoah National Park (Figure 14.11) is located in Virginia and was established as a National Park in 1935. This park, which is located close to Washington, D.C., is a favorite of hikers containing 101 miles of the Appalachian Trail, which runs from Northern Georgia to Maine. Search for 38 17 53.72N 78 40 26.42W and zoom out to an eye altitude of 25,000 feet.

Figure 14.11 | Shenandoah National Park in Virginia.
Author: Beau Considine
Source: Flickr
License: CC BY-SA 2.0

13. Look over the region and examine the mountains, note their shape, ground cover, and height. Then Search for 43 48 25.03N 110 50 26.19W to examine Grand Tetons National Park. How do the Shenandoah Mountains compare to the Rocky Mountains?

   a. The Shenandoah Mountains are shorter.

   b. The Shenandoah Mountains are more rounded.

   c. The Shenandoah Mountains have more vegetation.

   d. All of the above.

14. Based on these observations, we can conclude that the mountains in Shenandoah National Park have undergone _____ erosion and are _____ the Rocky Mountains.

   a. more, older than   b. less, younger than   c. the same amount of, the same age as

15. Shanandoah National Park is located in which of the following physiographic provinces?

   a. Blue Ridge   b. Piedmont   c. Appalachian Plateau

   d. Valley and Ridge   e. Adirondack

Black Canyon of the Gunnison National Park

Black Canyon of the Gunnison National Park (Figure 14.12) is located in Western Colorado and was established as a National Park in 1999. This park is often overshadowed by the Grand Canyon, but is striking in its own right. The Gunnison River has a high gradient, which has produced an incredibly steep canyon. In fact, it is called Black Canyon not because of the color of the rocks, but because of the dark shadows produced by the steep walls of the canyon. Search for 38 34 43.18N 107 43 43.74W and zoom out to an eye altitude of 30,000 feet.

Figure 14.12 | Black Canyon of the Gunnison National Park in Colorado.
Author: User "daveynin"
Source: Flickr
License: CC BY 2.0

16. What type of weathering is primarily responsible for the formation of this canyon?

    a. Frost wedging

    b. Chemical weathering

    c. Mechanical weathering from air

    d. Mechanical weathering from water

17. We can measure the rate of erosion in this canyon at about 0.01 inches/year. Based on this rate, when did the canyon start to erode? (Hint: Measure the difference in elevation from the bottom of the canyon at the latitude and longitude given above and the top, measured at the road on the edge of the canyon due south of the previous point. Then divide by the rate of erosion. Make sure to use the correct units.)

    a. 750,000 years   b. 1,200,000 years   c. 2,700,000 years   d. 3,500,000 years

18. Black Canyon of the Gunnison National Park is located in which of the following physiographic provinces?

    a. Blue Ridge   b. Colorado Plateau   c. Appalachian Plateau

    d. Rocky Mountains   e. Basin and Range   f. Columbia Plateau

14.4 GEOLOGY OF GEORGIA

Georgia is a wonderful natural laboratory for the study of geology. The rocks within this state span over a billion years of history and through this lens we can study all of the topics presented within this lab manual. Within the state we have mountains, coastlines, folds, faults, earthquakes, fossils, a diversity of rocks, and evidence for ancient volcanic eruptions. As would be expected with this geologic diversity, Georgia contains multiple physiographic provinces (Figure 14.13) that have been discussed above.

The northwestern portion of the state is within the Valley and Ridge Province and shows the characteristic sandstone ridges with folded and deformed shale within valleys. As you might expect, the shale is relatively soft (which is why they show more deformation) and erodes quickly underneath the massive sandstones. This causes the sandstones to break and tumble

Figure 14.13 | Physiographic provinces of Georgia.
Author: USGS
Source: Wikimedia Commons
License: Public Domain

downhill making the Rock Cities that are a tourist attraction surrounding Lookout Mountain. The sands are Pennsylvanian in age and interbedded within them are thick coal deposits of ancient forests. These deposits were mined in the past, but for the most part are not currently active. The northeastern corner of the state is in the Blue Ridge physiographic province, which contains mountains consisting of ancient igneous and metamorphic rock. This area of Georgia contains large protected areas that preserve its natural beauty (such as Chattahoochee National Forest) and contains spectacular waterfalls as seen at Tallulah Falls State Park (Figure 14.14).

Figure 14.14 | Tallulah Falls State Park in Northern Georgia.
Author: Stanislav Vitebskiy
Source: Flickr
License: CC BY-ND 2.0

The central portion of the state is within the Piedmont Province that consists of rolling hills of igneous and metamorphic rocks punctuated with large batholiths. This region has several important geologic resources. First, the granite within the Piedmont has been mined for buildings, monuments, and memorial stones. The granite mining industry is one of the leading producers of granite within the United State, which is centered in Elberton, Georgia. The Piedmont Province has also produced gold, which can be found associated with quartz veins within saprolite (a soft and porous rock often formed by the weathering of granite). Starting around 1830, the discovery of gold in Dahlonega and Villa Rica, Georgia, led to the first major, though short-lived, gold rush.

The largest physiographic province within Georgia is the **Coastal Plain** in the southern half of the state. The transition from the Piedmont into the Coastal Plain is striking in that the topography dramatically changes from rolling hills to flat terrain. The boundary of the Coastal Plain is referred to as the **Fall Line**, which is a line of waterfalls along the boundary between the provinces caused by differences in the rate of erosion in the two provinces. The rocks and sediments in the Coastal Plain vary from Mid-Mesozoic in the central portion of the state to Late Cenozoic in the southern portion, tracking a fall in sea level through time. The Coastal Plain province has several useful geological resources. Foremost, its organic-rich soil and flat topography makes this region ideal for agriculture thus allowing for the abundant production of Georgia's staples such as peanuts, pecans, onions, cotton, and peaches. The mineral Kaolin, which is a product of the chemical weathering of feldspar, is also mined in this physiographic province. This white clay mineral is used in the production of many products including ceramics, toothpaste, cosmetics, and glossy paper—just to name a few.

However, one of the most important resources across Georgia is water. As with many parts of the country, Georgia has experienced steady population growth as well as droughts that have made discussions about water reservoirs and usage particularly vital to the future of the state. Georgia contains multiple river basins (Figure 14.15), many of which have been dammed to build large reservoirs, such as Lake Lanier. Study Figure 14.15, and identify the river basins that contain Georgia's major cities (such as Atlanta, Macon, and Savannah) as well as the river basin in which you live.

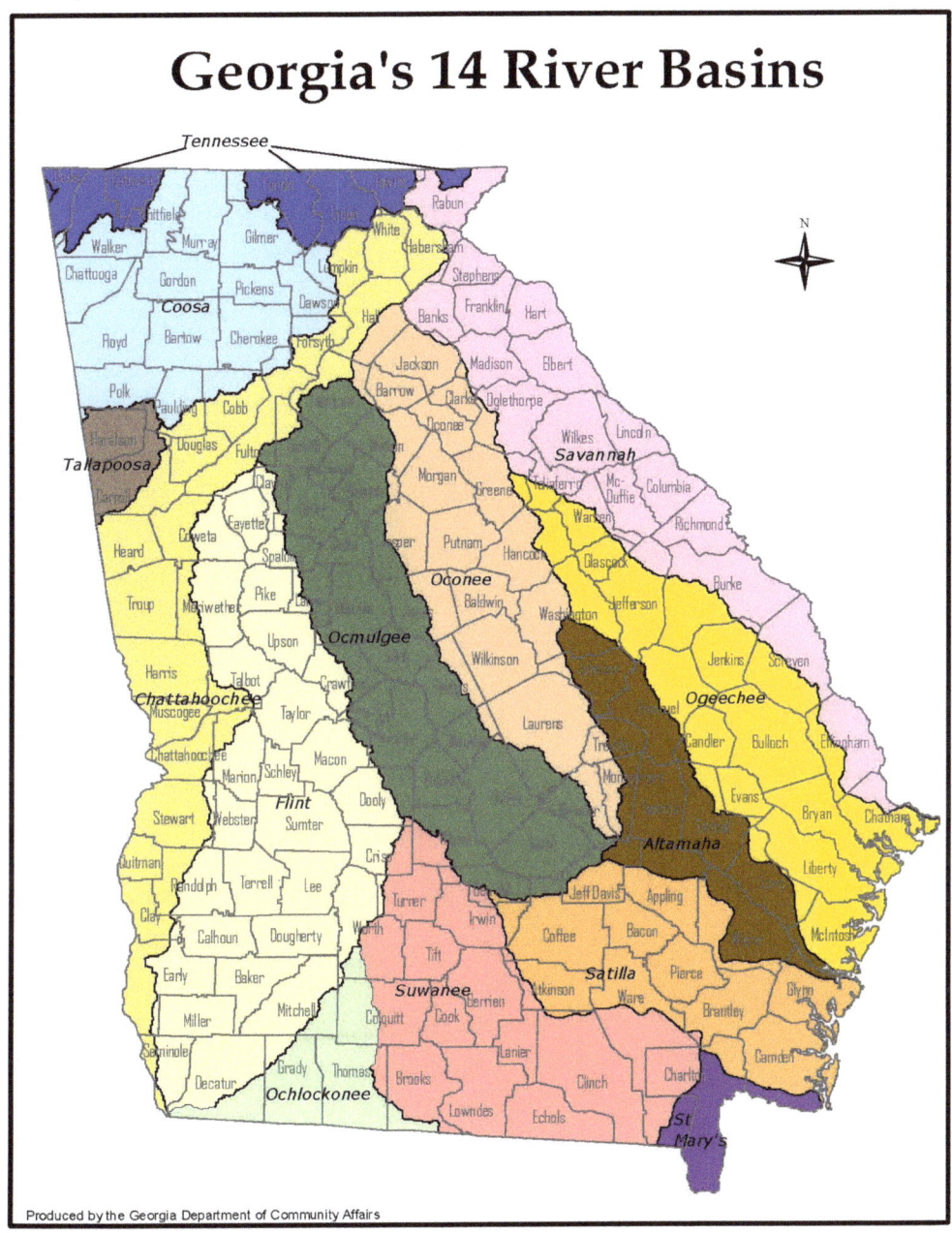

Figure 14.15 | The major river basins within Georgia.
Author: Georgia Department of Community Affairs
Source: Upper Oconee Watershed Network
License: Public Domain

14.5 LAB EXERCISE

Materials

Download the file "GA Geologic Cross Section N-S.gif" from your course website. This cross section runs north to south through the state of Georgia showing the subsurface geology across the Valley and Ridge, Piedmont, and Coastal Plain physiographic provinces. This cross section will be used to answer question 25-28. This cross-section was provided for use with permission from Geological Highway Map, Southeastern Region, American Association of Petroleum Geologists. Also, download the "Chart of Surface Time and Rock Units" legend that gives a key to the symbols and rock units included in the cross-section. We will examine the physiographic provinces of Georgia by examining a couple of Georgia's State Parks.

Part B – Georgia Provinces

Cloudland Canyon State Park

Cloudland Canyon State Park (Figure 14.16) is located in Dade County and was established as a park in 1938. Many of the roads and civil infrastructure within Dade County were built following the designation of this area as a state park in order to allow access to the park. Cloudland Canyon State Park has multiple waterfalls as well as some of the most scenic overlooks within Georgia, especially in autumn. Search for 34 50 44.57N 85 28 40.92W in google earth and zoom out to an eye altitude of 5,000 feet.

Figure 14.16 | Cloudland Canyon State Park, Dade County, Georgia.
Author: Moultrie Creek
Source: Flickr
License: CC BY-SA 2.0

19. Examine the walls of Cloudland Canyon, what type of rock makes up the rim of the canyon?

   a. Granite   b. Shale

   c. Sandstone   d. Gneiss

20. Zoom out to an eye altitude of 20 miles and examine the valley 6 miles to the southeast (34 43 10.37N 85 24 55.74W at the same eye altitude). What major geologic structure can be seen in this valley?

   a. Normal Fault   b. Dome   c. Horizontal Fold

   d. Reverse Fault   e. Plunging Fold

21. Cloudland Canyon State Park is located in which of the following physiographic provinces?

    a. Blue Ridge   b. Valley and Ridge   c. Piedmont   d. Coastal Plain

High Falls State Park

High Falls State Park (Figure 14.17) is a small park located in Butts County and was designated as a state park in 1966. This park, located between Macon and Atlanta, has the largest waterfalls in central Georgia and sits on the Fall Line. The Fall line is a line of waterfalls that is caused by differences in rock strength between two physiographic provinces. Search for 33 10 42.13N 84 01 01.11W zoom out to an eye altitude of 2,500 feet.

Figure 14.17 | High Falls State Park, Butts County, Georgia.
Author: B A Bowen Photography
Source: Flickr
License: CC BY 2.0

22. Notice that the Towaliga River Flows south, based on the position of the waterfalls, the rocks upstream are _____ to weathering while the rocks downstream are _____ to weathering.

    a. more resistant, less resistant   b. less resistant, more resistant

23. Which physiographic province is located upstream of these waterfalls?

    a. Blue Ridge   b. Valley and Ridge   c. Piedmont   d. Coastal Plain

24. Which physiographic province is located downstream of these waterfalls?

    a. Blue Ridge   b. Valley and Ridge

    c. Piedmont   d. Coastal Plain

Georgia North to South Cross-Section

Reminder: the following questions are based on the cross-section and legend provided on your courses website.

25. Examine the Cartersville Fault that is the boundary between the Valley and Ridge and Piedmont Geological Provinces. Which type of fault is the Cartersville Fault?

    a. Reverse   b. Normal   c. Strike-Slip

26. Which of the following statements regarding the four major faults within the Piedmont are true?

    a. The faults are all reverse like the Cartersville Fault.

    b. The faults are all normal like the Cartersville Fault.

    c. The faults are all strike-slip like the Cartersville Fault.

    d. With one exception, the faults are all similar in type to the Cartersville Fault.

    e. With two exceptions, the faults are all similar in type to the Cartersville Fault.

27. Which geologic province has abundant plutons?

    a. Valley and Ridge   b. Piedmont   c. Coastal Plain

28. What is the dip direction of the sedimentary layers in the Coastal Plain Province in Peach County?

    a. North   b. South   c. East   d. West

Part C – Georgia's Natural Resources

Search for 33 18 32.95N 82 26 08.55W and zoom to an eye altitude of 8,000 feet.

29. What natural resource is being mined in this area?

    a. Gold   b. Granite   c. Kaolin   d. Coal

30. This mine is located in which of the following physiographic province?

    a. Blue Ridge   b. Valley and Ridge   c. Piedmont   d. Coastal Plain

31. The mineral that was weathered to produce these large deposits likely originated from which of the following physiographic province?

    a. Blue Ridge   b. Valley and Ridge

    c. Piedmont   d. Coastal Plain

Search for 33 27 11.53N 84 56 12.56W and zoom out to an eye altitude of 5,000 feet. To answer the following questions also refer to Figure 14.15.

32. Rainfall in this forest would be part of which of the following Georgia River Basins?

    a. Flint   b. Tallapoosa   c. Coosa   d. Ocmulgee

    e. Oconee   f. Chattahoochee   g. Suwanee

33. Zoom out to at least an eye altitude of 15 miles and examine the major river in this drainage basin. What overall direction does this river flow?

    a. Northeast to Southwest   b. Southwest to Northeast

    c. Northwest to Southeast   d. Southeast to Northwest

34. Follow this river downstream, into what body of water does this basin ultimately flow?

    a. Atlantic Ocean   b. Lake Lanier   c. Lake Martin   d. Gulf of Mexico

35. How many times has this river been dammed to produce reservoirs that are most likely used for drinking water, power, or recreation?

    a. Once   b. Twice   c. Three times   d. Greater than three times

14.6 STUDENT RESPONSES

1. What type of volcano is Crater Lake?

   a. Composite Volcano   b. Shield Volcano

   c. Caldera   d. Cinder Cone

2. What is the origin of this volcano?

   a. Subduction of an oceanic plate at a Convergent Boundary

   b. Subduction of a continental plate at a Convergent Boundary

   c. Development of a Hotspot

   d. Continental Rifting

3. Crater Lake National Park is located in which of the following physiographic provinces?

   a. Rocky Mountains   b. Sierra Nevada

   c. Cascades   d. Columbia Plateau

   e. Colorado Plateau   f. Basin and Range

4. What is the prominent drainage pattern in this park?

   a. Radial   b. Trellis   c. Rectangular   d. Dendritic

5. One of the main attractions at this park is a forest of large petrified trees. These trees were preserved by being replaced with silica from ash layers within the Triassic (Mesozoic) sandstones. Based on the geologic history and features of the area, which of the following statements about these sedimentary rocks is TRUE?

   a. The clastic sand that forms these rocks was weathered and eroded from the Appalachian Mountains.

   b. The ash was produced from volcanoes associated with the Yellowstone Hotspot.

   c. The clastic sand that forms these rocks was weathered and eroded from the Rocky Mountains.

   d. The clastic sand that forms these rocks was weathered and eroded from the Sierra Nevada Mountains.

6. Theodore Roosevelt National Park is located in which of the following physiographic provinces?

    a. Rocky Mountains b. Colorado Plateau c. Interior Lowlands

    d. Great Plains e. Basin and Range f. Columbia Plateau

7. Based on history of the region containing Acadia National Park and the color of the rocks, what type of igneous rock occur in this area?

    a. Gabbro b. Granite c. Rhyolite d. Andesite e. Diorite

8. What type of igneous body does this structure represent?

    a. Stock b. Dike c. Sill

9. Acadia National Park is located in which of the following physiographic provinces?

    a. Interior Lowlands b. Adirondacks

    c. Great Plains d. Appalachian Plateau

    e. New England f. Piedmont

10. What is the maturity of the river in this area?

    a. Youthful b. Mature c. Old Age

11. Zoom out to an eye altitude of 500 miles to see the source of the sediment that is accumulating in this area. This sediment is _____ and if lithified would be called _____.

    a. Immature, Conglomerate b. Intermediate, Sandstone c. Mature, Shale

12. Congaree National Park is located in which of the following physiographic provinces?

    a. Interior Lowlands b. Blue Ridge

    c. Piedmont d. Coastal Plain

    e. New England

13. Look over the region and examine the mountains, note their shape, ground cover, and height. How do these mountains compare to the Rocky Mountains (search for 43 48 25.03N 110 50 26.19W to examine Grand Tetons National Park)?

    a. The Shenandoah mountains are shorter.

    b. The Shenandoah mountains are more rounded.

    c. The Shenandoah mountains have more vegetation.

    d. All of the above.

14. Based on these observations, we can conclude that the mountains in Shenandoah National Park have undergone _____ erosion and are _____ the Rocky Mountains.

    a. more, older than   b. less, younger than   c. the same amount, the same age as

15. Shenandoah National Park is located in which of the following physiographic provinces?

    a. Blue Ridge   b. Piedmont   c. Appalachian Plateau

    d. Valley and Ridge   e. Adirondack

16. What type of weathering is primarily responsible for the formation of this canyon?

    a. Frost wedging   b. Chemical weathering

    c. Mechanical weathering from air   d. Mechanical weathering from water

17. We can measure the rate of erosion in this canyon at about 0.01 inches/year. Based on this rate, when did the canyon start to erode? (Hint: Measure the difference in elevation from the bottom of the canyon at the latitude and longitude given above and the top, measured at the road on the edge of the canyon due south of the previous point. Then divide by the rate of erosion. Make sure to use the correct units.)

    a. 750,000 years   b. 1,200,000 years   c. 2,700,000 years   d. 3,500,000 years

18. Black Canyon of the Gunnison National Park is located in which of the following physiographic provinces?

    a. Blue Ridge   b. Colorado Plateau   c. Appalachian Plateau

    d. Rocky Mountains   e. Basin and Range   f. Columbia Plateau

19. Examine the walls of Cloudland Canyon, what type of rock makes up the rim of the canyon?

    a. Granite   b. Shale   c. Sandstone   d. Gneiss

20. Zoom out to an eye altitude of 20 miles and examine the valley 6 miles to the southeast (34 43 10.37N 85 24 55.74W at an eye altitude of 20 miles). What major geologic structure can be seen in this valley?

    a. Normal Fault   b. Dome

    c. Horizontal Fold   d. Reverse Fault

    e. Plunging Fold

21. Cloudland Canyon State Park is located in which of the following physiographic provinces?

    a. Blue Ridge   b. Valley and Ridge   c. Piedmont   d. Coastal Plain

22. Notice that the Towaliga River Flows south, based on the position of the waterfalls, the rocks upstream are _____ to weathering while the rocks downstream are _____ to weathering.

    a. more resistant, less resistant   b. less resistant, more resistant

23. Which physiographic province is located upstream of these waterfalls?

    a. Blue Ridge   b. Valley and Ridge   c. Piedmont   d. Coastal Plain

24. Which physiographic province is located downstream of these waterfalls?

    a. Blue Ridge   b. Valley and Ridge   c. Piedmont   d. Coastal Plain

25. Examine the Cartersville Fault that is the boundary between the Valley and Ridge and Piedmont Geological Provinces. Which type of fault is the Cartersville Fault?

    a. Reverse   b. Normal   c. Strike-Slip

26. Which of the following statements regarding the four major faults within the Piedmont are true?

   a. The faults are all reverse like the Cartersville Fault

   b. The faults are all normal like the Cartersville Fault

   c. The faults are all strike slip like the Cartersville Fault

   d. With one exception, the faults are all similar in type to the Cartersville Fault

   e. With two exceptions, the faults are all similar in type to the Cartersville Fault

27. Which geologic province has abundant plutons?

   a. Valley and Ridge   b. Piedmont   c. Coastal Plain

28. What is the dip direction of the sedimentary layers in the Coastal Plain Province in Peach County?

   a. North   b. South   c. East   d. West

29. What natural resource is being mined in this area?

   a. Gold   b. Granite   c. Kaolin   d. Coal

30. This mine is located in which of the following physiographic province?

   a. Blue Ridge   b. Valley and Ridge   c. Piedmont   d. Coastal Plain

31. The mineral that was weathered to produce these large deposits likely originated from which of the following physiographic province?

   a. Blue Ridge   b. Valley and Ridge   c. Piedmont   d. Coastal Plain

32. Rainfall in this forest would be part of which of the following Georgia River Basins?

   a. Flint   b. Tallapoosa   c. Coosa   d. Ocmulgee

   e. Oconee   f. Chattahoochee   g. Suwanee

33. Zoom out to at least an eye altitude of 15 miles and examine the major river in this drainage basin. What overall direction does this river flow?

    a. Northeast to Southwest   b. Southwest to Northeast

    c. Northwest to Southeast   d. Southeast to Northwest

34. Follow this river downstream, into what body of water does this basin ultimately flow?

    a. Atlantic Ocean   b. Lake Lanier   c. Lake Martin   d. Gulf of Mexico

35. How many times has this river been dammed to produce reservoirs that are most likely used for drinking water, power, or recreation?

    a. Once   b. Twice   c. Three times   d. Greater than three times

CPSIA information can be obtained
at www.ICGtesting.com
Printed in the USA
LVHW071554010920
PP16189700001B/13